Barron's Review Course Series

Let's Review:

Integrated Algebra

Second Edition

Lawrence S. Leff, M.S.
Former Assistant Principal, Mathematics Supervision
Franklin D. Roosevelt High School
Brooklyn, New York

Dedication

To Rhona . . .
For the understanding,
for the sacrifices,
for the love,
. . . and with love.

All inquiries should be addressed to:
Barron's Educational Series, Inc.
250 Wireless Boulevard
Hauppauge, New York 11788
www.barronseduc.com

ISBN: 978-1-4380-0017-6

ISSN: 2166-6008

PRINTED IN THE UNITED STATES OF AMERICA
9 8 7 6 5 4 3 2 1

TABLE OF CONTENTS

PREFACE

This book organizes all of the various topics, concepts, and skills required for the NYS Integrated Algebra Regents Examination in a way that is easy for students to understand and convenient for teachers to use when planning their daily lessons.

Who Should Use This Book?

- *Students Who Want to Raise Their Grades*
  Students enrolled in Regents-level Integrated Algebra will find this book helpful when they either need additional explanation and practice on a troublesome topic being studied in class or want to review specific topics before a classroom test or the Regents examination.
- *Teachers Who Want an Additional Resource When Lesson Planning*
  Because this book is designed to be compatible with all styles of classroom instruction and curriculum organization, classroom teachers will want to include *Let's Review: Integrated Algebra* in their personal, departmental, and school libraries. Teachers will find it a valuable lesson planning aid as well as a helpful source of classroom exercises, homework problems, and test questions.
- *School Districts and Mathematics Departments*
  School districts and mathematics departments that want to align their mathematics curricula with the set of prescribed topics and performance indicators of the Integrated Algebra course will find this book particularly helpful. The table of contents shows at a glance how the various topics, skills, and concepts required by Integrated Algebra can be organized into a logical and cohesive set of lessons.

What Special Features Does This Book Have?

- *Reflects Core Curriculum*
  This book offers complete topic coverage of Integrated Algebra as described in the 2005 revision of the New York State Core Curriculum for Mathematics.
- *A Compact Format Designed for Self-Study and Rapid Learning*
  The clear writing style quickly identifies essential ideas while avoiding unnecessary details. Helpful diagrams, convenient Math Fact summaries, and numerous step-by-step demonstration examples will be appreciated by students who need an easy-to-read book that provides

complete and systematic preparation for both classroom and Integrated Algebra Regents tests.

- *Graphing Calculator Approaches*
  Graphing calculators are required for the Integrated Algebra Regents Examination. The key features of the TI-83/84 graphing calculator are introduced and used throughout the book to help support and extend algebraic approaches. Descriptions of keystroke sequences as well as actual calculator screen displays are included.
- *Practice Exercises with Answers*
  Comprehensive sets of practice exercises include questions at different levels of difficulty that are designed to clinch understanding while building skill and confidence. Exercise sections feature sample multiple-choice and extended-response Regents questions that will help students prepare for both classroom tests and the Integrated Algebra Regents Examination. Answers to the practice exercises are intended to give students valuable feedback that will lead to greater understanding and higher test grades.

LAWRENCE S. LEFF

CHAPTER 1

SETS, OPERATIONS, AND ALGEBRAIC LANGUAGE

1.1 NUMBERS, VARIABLES, AND SYMBOLS

KEY IDEAS

A **constant** is a quantity that does not change such as the number of days in a week. A quantity that can change such as the weight of a person is a **variable**. It is customary to represent variables using lowercase letters such as x and y. Algebra is a branch of mathematics that uses variables and special symbols to make general statements about how numbers are related.

Equations and Inequations

An **equation** is a mathematical statement that two expressions have the same value. An equal sign, =, is used to separate the two expressions, as in $5 + 3 = 2 \times 4$, thereby indicating the left side has the same value as the right side. The equal sign is read "is equal to." An **inequation** is a mathematical statement that two expressions do not have the same value, as in $1 + 1 \neq 3$ where the symbol $\neq$ is read "is not equal to." The symbol $\approx$ is read, "is approximately equal to." For example, $\frac{1}{9} \approx 0.111$.

Sets and Subsets

A **set** is a collection of objects. The members of a set are called **elements**. A set is usually named by a capital letter such as $A, B, C,$ and so forth.

- A set can be described by listing its elements within braces, { }. If A is the set of all even numbers from 2 to 10, then

$$A = \{2, 4, 6, 8, 10\}$$

- The order in which the elements of a set appear within the braces does not matter. If S is the set of the first three lowercase letters of the English alphabet, then

$$S = \{a, b, c\} = \{b, a, c\} = \{c, a, b\} \text{ and so on}$$

- If each element of set B is also an element of set A, then B is a **subset** of A. If $B = \{2, 4\}$ and $A = \{1, \mathbf{2}, 3, \mathbf{4}, 5\}$, then B is a subset of A.

Variables and Replacement Sets

The students who attend Atlantic Beach High School range in age from 14 to 19 years old. If x represents the age, in years, of a student who attends the school, then *variable* x can take on any value from $\{14,15,16,17,18,19\}$, which is called the *replacement set*.

- A **variable** is a symbol that represents an unspecified member of a given set called the **replacement set** or **domain**.
- Each member of the replacement set is a possible **value** of the variable. Thus, 14, 15, 16, 17, 18, and 19 are the possible *values* of x. If the replacement set consists of only one element, then the variable is a **constant**.

Different Uses of Variables

Variables serve different purposes. In algebra, variables are used:

- as placeholders for numbers. In the equation $x + 5 = 7$, variable x serves as a placeholder for the number 2 since $2 + 5 = 7$.
- to represent quantities that change. In the equation $y = x + 5$, y changes when x changes. When $x = 1$, $y = 6$; when $x = 2$, $y = 7$; and so forth.
- to represent quantities related by a formula such as $A = L \times W$, where A, L, and W represent the area, length, and width of a rectangle.
- to make general statements about how numbers behave. You know that $2 + 3 = 3 + 2$. To generalize that the order in which *any* two numbers are added together does not matter, you can write that "$x + y = y + x$" where variables x and y are placeholders for *any* two numbers.

Signed Numbers

Numbers with positive (+) or negative (–) signs are called **signed numbers**. Positive 3 and negative 3 are *opposites* in the same sense that winning 3 games (+3) and losing 3 games (–3) are opposite results. The number 0 is neither positive nor negative. The sum of a number and its opposite is always 0. For example, $(+3) + (-3) = 0$. The opposite of 0 is 0.

- A nonzero number that is written without a sign is assumed to be positive. Thus, 5 and +5 represent the same quantity.
- A negative sign in front of a number that is enclosed by parentheses means to take the opposite of that number. For example,

$$-(+4) = -4 \quad \text{and} \quad -(-4) = +4 \text{ or } 4$$

- A negative sign in front of a variable means take the opposite of the value of that variable. For example,

$$\text{if } x = 5, \text{ then } -x = -5$$

$$\text{if } y = -8, \text{ then } -y = 8$$

Representing Multiplication

Multiplication is indicated when:

- a period is centered between two quantities:

$4 \cdot y$ means 4 times y

- parentheses are placed around quantities written next to each other:

$(+2)(-3)$ indicates $+2$ times -3

$5(x + y)$ represents 5 times the sum of x and y

- numbers and variables are written consecutively:

$3x$ means 3 times x

$2xy$ is 2 times x times y

Inequality Relations

The inequality symbols $<$ and $>$ can be used to compare two numbers:

- $8 > 2$ is read "8 *is greater than* 2."
- $3 < 5$ is read "3 *is less than* 5."

An equal sign can be combined with an inequality symbol:

- $x \leq 3$ is read "x is less than *or* equal to 3." Hence, x is *at most* 3.
- $y \geq 6$ is read "y is greater than *or* equal to 6." Hence, y is *at least* 6.

Algebraic Expressions Versus Equations

An **algebraic expression** is any combination of numbers and variables connected by one or more arithmetic operations, as in

$$6x + 1 \qquad x - \frac{1}{2}y + z \qquad \frac{x}{3y} - 7xy \qquad 3a \times 2b \div c$$

The **terms** of an algebraic expression are the parts of the expression that are separated by addition or subtraction signs. The terms of $2ab - 5bc + 3ac$ are $2ab$, $5bc$, and $3ac$. The number that multiplies the variable part of a term is its **numerical coefficient**. The numerical coefficient of $-4ab$ is -4. The numerical coefficient of a single variable or a product of variables such as xy is understood to be 1.

Since an algebraic expression is not a mathematical sentence, it *never* includes an equal sign or an inequality symbol. In an *equation*, an equal sign separates two expressions, thereby creating left and right sides of the equation. The left side of the equation $3x + 1 = 5x - 9$ is $3x + 1$, and the right side is $5x - 9$. Much of the study of algebra involves manipulating algebraic expressions and determining the values of variables that make equations true statements.

Divisibility and Factors

An integer is **evenly divisible** by a nonzero integer if their quotient is an integer with a remainder of 0. For example, 12 is evenly divisible by 4 (or 4 is an *exact divisor* of 12) because their quotient is 3 with a remainder of 0. Because 12 is evenly divisible by 4, 4 is a *factor* of 12. A **factor** of a term is an exact divisor of it. Since $12 \div 7 = 1$ with a remainder of 5, 12 is *not* evenly divisible by 7 so 7 is *not* a factor of 12.

Example 1

In a softball league, 92 players are on 8 different teams. If each team has at least 11 players, what is the largest possible number of players on any one team?

Solution: Since $92 \div 11 = 8$ teams with a remainder of 4 players, there are 4 extra players in the softball league. If all 4 players are on the same team, then that team has $11 + 4 =$ **15** players, which is, therefore, the largest possible number of players on any one team.

Prime and Composite Numbers

A **prime number** is an integer greater than 1 that is divisible only by itself and 1. The numbers 3, 5, 7, 11, 13, 17, 19, and 23 are examples of prime numbers. The number 2 is the only even number that is a prime number. Positive integers greater than 1 that are *not* prime are called **composite numbers**.

Inequalities That Describe an Interval

The inequality $1 \le x \le 5$ describes an interval in which x can be any number from 1 to 5, including 1 and 5, as illustrated in Figure 1.1.

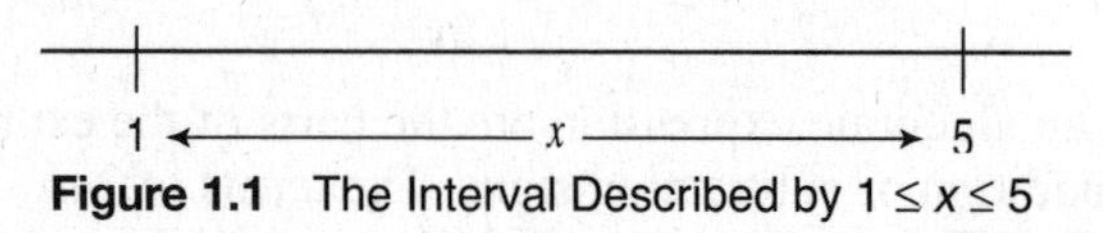

Figure 1.1 The Interval Described by $1 \le x \le 5$

The inequality $1 \le x \le 5$ is read "1 is less than or equal to x *and* x is less than or equal to 5." An interval need not include its left or right boundary points.

- $1 \le x < 5$ is the interval of numbers from 1 to 5, including 1 but *not* 5. If x is a whole number, then x may be equal to 1, 2, 3, or 4.
- $1 < x \le 5$ is the interval of numbers from 1 to 5, including 5, but *not* 1. If x is a whole number, then x may be equal to 2, 3, 4, or 5.
- $1 < x < 5$ is the interval of numbers from 1 to 5, but does *not* include 1 or 5. If x is a whole number, then x may be equal to 2, 3, or 4.

Interval Notation

Instead of using an inequality to describe an interval, a shorthand notation can be used in which the endpoints of the interval are written as an ordered pair. For example, the interval $1 < x \leq 5$ may also be represented as $(1, 5]$, where the bracket next to 5 indicates that the interval includes 5. An interval that includes an endpoint is **closed** on that side. The left parenthesis next to 1 means that the interval does *not* include 1. An interval that does not include an endpoint is **open** on that side. Table 1.1 compares inequality and interval notations for the interval that has 1 and 5 as endpoints.

TABLE 1.1 INTERVAL NOTATION

Type of Interval	Inequality Notation	Interval Notation	Verbal Description
Open interval: both endpoints are not included	$1 < x < 5$	$(1, 5)$	All real numbers between 1 and 5, *not* including 1 and 5
Half-open interval: only one endpoint is included	$1 \leq x < 5$	$[1, 5)$	All real numbers between 1 and 5, including 1 but *not* 5
	$1 < x \leq 5$	$(1, 5]$	All real numbers between 1 and 5, including 5 but *not* 1
Closed interval: both endpoints are included	$1 \leq x \leq 5$	$[1, 5]$	All real numbers between 1 and 5, including both 1 and 5

Example 1

Which interval notation represents the set of all real numbers greater than or equal to –3 and less than 17?

(1) $(-3, 17)$ (2) $(-3, 17]$ (3) $[-3, 17)$ (4) $[-3, 17]$

Solution: Write the endpoints –3 and 17 as an ordered pair: –3, 17. Since the interval contains –3, a left bracket is needed next to –3. The interval does not include 17, so a right parenthesis is written next to 17. Hence, the interval notation $[-3, 17)$ represents the set of all real numbers greater than or equal to –3 and less than 17. The correct choice is **(3)**.

Check Your Understanding of Section 1.1

A. *Multiple Choice.*

1. Which inequality could be used to represent the statement, "x is *at least* 4"?
(1) $x < 4$ (2) $x \leq 4$ (3) $x > 4$ (4) $x \geq 4$

2. Which inequality could be used to represent the statement, "x is *at most* 7"?
(1) $x < 7$ (2) $x \leq 7$ (3) $x > 7$ (4) $x \geq 7$

3. Which number is a factor of both 56 and 105?
(1) 5 (2) 2 (3) 3 (4) 7

4. If $A = \{0, 1, 2\}$ and $B = \{0, -1, -2\}$, which set is a subset of both A and B?
(1) $\{1, 2\}$ (2) $\{-2, -1, 0, 1, 2\}$ (3) $\{0\}$ (4) $\{-1, -2, 1, 2\}$

5. A class of 43 students is to be divided into committees so that each student serves on exactly one committee. Each committee must have at least three members and at most five members. Which inequality represents the possible number of subcommittees, x, that could be formed?
(1) $8 \leq x \leq 14$ (2) $8 \leq x \leq 15$ (3) $9 \leq x \leq 14$ (4) $9 \leq x \leq 15$

6. There are 461 students and 20 teachers taking buses on a trip to a museum. Each bus can seat a maximum of 52. What is the least number of buses needed for the trip?
(1) 8 (2) 9 (3) 10 (4) 11

7. A jar contains between 40 and 50 marbles. If the marbles are taken out of the jar three at a time, two marbles will be left in the jar. If the marbles are taken out of the jar five at a time, four marbles will be left in the jar. How many marbles are in the jar?
(1) 41 (2) 43 (3) 44 (4) 47

8. Which interval includes the greatest number of prime numbers?
(1) $1 \leq x \leq 20$ (2) $21 \leq x \leq 45$ (3) $46 \leq x \leq 74$ (4) $75 \leq x \leq 100$

9. Paula's weight when attending high school was never less than 115 lbs or more than 128 lbs. Which interval notation represents the set of Paula's possible weights when she attended high school?
(1) (115, 128) (2) (115, 128] (3) [115, 128) (4) [115, 128]

10. During July and August, the temperature of a certain city never falls below 60° Fahrenheit and is always less than 95° Fahrenheit. Which interval notation represents the set of possible temperatures in this city during July and August?
(1) (60, 95) (2) (60, 95] (3) [60, 95) (4) [60, 95]

B. *Show how you arrived at your answer.*

11. The manufacturer of Ron's car recommends that the tire pressure be at least 26 pounds per square inch and less than 35 pounds per square inch. If x is the number of pounds per square inch in the recommended tire pressure, write an inequality that represents the possible values of x.

12. In a softball league, 92 players are on 8 different teams. If each team has at least 11 players, what is the largest possible number of players on any one team?

13. If all of the books on a shelf with fewer than 45 books were put into piles of five books each, no books would remain. If the same set of books were put into piles of seven books each, two books would remain. What is the greatest number of books that could be on the shelf?

1.2 CLASSIFYING REAL NUMBERS

KEY IDEAS

Each distinct point on a number line corresponds to a different number. The set of all such numbers forms the set of real numbers.

The Real Numbers and Its Subsets

As pictured in Figure 1.2, the set of **real numbers** is the set of numbers comprised of these subsets:

- Natural or counting numbers = {1, 2, 3, . . . }.
- Whole numbers = {0, 1, 2, 3, . . . }. The set of whole numbers adds 0 to the set of natural numbers.
- Integers = { . . . , −3, −2, −1, 0, 1, 2, 3, . . . }. The set of integers includes all whole numbers and their opposites.
- Rational numbers. A **rational number** is the quotient of two integers provided the denominator is not 0, as in $\frac{-3}{5}$. Because integers can be written as fractions with a denominator of 1, as in $2 = \frac{2}{1}$, the set of

rational numbers includes the set of integers. Rational numbers when written in decimal form either terminate or never end. For example, $\frac{3}{4} = 0.75$, while $\frac{4}{11} = 0.363636\ldots$, where the three trailing periods indicate that this is a non-ending decimal in which the digits 3 and 6 repeat endlessly. A repeating decimal such as 0.363636 . . . can be written in a more compact form by placing a bar over the first appearance of the repeating digits, as in $0.\overline{36}$.

- Irrational numbers. Not all real numbers are rational. A non-ending decimal number that does *not* have a repeating set of digits, such as 0.210210021000 . . . , is an **irrational number**. The number π ($\approx 3.141592654\ldots$) that arises when working with circles is irrational. Irrational numbers do *not* have exact decimal equivalents. Every real number is either rational or irrational.

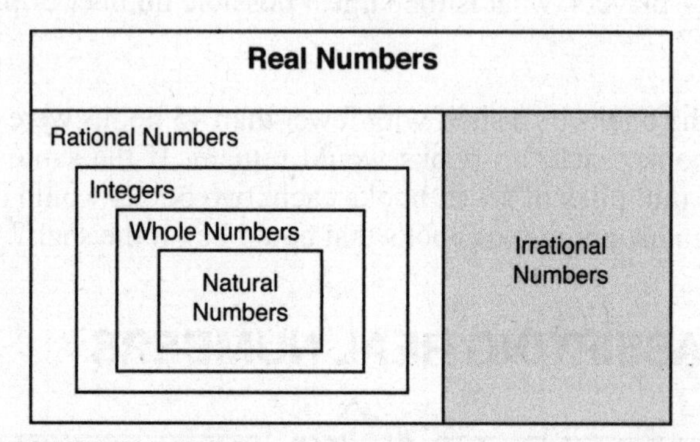

Figure 1.2 The Set of Real Numbers and Its Subsets

The Real Number Line

A **number line** is a ruler-like graph used to visually represent the set of real numbers, as in Figure 1.3. Any point on the number line may be designated the **origin** and labeled 0. Positive numbers are placed to the right of the origin, and negative numbers are located to the left of the origin, the same distance from 0 as their positive counterparts.

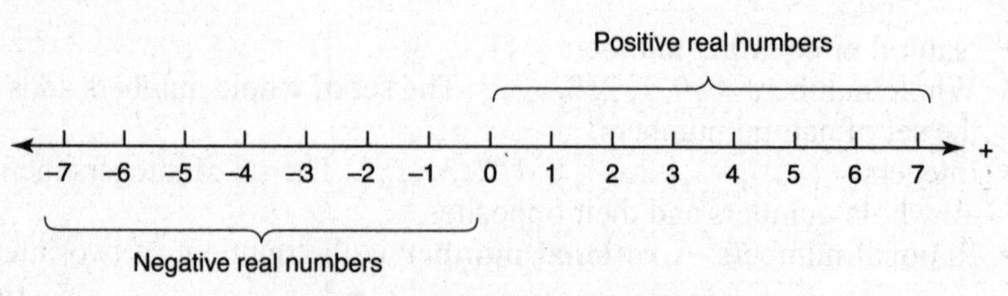

Figure 1.3 The Real Number Line

- The set of *all* points on the number line, including those between consecutive integers, corresponds to the set of **real numbers**.
- When moving along the real number line from left to right, the real numbers increase so that each real number is greater than any real number to its left. For example, 1 is greater than –2.

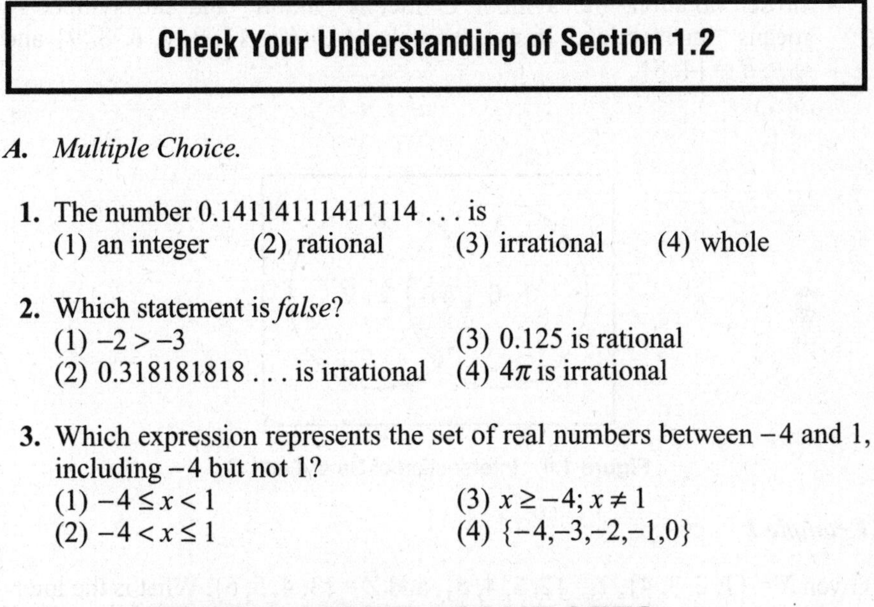

Check Your Understanding of Section 1.2

A. Multiple Choice.

1. The number 0.14114111411114 . . . is
(1) an integer (2) rational (3) irrational (4) whole

2. Which statement is *false*?
(1) $-2 > -3$
(2) 0.318181818 . . . is irrational
(3) 0.125 is rational
(4) 4π is irrational

3. Which expression represents the set of real numbers between -4 and 1, including -4 but not 1?
(1) $-4 \leq x < 1$
(2) $-4 < x \leq 1$
(3) $x \geq -4; x \neq 1$
(4) $\{-4,-3,-2,-1,0\}$

1.3 LEARNING MORE ABOUT SETS

KEY IDEAS

Special set notation is used to indicate how sets are related. The members of a set can be described in **roster-form** by listing them one by one within braces or by using **set-builder notation**, which places within braces a rule for determining whether or not a particular element belongs to the set.

Union and Intersection of Sets

The elements of two sets can be combined to form a new set.

If $A = \{3, 4, 6, 8\}$ and $B = \{2, 4, 8, 9\}$, then:

- The union of the two sets is $\{2, 3, 4, 6, 8, 9\}$. Each element of this set is in set A, set B, or in both sets. The union of two or more sets is the set whose elements are in at least one of the sets. Although 4 and 8 are in both set A and set B, each of these elements is written only one time in the intersection of the two sets.

- The *intersection* of the two sets is {4, 8}. Each element of this set is in set A and in set B. The **intersection** of two or more sets is the set whose members are common to all of the sets. The shaded region in the accompanying Venn diagram represents the intersection of sets A and B.
- In set notation, the symbol $\cup$ means "union" and the symbol $\cap$ means "intersection." In our example, $A \cup B = \{2, 3, 4, 6, 8, 9\}$ and $A \cap B = \{4, 8\}$.

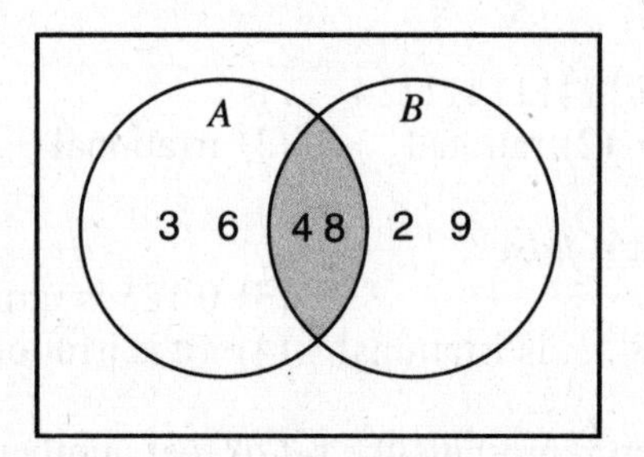

Figure 1.4 Intersection of Sets *A* and *B*.

Example 1

Given $X = \{1, 2, 3, 4\}$, $Y = \{2, 3, 4, 5\}$, and $Z = \{3, 4, 5, 6\}$. What is the intersection of sets X, Y, and Z?

(1) {3, 4}
(2) {4, 5}
(3) {3, 4, 5}
(4) {1, 2, 3, 4, 5, 6}

Solution: Compare the elements of set X, one by one, to the members of the other two sets. Draw a diagonal line through any element of set X that is *not* also contained in both set Y and set Z:

$$X = \{\not{1}, \not{2}, \mathbf{3}, \mathbf{4}\}$$
$$Y = \{2, \mathbf{3}, \mathbf{4}, 5\}$$
$$Z = \{\mathbf{3}, \mathbf{4}, 5, 6\}$$

Because 3 and 4 are the only elements common to all three sets, the intersection of sets X, Y, and Z is {3, 4}. The shaded region in the accompanying Venn diagram represents the intersection of the three sets.

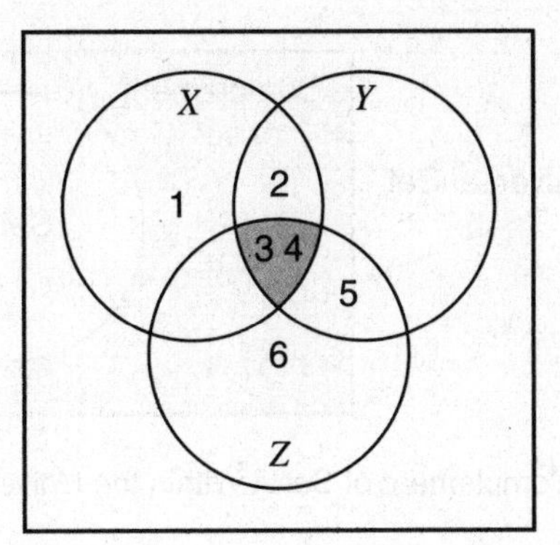

The correct choice is **(1)**.

Number of Elements in a Set

A set may have a specific number of elements, an uncountable number of elements, or no elements.

- A **finite set** has a countable number of distinct elements such as the set of all teachers in your school.
- An **infinite set** has a nonending, uncountable number of elements such as the set of all positive numbers.
- The **empty set** is a set having no elements. For example, the set of elephants that can fly is the empty set. The intersection of the sets {1, 3} and {2, 4, 6} is the empty set.

The Complement of a Set

The **universal set** is the set of all objects under consideration. In Figure 1.5, the universal set is represented by the entire rectangular region. Set *A*, pictured as a circular region inside the rectangle, is any subset of the universal set. The shaded region of the rectangle represents the set of all elements within the universal set *U* that are *not* in set *A*. This set is called the **complement** of set *A*. The sum of the number of elements in set *A* and its complement must be equal to the number of elements in the universal set. For example, if the universal set is $U = \{1, 2, 3, 4, 5\}$ and $A = \{3, 5\}$:

- The complement of set *A* within the universe of set *U* is {1, 2, 4}.
- 5 elements are in set *U*, 2 elements in set *A*, and 3 elements in the complement of set *A*. Thus, $2 + 3 = 5$.

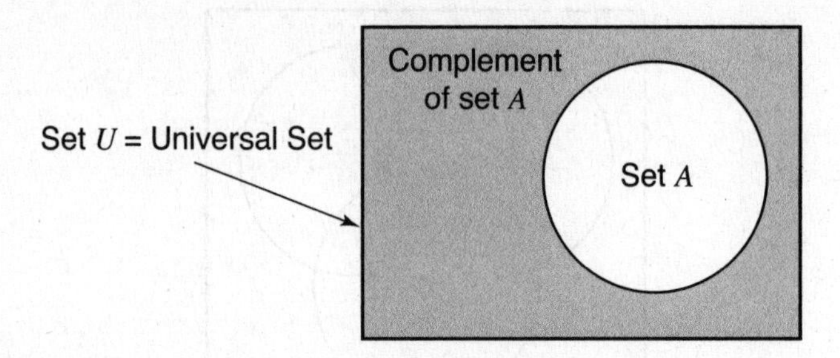

Figure 1.5 Complement of Set *A* Within the Universe of Set *U*.

Example 2

Given set $U = \{T, R, I, A, N, G, L, E\}$ and set $B = \{T, A, N, G\}$, what is the complement of set B within the universe of set U?

Solution: To help determine those elements of U that are not in B, draw a diagonal line through each element that is contained in both sets:

$$\text{set } U = \{\cancel{T}, R, I, \cancel{A}, \cancel{N}, \cancel{G}, L, E\}$$
$$\text{set } B = \{\cancel{T}, \cancel{A}, \cancel{N}, \cancel{G}\}$$

The set of elements in set U that do not have diagonal lines represents the complement of set B. Thus, the complement of set B within the universe of set U is $\{\boldsymbol{R, I, L, E}\}$.

Example 3

Given:

$A = \{$All odd integers from 5 to 21, inclusive$\}$ and $B = \{9, 11, 13, 17, 19\}$. What is the complement of set B within the universe of set A?

Solution:

- List the elements of universal set A:

$$A = \{5, 7, 9, 11, 13, 15, 17, 19, 21\}$$

- Draw a diagonal line through each element that is contained in both sets:

$$A = \{5, 7, \cancel{9}, \cancel{11}, \cancel{13}, 15, \cancel{17}, \cancel{19}, 21\}$$
$$B = \{\cancel{9}, \cancel{11}, \cancel{13}, \cancel{17}, \cancel{19}\}$$

- The set of elements in set A that do not have diagonal lines represents the complement of set B within the universe of set A.

The complement of set B within the universe of set A is $\{\mathbf{5, 7, 15, 21}\}$.

Set-Builder Notation

A set is written in **roster-form** when its members are listed individually within braces, as in $A = \{7, 8, 9, 10, 11\}$. **Set-builder notation** places inside braces a general rule for determining whether or not a particular number is a member of the set. Using set-builder notation,

$$A=\{x \mid \underbrace{7 \le x \le 11}_{\text{rule}};\ \underbrace{x \text{ is a positive integer}}_{\text{replacement set for } x}\}$$

Set A is read, "the set of all x such that x is greater than or equal to 7 and less than or equal to 11, where x is a positive integer." Sometimes a colon is used instead of a vertical bar, as in $A = \{x : 7 \le x \le 11; x \text{ is a positive integer}\}$. The colon and vertical bar are each translated as, "such that." If the replacement set for x is not indicated within the braces, assume it is the largest possible set of real numbers. For example, $\{x \mid x > 3\}$ represents the set of all *real numbers* greater than 3.

Example 4

Represent each set in roster-form.

a. $\{x : x + 2 = 9\}$

b. $\{x \mid -4 \le x < \frac{1}{2}; x \text{ is an integer}\}$

c. $\{x \mid -2 < x \le 3; x \text{ is an integer}\}$

Solution:

a. Since $7 + 2 = 9$, x is a placeholder for 7. Hence,

$$\{x : x + 2 = 9\} = \mathbf{\{7\}}$$

b. The set includes the integers between -4 and $\frac{1}{2}$, including -4:

$$\{x \mid -4 \le x < \tfrac{1}{2}, \text{ where } x \text{ is an integer}\} = \mathbf{\{-4, -3, -2, -1, 0\}}$$

c. The set includes the integers from -2 to 3, including 3 but not -2:

$$\{x \mid -2 < x \le 3, \text{ where } x \text{ is an integer}\} = \mathbf{\{-1, 0, 1, 2, 3\}}$$

Example 5

Which set-builder notation describes $\{-4, -3, -2, -1, 0, 1\}$?
(1) $\{x \mid -4 \le x < 1$, where x is an integer$\}$
(2) $\{x \mid -4 < x \le 1$, where x is an integer$\}$
(3) $\{x \mid -4 < x < 1$, where x is an integer$\}$
(4) $\{x \mid -4 \le x \le 1$, where x is an integer$\}$

Solution: Since the inequality must include both –4 and 1, the correct choice is **(4)**.

Comparing Set Notations

Some sets, such as $A = \{7, 8, 9, 10, 11\}$, can be represented without difficulty in roster-form or in set-builder notation. A set of unrelated objects whose members do not fit a pattern, such as $\{\frac{1}{2}$, ♫, W, 13, $\Delta\}$, cannot be represented in set-builder notation. Set-builder notation is especially useful when describing sets containing a large number of elements that are related by a rule. The members of the set of negative integers are related by the rule that each integer in the set is less than 0. Thus, $\{x \mid x < 0$, where x is an integer$\}$ represents the set of all *negative* integers.

Check Your Understanding of Section 1.3

A. *Multiple Choice.*

1. Which set could represent the set of nonnegative real numbers?
(1) $\{x \mid x > 0\}$ (2) $\{x \mid x \ge 0\}$ (3) $\{x \mid x < 0\}$ (4) $\{x \mid x \le 0\}$

2. Which of the following is an infinite set?
(1) $\{x \mid x$ is a United States Senator$\}$
(2) $\{x \mid x$ is an odd integer$\}$
(3) $\{x \mid x$ is a positive integer less than 1 million$\}$
(4) $\{x \mid x$ is a human being born in 1982$\}$

3. If $C = \{y \mid y < -2\}$and y is an integer, which set is a subset of C?
(1) $\{-2, -3\}$ (2) $\{-2, -1\}$ (3) $\{-3\}$ (4) $\{0\}$

4. The set $\{1, 2, 3, 4\}$ is equivalent to
(1) $\{x \mid 1 < x < 4$, where x is a whole number$\}$
(2) $\{x \mid 0 < x < 4$, where x is a whole number$\}$
(3) $\{x \mid 0 < x \le 4$, where x is a whole number$\}$
(4) $\{x \mid 1 < x \le 4$, where x is a whole number$\}$

5. Consider a set U whose members are those integers greater than –3 and less than 7. A subset of set U is the set of positive factors of 6. What is the complement of this subset within the universe of set U?
(1) $\{0, 4, 5\}$
(2) $\{-2, -1, 0, 4, 5\}$
(3) $\{0, 4, 5, 6\}$
(4) $\{-2, -1, 4, 5\}$

6. Given: $Q = \{0, 2, 4, 6\}$, $W = \{0, 1, 2, 3\}$, and $Z = \{1, 2, 3, 4\}$. What is the intersection of sets Q, W, and Z?
(1) $\{2\}$
(2) $\{0, 2\}$
(3) $\{1, 2, 3\}$
(4) $\{0, 1, 2, 3, 4, 6\}$

B. *Show how you arrived at your answer.*

7–9. *Represent the members of each set using the roster method.*

7. $\{x \mid 4 < x \leq 9$, where x is an integer$\}$

8. $\{y \mid -3 \leq y \leq 2$, where y is an integer$\}$

9. $\{x : x = 2n + 1$, where n is a whole number less than $10\}$

10. Given: $U = \{x \mid -3 \leq x < 4$, where x is an integer$\}$ and $A = \{-1, 0, 1\}$. What is the complement of A within the universe of set U?

11. Consider a set U whose members are the positive factors of 12 and a set A whose members are the positive factors of 4. What is the complement of set A within the universe of set U?

12. Given set U whose members are those positive integers that are at most 20. A subset of U are those prime numbers that are less than 20. What is the complement of this subset within the universe of set U?

13. If $A = \{2, 6, 8\}$, $B = \{2, 3, 5, 8\}$, and $C = \{7, 8, 9\}$. Find the intersection of sets A, B, and C.

14. If given: $S = \{n \mid 30 \leq n \leq 50$; n is a prime number$\}$. Rewrite S in roster-form.

15. Given: The members of set A are those odd positive integers less than 20, the members of set B are the positive factors of 15, and the members of set C are the prime numbers less than 10.
a. What is the union of sets B and C?
b. What is the intersection of sets B and C?
c. What is the complement of set B within the universe of set A?

16. The accompanying diagram shows the results of a survey asking which sports the members of the Key Club watch on television.

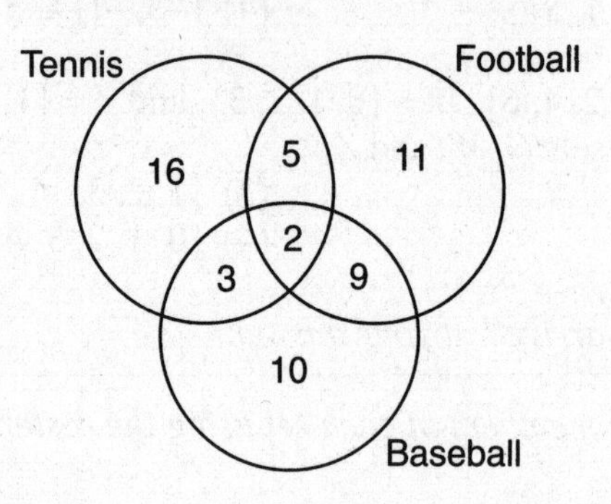

Based on the diagram, how many club members surveyed watch:

a. baseball?
b. both tennis and football?
c. exactly one of the three sports?
d. all three sports?

17. The accompanying diagram shows the elements of sets A, B, and C.

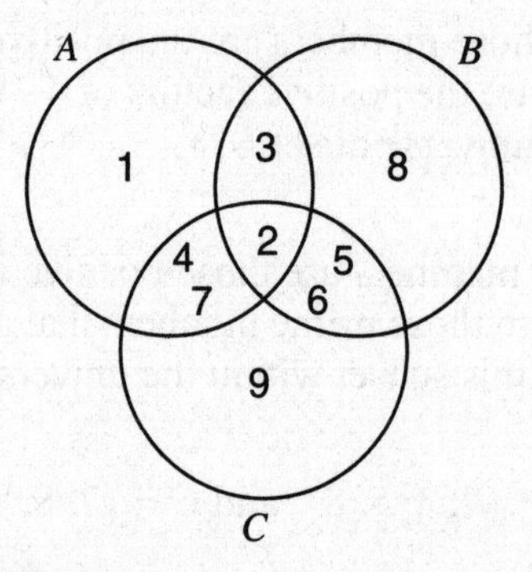

a. Write the set that represents the intersection of sets A and B.
b. Write the set that represents the intersection of sets B and C.
c. Write the set that represents the intersection of sets A, B, and C.
d. If set U is the set of all positive integers that are at most 10, what is the complement of set C within the universe of set U?

1.4 OPERATIONS WITH SIGNED NUMBERS

KEY IDEAS

The **absolute value** of a real number refers to its magnitude or numerical value without regard to its sign. Special rules are followed when performing arithmetic operations with signed numbers.

Absolute Value

The **absolute value** of a real number x is written as $|x|$. The absolute value of a real number is never negative as it depends only on its *distance* from 0 and not its *direction* from 0.

- If x is nonnegative, $|x|$ is simply x. Thus, $|+2| = 2$ and $|0| = 0$.
- If x is negative, $|x|$ is the opposite of x. Thus, $|-2| = 2$.

Multiplying and Dividing Signed Numbers

To multiply or divide two signed numbers, first multiply or divide the numbers while ignoring their signs.

- If the two numbers have the same sign, the answer is positive:

$$(+2)\times(+7) = +14 \qquad \frac{+24}{+3} = +8$$

and

$$(-2)\times(-7) = +14 \qquad \frac{-24}{-3} = +8$$

- If the two numbers have opposite signs, the answer is negative:

$$(-2)\times(+7) = -14 \text{ and } \frac{-24}{+3} = -8$$

Adding Signed Numbers

To add two signed numbers, follow these rules:

- If the two numbers have the *same* sign, *add* the numbers without their signs. Write the sum using their common sign, as in

$$(+2) + (+7) = +9 \quad \text{and} \quad (-2) + (-7) = -9$$

- If the two numbers have *different* signs, *subtract* the numbers while ignoring their signs. Write the difference with the sign of the number having the larger absolute value, as in

$$(-2) + (+7) = +5 \quad \text{and} \quad (+2) + (-7) = -5$$

Subtracting Signed Numbers

To subtract signed numbers, change the sign of the number that is being subtracted to its opposite and then add, as in

$$(+2) - (-7) = (+2) + (+7) = +9 \quad \text{and} \quad (+2) - (+7) = (+2) + (-7) = -5$$

Example 1

Find the pair of integers whose product is -20 and whose sum is $+8$.

Solution: Make a chart that includes all pairs of integers whose product is -20. For each such pair of integers, determine if their sum is $+8$.

Product = −20	**Sum**
−1 and +20	+19
+1 and −20	−19
−4 and +5	+1
+4 and −5	−1
−2 and +10	+8 ✓
+2 and −10	−8

The two integers are **−2** and **+10**.

Example 2

Simplify: a. $-3|2-6|$ b. $1-|-3|$ c. $\dfrac{4|-21|}{-6}$

Solution:

a. $-3|2-6| = -3|-4|$
$= -3 \times 4$
$= -12$

b. $1-|-3| = 1-3$
$= -2$

c. $\dfrac{4|-21|}{-6} = \dfrac{4 \times 21}{-6}$
$= \dfrac{84}{-6}$
$= -14$

Evaluating Algebraic Expressions

An algebraic expression can be evaluated when each of its variables is given a numerical value.

Example 3

Evaluate each expression when $m = -8$ and $n = +4$.

a. $\frac{m-n}{m+n}$ b. $-2(n - m)$

Solution: Replace each variable with its assigned value, and then perform the indicated operations.

a. $\frac{m-n}{m+n} = \frac{-8-4}{-8+4}$

$= \frac{-12}{-4}$

$= \mathbf{+3}$

b. $-2(n - m) = -2(+4 - (-8))$

$= -2(+4 + 8)$

$= -2(+12)$

$= \mathbf{-24}$

Check Your Understanding of Section 1.4

A. *Multiple Choice.*

1. Which statement is always true?
(1) $-2 < -1$ (2) $|-2| < |-1|$ (3) $|-2| < 1$ (4) $|-1| < 0 < |-2|$

2. If $-5 < |x| < 2$, what is a possible value of x?
(1) -6 (2) -4 (3) 3 (4) -1

3. What is the value of $1 - |-5 + 2|$?
(1) -2 (2) 2 (3) -4 (4) 4

B. *Show how you arrived at your answer.*

4–7. *Find each sum.*

4. $(-6) + (-1)$ **5.** $-3 + 8$ **6.** $-6.4 + (+3.9)$ **7.** $\left(+\frac{1}{2}\right) + \left(-\frac{1}{3}\right)$

8–11. *Find each difference.*

8. $(-7) - (-3)$ **9.** $(-2) - (-6)$ **10.** $5 - 7$ **11.** $-(-8) - (-5)$

12–23. Perform the indicated operation(s).

12. $(-4)(-3)$

13. $(-5)(+4)(-3)$

14. $\frac{-21}{+3}+5$

15. $\frac{-18}{-6}-4$

16. $\frac{2.4}{-0.6}$

17. $(+1.5)(-0.3)$

18. $\left(-\frac{3}{5}\right)\left(-\frac{10}{3}\right)$

19. $\frac{(-14)(-9)}{-6}$

20. $\frac{-18-(-6)}{-1+5}$

21. $\frac{(-14)-(-2)}{-3}$

22. $\left(\frac{-56}{+8}\right)+\left(\frac{-21}{-7}\right)$

23. $\left(\frac{-45}{+5}\right)-\left(\frac{+12}{-3}\right)$

24. Represent $\{x \mid x+3=-1;\ x \text{ is an integer}\}$ in roster-form.

25. The product of two integers is −40. If the same two integers are added together, the sum is −3. What is the *smaller* of the two integers?

26. The product of two integers is 24. If the same two integers are added together, the sum is −25. What is the *smaller* of the two integers?

27. The product of two integers is −56. If the same two integers are added together, the sum is 1. What is the *larger* of the two integers?

***28–31.** Find the value of each algebraic expression when $x = 12$ and $y = -3$.*

28. $\frac{1}{2}xy$

29. $2(x-y)$

30. $\frac{y-x}{2-y}$

31. $\frac{xy}{x+y}$

1.5 PROPERTIES OF REAL NUMBERS

KEY IDEAS

Real numbers behave in predictable ways. For example, the order in which any two real numbers are added or multiplied does not matter.

Properties of Addition and Multiplication

In order to be able to work with real numbers, basic assumptions are made about how real numbers behave.

- **Closure Property**. A set of numbers is **closed** under an operation when the result of performing that operation is a number in the same set. The set of real numbers is *closed* under the operations of addition and multiplication. The set of integers is *not* closed under division since the quotient of two integers is not necessarily equal to another integer, as in $3 \div 4$.
- **Commutative Properties**. An operation with two numbers is **commutative** if the order in which the two numbers are taken does not matter. Addition and multiplication of real numbers are commutative operations. For instance, $2 + 3 = 3 + 2 = 5$ and $2 \cdot 3 = 3 \cdot 2 = 6$. Division and subtraction are *not* commutative operations. For example, $3 \div 4$, is not the same as $4 \div 3$, and $4 - 3$ is different than $3 - 4$.
- **Associative Properties**. When finding the sum or product of three numbers, only two numbers can be operated on at a time. Addition and multiplication of real numbers are associative operations because it does not matter which two of the three numbers are grouped together and operated on first, as illustrated by:

$$(2 + 3) + 4 = 5 + 4 = 9 \text{ and } 2 + (3 + 4) = 2 + 7 = 9$$

$$(2 \times 3) \times 4 = 6 \times 4 = 24 \text{ and } 2 \times (3 \times 4) = 2 \times 12 = 24$$

- **Identity Properties**. Because the sum of any real number and 0 is that same real number, 0 is called the **identity element for addition**. For example, $4 + 0 = 4$ and $0 + 4 = 4$. The product of any real number and 1 is always that same real number, as in $4 \times 1 = 4$ and $1 \times 4 = 4$. Hence, 1 is the **identity element for multiplication**.
- **Inverse Properties**. Each real number has an opposite, called its **additive inverse,** such that their sum is 0, the identity element for addition. For example, the additive inverse of $+4$ is -4 since $(+4) + (-4) = 0$. Each nonzero real number a has a reciprocal $\frac{1}{a}$, called its **multiplicative inverse,** such that their product is 1, the identity element for multiplication. For example, the multiplicative inverse of 4 is its reciprocal, $\frac{1}{4}$, because $4 \times \frac{1}{4} = 1$.
- **Distributive Property**. The distributive property of multiplication over addition states that multiplying a number by a sum, as in $3(2 + 4) = 3 \times 6 = 18$, is the same as multiplying each addend separately by that number and then adding the two products:

$$3(2 + 4) = (3 \times 2) + (3 \times 4) = 6 + 12 = 18$$

Example 1

Remove the parentheses by applying the distributive property:

a. $5(x + 2)$ b. $6(2y - 3)$ c. $-2(x - 4)$

Solution:

a. $5(x + 2) = 5 \cdot x + 5 \cdot 2 = \mathbf{5x + 10}$

b. $6(2y - 3) = 6(2y + (-3)) = 6(2y) + 6(-3) = \mathbf{12y - 18}$

c. $-2(x - 4) = -2(x + (-4)) = -2x + (-2)(-4) = \mathbf{-2x + 8}$

Combining Like Terms

Like terms are terms such as $4x$ and $5x$ that differ only in their numerical coefficients. To add or subtract like terms, simply combine their numerical coefficients:

- $4x + 5x = 9x$
- $9y - 7y = 2y$
- $7b - b = 7b - 1b = 6b$
- $2xy + 3xy + 4xy = 9xy$

Example 2

Simplify: $2(5x + 3) - 9(x + 1)$

Solution: Remove the parentheses using the distributive property. Then combine like terms.

Apply the distributive property: $2(5x + 3) - 9(x + 1) = 10x + 6 - 9x - 9$

Group like terms together: $= (10x - 9x) + (6 - 9)$

Combine like terms: $= \mathbf{x - 3}$

Check Your Understanding of Section 1.5

A. Multiple Choice.

1. Which set of numbers is *not* closed under multiplication?

(1) {odd integers}
(2) {even integers}
(3) {prime numbers}
(4) {rational numbers}

2. The statement $2x + 2y = 2(x + y)$ uses which of the following properties of real numbers?
(1) commutative property
(2) associative property
(3) distributive property
(4) closure property

3. Which set has the property of closure under the operation of addition?
(1) $\{0, 1, 2\}$
(2) $\{x \mid x \text{ is an odd integer}\}$
(3) $\{0, 1, 2, 3, \ldots, 10\}$
(4) $\{x \mid x \text{ is an even integer}\}$

4. Which property is illustrated by the equation $\heartsuit + (\square + 0) = (\heartsuit + \square) + 0$?
(1) distributive
(2) associative for addition
(3) commutative for addition
(4) additive inverse

5. The statement $(x + y) + z = z + (x + y)$ illustrates which property of real numbers?
(1) commutative (2) associative (3) distributive (4) closure

6. Which equation illustrates the additive inverse property for real numbers?
(1) $a + (-a) = 0$ (2) $a + 0 = a$ (3) $a + (-a) = -1$ (4) $a \cdot \frac{1}{a} = 1$

7. Which equation illustrates the multiplicative identity property for real numbers?
(1) $x + 0 = x$ (2) $x \cdot 1 = x$ (3) $x \cdot \frac{1}{x} = 1$ (4) $x \cdot 0 = 0$

8. Which equation illustrates the multiplicative inverse property?
(1) $b \cdot 0 = 0$ (2) $b + (-b) = 0$ (3) $b + 0 = b$ (4) $b \cdot \frac{1}{b} = 1$

9. Which set of numbers is closed under subtraction?
(1) odd integers
(2) rational
(3) counting numbers
(4) prime numbers

10. What is the additive inverse of $-4a$?
(1) $\frac{a}{4}$ (2) $4a$ (3) $-\frac{4}{a}$ (4) $-\frac{1}{4a}$

11. What is the multiplicative inverse of $\frac{x}{2}$ $(x \neq 0)$?
(1) 1 (2) $\frac{2}{x}$ (3) $-\frac{x}{2}$ (4) $2x$

12. To add $\frac{1}{2}+\frac{1}{3}$, the first step is to change each fraction into an equivalent fraction with the least common denominator as its denominator by writing

$$\frac{1}{2}+\frac{1}{3}=\frac{1}{2}\times\left(\frac{3}{3}\right)+\frac{1}{3}\times\left(\frac{2}{2}\right)$$

This step can be justified by which of the following properties of real numbers?
(1) commutative property
(2) existence of an additive identity
(3) existence of a multiplicative identity
(4) distributive property

B. *Show how you arrived at your answer.*

13–18. *Simplify.*

13. $3(x+1)-4x$ **15.** $5(x-2)+9(x+1)$ **17.** $7-4(x-1)$

14. $2(n-3)+7$ **16.** $2(3-4y)+5(3y-2)$ **18.** $8(m+1)-(3-m)$

1.6 EXPONENTS AND SCIENTIFIC NOTATION

KEY IDEAS

An **exponent** tells how many times a number, called the **base**, is repeated in a product. For example, $3\times3\times3\times3\times3=3^5$, where 3 is the base and 5 is the exponent. When an exponent is not written, assume it is 1. Thus, $4=4^1$ and $x=x^1$.

Powers of Bases

A base with its exponent is called a **power** of that base. Thus, 3^5 is read, "3 raised to the fifth power." If an exponent is 2, as in x^2, the expression is read, "x raised to the second power," the square of x," or "x squared." Similarly, if an exponent is 3, as in x^3, the expression is read, "x raised to the third power," "the cube of x," or "x cubed."

Example 1

Evaluate: a. $(-4)^2$ b. -4^2 c. $3^2 \cdot (-2)^3$

Solution:

a. $(-4)^2 = (-4) \times (-4)$
$= \mathbf{16}$

b. $-4^2 = -(4^2)$
$= -(4 \times 4)$
$= \mathbf{-16}$

c. $3^2 \cdot (-2)^3 = \underbrace{(3)(3)}_{9}\underbrace{(-2)(-2)(-2)}_{-8}$
$= (9) \times (-8)$
$= \mathbf{-72}$

Positive Integer Exponents

The accompanying table reviews the rules for working with positive integers exponents.

Exponent Law	Rule	Example
Product	To multiply powers of the same base, *add* the exponents: $x^a \cdot x^b = x^{a+b}$	$n^5 \cdot n^2 = n^{5+2} = n^7$
Quotient	To divide powers of the same base, *subtract* the exponents: $\dfrac{x^a}{x^b} = x^{a-b}$	$\dfrac{n^5}{n^2} = n^{5-2} = n^3$
Power of a Power	To raise a power to another power, *multiply* the exponents: $(x^a)^b = x^{ab}$	$(n^2)^5 = n^{2\times 5} = n^{10}$

Example 2

The expression $2^4 \cdot 4^3$ is equivalent to

(1) 8^{12} (2) 8^7 (3) 2^{10} (4) 2^9

Solution:

Rewrite 4^3 as a power of 2: $2^4 \cdot 4^3 = 2^4 \cdot (2^2)^3$
Raise a power to a power: $= 2^4 \cdot 2^6$
Multiply by adding exponents: $= 2^{10}$
The correct choice is **(3)**.

Example 3

Simplify $(x^4y^2)^3$.

Solution: $(x^4y^2)^3 = (x^4)^3 \cdot (y^2)^3 = \mathbf{x^{12}y^6}$

Zero Exponent

The quotient $\frac{2^3}{2^3}$ can be evaluated in two different ways:

- Use the quotient law of exponents,

$$\frac{2^3}{2^3} = 2^{3-3} = 2^0$$

- Because any nonzero quantity divided by itself is 1, $\frac{2^3}{2^3} = 1$.

Since the two answers must be equivalent, $2^0 = 1$. In general, any nonzero quantity raised to the 0 power is 1.

Example 4

If $y \neq 0$, evaluate: **a.** $2y^0$ **b.** $(2y)^0$

Solution:

a. $2y^0 = 2 \times y^0 = 2 \times 1 = \mathbf{2}$.
b. Because any nonzero quantity raised to the 0 power is 1, $(2y)^0 = \mathbf{1}$.

Negative Integer Exponents

The quotient $\frac{3^4}{3^6}$ can be evaluated in two different ways:

- Use the quotient law of exponents:

$$\frac{3^4}{3^6} = 3^{4-6} = 3^{-2}$$

- Expand the powers in the numerator and denominator and then simplify:

$$\frac{3^4}{3^6} = \frac{\cancel{3 \times 3 \times 3 \times 3}}{\cancel{3 \times 3 \times 3 \times 3} \times 3 \times 3} = \frac{1}{3^2}$$

Because the two answers must be the equivalent, $3^{-2} = \frac{1}{3^2}$, which means that 3^{-2} and 3^2 are reciprocals.

Here are a few more examples:

$$2m^{-5} = \frac{2}{m^5},\ \frac{x}{y^{-3}} = xy^3, \text{ and } \left(\frac{2x}{3y}\right)^{-4} = \frac{(3y)^4}{(2x)^4} = \frac{81y^4}{16x^4}$$

Example 5

What is the value of $(b - b^0)^{a+1}$ when $a = -4$ and $b = 3$?

(1) $\frac{1}{4}$ (2) $\frac{1}{8}$ (3) 8 (4) 4

Solution: Since $b^0 = 1$, $(b - b^0)^{a+1} = (b - 1)^{a+1}$. Substitute 3 for b and -4 for a:

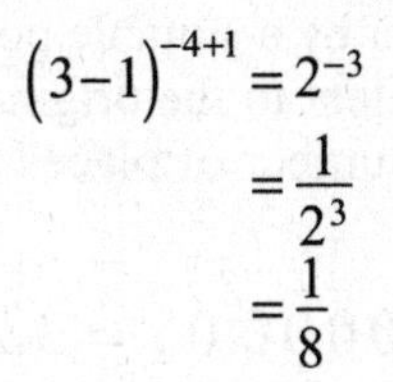

$$(3-1)^{-4+1} = 2^{-3} = \frac{1}{2^3} = \frac{1}{8}$$

The correct choice is **(2)**.

MATH FACTS

Zero and Negative Exponent Laws

- Any nonzero base raised to the 0 power is 1:

$$x^0 = 1, \text{ provided } x \neq 0$$

- For any nonzero base x with an integer exponent n, x^{-n} and x^n are reciprocals:

$$x^{-n} = \frac{1}{x^n} \text{ and } \frac{1}{x^{-n}} = x^n$$

Using Technology: Evaluating Powers

To use a graphing calculator to find 3^7, press

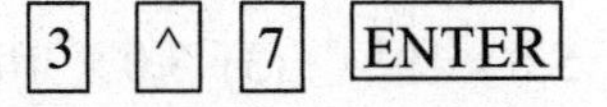

The calculator display should read 2187. Pressing the exponent key ^ tells the calculator to treat the number entered immediately before it as the base and the number entered immediately after it as the exponent of that base.

Scientific Notation

Very large and very small numbers are sometimes easier to read and to work with if they are expressed in scientific notation. When written in scientific notation, 32,000,000 becomes 3.2×10^7 and 0.0000608 becomes 6.08×10^{-5}. A number is in scientific notation when it is written as a number greater than or equal to 1 but less than 10 that is times a power of 10. To write a number in scientific notation, follow these steps:

STEP 1 Rewrite the digits of the original number but with its decimal point placed to the right of the first digit so that the resulting number is greater than or equal to 1 but less than 10. For example, rewrite 32,000,000 without any trailing zeros as 3.2.

STEP 2 Multiply this number by a suitable power of 10 so that the resulting expression is equivalent to the original number. The power of 10 is determined by the number of places the decimal point gets shifted in step 1:

$$32000000. = 3.2 \times 10^7$$

7 places

$$0.0000608 = 6.08 \times 10^{-5}$$

5 places

Notice that the power of 10 is positive if the original number is greater than 10 and is negative if the original number is less than 1.

You should always confirm that the scientific notation representation of a number is equivalent to the original number by actually multiplying the decimal number by the power of 10. The decimal point gets shifted to the right when the exponent of 10 is positive and shifted to the left if the exponent of 10 is negative.

Example 6

What is the quotient of 8.4×10^8 and 2.8×10^2 expressed in scientific notation?

(1) 3.0×10^4 (2) 5.6×10^4 (3) 3.0×10^6 (4) 5.6×10^6

Solution: Divide the numerical coefficients, and then divide the powers of 10 by subtracting their exponents:

$$8.4 \times 10^8 \div 2.8 \times 10^2 = \frac{8.4 \times 10^8}{2.8 \times 10^2}$$
$$= \frac{8.4}{2.8} \times 10^{8-2}$$
$$= 3 \times 10^6$$

The correct choice is **(3)**.

Example 7

What is the product of 8.4×10^3, 5.0×10^{-2}, and 7.5×10^6 expressed in scientific notation?

(1) 315×10^7 (2) 3.15×10^9 (3) 315×10^8 (4) 3.15×10^{11}

Solution: Multiply the numerical coefficients, and multiply the powers of 10 by adding their exponents. Write the product in scientific notation:

$$(8.4 \times 10^3) \times (5.0 \times 10^{-2}) \times (7.5 \times 10^6) = (8.4 \times 5.0 \times 7.5) \times (10^3 \times 10^{-2} \times 10^6)$$
$$= (315) \times (10^{3-2+6})$$
$$= (315) \times (10^7)$$
$$= (3.15 \times 10^2) \times (10^7)$$
$$= 3.15 \times (10^2 \times 10^7)$$
$$= 3.15 \times 10^9$$

The correct choice is **(2)**.

Check Your Understanding of Section 1.6

A. Multiple Choice.

1. If $0.0000603 = 6.03 \times 10^n$, what is the value of n?
(1) -5 (2) 5 (3) -4 (4) 4

2. If $x = 2^5$ and $y = 4^3$, then $\frac{x}{y} =$
(1) $\frac{1}{2}$ (2) 2 (3) 8 (4) 4

3. What is the value of $3^0 + 3^{-2}$?
(1) 0 (2) $\frac{1}{9}$ (3) $1\frac{1}{9}$ (4) 6

4. The distance from the Sun to the planet Neptune is about 2,790,000,000 miles. Expressed in scientific notation, this distance in miles is
(1) 2.79×10^9 (2) 2.79×10^{-9} (3) 27.9×10^7 (4) 27.9×10^{-7}

5. If $4^x = 64$, what is the value of x^4?
(1) $\frac{1}{64}$ (2) 12 (3) 16 (4) 81

6. Expressed in scientific notation, 0.003146 is equivalent to
(1) 31.46×10^4 (2) 3.146×10^{-3} (3) 31.46×10^3 (4) 3.146×10^{-2}

7. The distance from the Earth to the Sun is approximately 93 million miles. A scientist would write that number as
(1) 9.3×10^6 (2) 9.3×10^7 (3) 93×10^7 (4) 93×10^{10}

8. The expression $3^4 \cdot 9^3$ is equivalent to
(1) 3^{10} (2) 3^{24} (3) 27^7 (4) 27^{12}

9. Which expression is equivalent to 4.08×10^{20}?
(1) 0.0408×10^{18} (2) 40.8×10^{18} (3) 408×10^{18} (4) 4080×10^{18}

10. Which inequality is true if $x = \frac{3.04}{1.48}$, $y = 1.99 + 0.33$, and $z = (1.3)^3$?
(1) $y < z < x$ (2) $y < x < z$ (3) $x < z < y$ (4) $x < y < z$

11. Which expression is equivalent to 8.12×10^{-9}?
(1) 812×10^{-7} (2) 0.0812×10^{-7} (3) 812×10^{-11} (4) 0.0812×10^{-11}

12. What is the sum of 6.0×10^3 and 3.0×10^2?
(1) 6.3×10^3 (2) 9.0×10^5 (3) 9.0×10^6 (4) 3.6×10^3

13. If $10^k = \frac{1}{2}$, what is the value of 10^{k+3}?
(1) 1.5 (2) 3.5 (3) 500 (4) 1000

14. Which expression represents 72 kilometers per hour as meters per hour?
(1) 7.2×10^{-2} (2) 7.2×10^2 (3) 7.2×10^{-3} (4) 7.2×10^4

15. If the mass of a proton is 1.67×10^{-24} gram, what is the number of grams in the mass of 1,000 protons?
(1) 1.67×10^{-27} (2) 1.67×10^{-23} (3) 1.67×10^{-22} (4) 1.67×10^{-21}

16. A star constellation is approximately 3.1×10^4 light years from Earth. One light year is about 5.9×10^{12} miles. What is the approximate distance, in miles, between Earth and the star constellation?
(1) 1.83×10^{17} (2) 9.0×10^{49} (3) 1.9×10^8 (4) 9.0×10^{16}

B. *Show how you arrived at your answer.*

17–19. *Express each product in scientific notation.*

17. $(2.5 \times 10^3)(4.2 \times 10^8)$ **18.** $(3.1 \times 10^{-5})(4.7 \times 10^{-4})$ **19.** $(0.000050)^3$

20–22. *Express each quotient in scientific notation.*

20. $\dfrac{6.0\times10^{12}}{1.5\times10^4}$ **21.** $\dfrac{9.0\times10^3}{0.75\times10^8}$ **22.** $\dfrac{1.8\times10^{-4}}{2.4\times10^{-9}}$

23. What is the value of $2 \cdot 4^{-5} \cdot 4^8$?

24. Simplify $(-2x^2y^3)^5$.

25. What is the value of $(3n - 12)^0$ when $n \neq 4$?

26. If $y = 3(8^{x-2}) - x^{-2}$, what is the value of y when $x = 2$?

27. Rewrite the set $\{x : 2^x < 33; x \text{ is a positive integer}\}$ in roster-form.

28. What is the value of $\left(1+\frac{b}{a}\right)^{-2}$ when $a = -4$ and $b = 3$?

1.7 ORDER OF OPERATIONS

KEY IDEAS

Arithmetic operations, as in $8 + 5 \times 2$, are not necessarily performed from left to right in the order in which they appear. Instead, they are performed according to an accepted set of rules called the **order of operations**. Since multiplication and division are performed before addition and subtraction, $8 + 5 \times 2 = 8 + 10 = 18$. Parentheses can be used to group an arithmetic operation that is to be performed first, as in $(8 + 5) \times 2 = 13 \times 2 = 26$.

Rules for Order of Operations

Complete arithmetic operations in the following order:

1. Performing operations inside parentheses or other grouping symbols.
2. Raising to powers.
3. Multiplying and dividing from left to right.
4. Adding and subtracting, from left to right.

To evaluate $(11 + 3 \times 7) \div 2^3 - 1$:

- Evaluate inside parentheses: $(11 + 3 \times 7) \div 2^3 - 1 = 32 \div 2^3 - 1$
- Do powers: $= 32 \div 8 - 1$
- Multiply or divide, left to right: $= 4 - 1$
- Add or subtract, left to right: $= 3$

Example 1

Evaluate $7 - 2^4 \div 48 \times 6$.

Solution: Evaluate the power first. Then follow the order of operations, working from left to right:

$$7 - 2^4 \div 48 \times 6 = 7 - 16 \div 48 \times 6$$

Divide: $= 7 - \frac{1}{3} \times 6$

Multiply: $= 7 - 2$

Subtract: $= \mathbf{5}$

Evaluating Algebraic Expressions

The rules for order of operations may need to be applied when evaluating an algebraic expression.

Example 2

Find the value of $-x^2+\frac{1}{2}xy$ when $x=-3$ and $y=-10$.

Solution:

$$\begin{aligned}-x^2+\frac{1}{2}xy &= -(-3)^2+\frac{1}{2}(-3)(-10)\\ &= -9+(-1.5)(-10)\\ &= -9+15\\ &= \mathbf{6}\end{aligned}$$

Example 3

If $y = 5x^3 - 7x - 20$, find the value of y when $x = -2$.

Solution:

$$\begin{aligned}y &= 5x^3 - 7x - 20\\ &= 5(-2)^3 - 7(-2) - 20\\ &= 5(-8) + 14 - 20\\ &= -40 + 14 - 20\\ &= \mathbf{-46}\end{aligned}$$

Example 4

What is the value of the expression $(a^3 + b^0)^2$ when $a = -2$ and $b = 5$?

Solution: Replace a with -2 and b with 5:

$$((-2)^3 + 5^0)^2$$
$$(-8 + 1)^2$$
$$(-7)^2$$
$$49$$

The value of the given expression is 49.

Check Your Understanding of Section 1.7

A. Multiple Choice.

1. When $x = 2$ and $y = 0.5$, which expression has the greatest value?
(1) $x + y^2$ (2) $\frac{y}{x}$ (3) $\frac{x}{y} - 1$ (4) $xy + 1$

2. What is the first step in simplifying the expression $(2 - 3 \times 4 + 5)^2$?
(1) square 5
(2) add 4 and 5
(3) subtract 3 from 2
(4) multiply 3 by 4

3. What is the value of $5xy^2$ if $x = -2$ and $y = -3$?
(1) -90 (2) 90 (3) -180 (4) 180

4. What is the value of $-3a^2b$ if $a = -2$ and $b = -4$?
(1) -48 (2) 48 (3) -144 (4) 144

5. If $y = -\frac{1}{4}$ and $z = 8$, what is the value of $\frac{1}{2}yz^2$?
(1) 8 (2) 2 (3) -8 (4) 4

6. If $x = -2$, $y = 6$, and $z = -4$, what is the value of $\frac{x^3y}{z}$?
(1) -12 (2) 12 (3) -9 (4) 9

7. What is the value of $n^4 \div 3n$ when $n = 3$?
(1) 9 (2) 27 (3) 36 (4) 81

8. If $y = 2x^2 - 5x + 6$, what is the value of y when $x = -2$?
(1) 32 (2) -24 (3) 24 (4) 4

***B.** Show how you arrived at your answer.*

9. What is the value of the expression $(a^3 - b^0)^2$ when $a = -2$ and $b = 3$?

10. What is the value of the expression $b^2 - |a - b|$ when $a = 7$ and $b = -3$?

11. What is the value of the expression $2x^0 - xy^3$ when $x = \frac{1}{2}$ and $y = -2$?

1.8 TRANSLATING BETWEEN ENGLISH AND ALGEBRA

KEY IDEAS

To use algebra to represent or "model" problem situations, you must know how to translate commonly encountered verbal phrases and sentences into algebraic expressions and equations.

Translating Algebraic Expressions

When translating an algebraic expression into a verbal phrase, pay attention to the arithmetic operation that connects the variable to one or more numbers. For example:

- $n + 5$ represents the *sum* of n and 5, or equivalently, 5 more than n.
- $n - 3$ represents 3 *subtracted* from n, 3 less than n, or n diminished by 3.
- $\frac{n}{2}$ represents n *divided* by 2 or one-half of n.
- $4n + 1$ represents 4 *times n increased* by 1 or, equivalently, 1 more than the product of 4 and n.
- $2(n + 3) + 1$ represents 2 times the sum of n and 3 increased by 1.

Representing One Quantity in Terms of Another

The same variable can be used when comparing two variable quantities.

- Tim's weight exceeds Sue's weight by 13 pounds. If Sue weighs x pounds, then Tim weighs $x + 13$ pounds.
- The number of dimes exceeds 3 times the number of pennies by 2. If there are x pennies, then there are $3x + 2$ dimes.
- Bill has 7 fewer dollars in his wallet than three times the number Roger has in his wallet. If Roger has x dollars, then Bill has $3x - 7$ dollars.
- The sum of two numbers is 54. If x represents one of these numbers, then $54 - x$ represents the other number.
- Victor is twice John's age 8 years ago. If y is John's current age, then $y - 8$ represents John's age 8 years ago so Victor's current age is $2(y - 8)$.

Translating Verbal Sentences

An **inequality** is a mathematical statement that two quantities are not equal, as in $2x + 1 < 7$. To translate a verbal sentence into an equation or an inequality, identify key phrases that can be translated directly into mathematical terms:

- $\underbrace{\text{Two times a number } n}_{2n}\ \underbrace{\text{increased by 5}}_{+5}\ \underbrace{\text{is 17}}_{=17}$.

  The equation is $2n + 5 = 17$.

- $\underbrace{\text{Five times a number } n}_{5n}\ \underbrace{\text{exceeds}}_{=}\ \underbrace{\text{2 times that number}}_{2n}\ \underbrace{\text{by 21}}_{+21}$.

  The equation is $5n = 2n + 21$.

- $\underbrace{\text{When 2 is } \textit{subtracted from} \text{ a number } n}_{n-2}\ \underbrace{\text{the result is } \textit{at most}}_{\leq}\ \underbrace{5}_{5}$.

  The inequality is $n - 2 \leq 5$.

- $\underbrace{\text{When twice a number } x}_{2x}\ \underbrace{\textit{is increased by } 5}_{+5}\ \underbrace{\text{the result is } \textit{at least}}_{\geq}\ \underbrace{27}_{27}$.

  The inequality is $2x + 5 \geq 27$.

Example 1

Marla sells jewelry on a web site that charges a flat fee of \$42 a month. Each piece of jewelry costs \$18 to make. Marla sells each piece of jewelry for \$25. Which inequality represents the number of pieces of jewelry, j, Marla must sell in a month to make a profit of at least \$150 for that month?

(1) $43j \leq 150$ (2) $43j \geq 150$ (3) $7j - 42 \geq 150$ (4) $7j + 42 \leq 150$

Solution: Use the relationship, monthly profit = total money earned − total costs.

- If Marla sells j pieces of jewelry in a month, she earns a total of $25j$ dollars.
- It costs Marla $18j$ dollars to make j pieces of jewelry. The monthly charge for the web site is \$42, so Marla's total monthly costs are $18j + 42$.
- Marla's monthly profit is $25j - (18j + 42) = 7j - 42$. If Marla's profit for the month is *at least* \$150, then $7j - 42 \geq 150$.

The correct choice is **(3)**.

Check Your Understanding of Section 1.8

A. Multiple Choice.

1. This year Danielle is three times as old as Amy. If x represents Amy's age now, which expression represents Danielle's age two years ago?
(1) $2x - 3$ (2) $3x - 2$ (3) $\frac{1}{3}x - 2$ (4) $3x + 2$

2. Lauren is twice as old as Aisha will be one year from now. If x represents Aisha's present age now, which expression represents Lauren's present age?
(1) $2(x - 1)$ (2) $2(x + 1)$ (3) $2x + 1$ (4) $2x - 1$

3. The product of 4 and n diminished by 1 is less than or equal to n divided by 2. Which inequality represents this relationship?
(1) $4(n-1) \leq \frac{n}{2}$ (2) $4(n-1) \geq \frac{n}{2}$ (3) $4n-1 \leq \frac{n}{2}$ (4) $4n-1 \geq \frac{n}{2}$

4. Jean's car has a 16-gallon gas tank. Before filling her tank with gas, the gas gauge of her car indicates that g gallons of gas are in the tank. If gas costs $\$p$ per gallon and Jean fills the tank to capacity, which expression represents how much Jean paid for the gas?
(1) $p(16 - g)$ (2) $pg - 16$ (3) $16g - p$ (4) $(p - g)16$

5. Tara buys three items that cost d dollars each. When she pays the cashier for the items, she receives \$8 in change. Which expression represents the amount of money Tara gave the cashier?
(1) $d + 8$ (2) $3d + 8$ (3) $3(8 - d)$ (4) $3d - 8$

6. Chris rents a booth at a flea market for \$120 for one day. At the flea market Chris sells picture frames each of which costs him \$5.00. If Chris sells each picture frame for \$11, write an inequality that represents the number of picture frames, p, he must sell to make a profit of at least \$275 for that day.
(1) $6p + 120 \leq 275$
(2) $6p \geq 275$
(3) $11p - 120 \geq 275$
(4) $6p - 120 \geq 275$

7. A store advertises that during its Labor Day sale \$15 will be deducted from every purchase over \$100. Furthermore, after the deduction is taken, the store offers an early-bird discount of 20% to any person who makes the purchase before 10:00 A.M. If Hakeem makes a purchase of x dollars where $x > 100$, at 8:00 A.M., what is the cost of Hakeem's purchase?
(1) $0.20x - 15$ (2) $0.20x - 3$ (3) $0.85x - 20$ (4) $0.80x - 12$

8. A certain type of hat costs \$4.50 to manufacture. Which inequality represents the number of dollars, x, the manufacturer must charge for each hat in order to make a profit of at least \$1,700 when h hats are sold?
(1) $(x - 4.50)h \geq 1{,}700$
(2) $(h - 4.50)\,x \geq 1{,}700$
(3) $hx - 4.50 \leq 1{,}700$
(4) $(h - 4.50)x \leq 1{,}700$

9. Kim wants to earn at least \$175 for a charity by participating in a walkathon. She will earn \$6 for each mile she walks plus \$5 from each person who pledges to support her walk. If 20 people have pledged to support her walk and m represents the number of miles she will walk, which inequality represents the amount of money Kim wants to earn?
(1) $100 + 6m \geq 175$
(2) $120 + 5m \geq 175$
(3) $100 + 6m \leq 175$
(4) $120 + 5m \leq 175$

10. A telephone call costs c cents for the first 3 minutes and m cents for each additional minute. What is the cost, in terms of c and m, of an 8-minute call?
(1) $c + 5m$ (2) $c + 8m$ (3) $3c + 5m$ (4) $3c + 8m$

B. *Represent the given information as an algebraic equation or inequality using x to represent the unknown quantity.*

11. When Jennifer celebrates her 18-year-old birthday, she will be twice as old as Mara was 5 years before that day.

12. Twice an unknown number diminished by 4 is less than 11.

13. In six years, three times Nicole's current age will be, at most, 39.

14. Twice the sum of a number and 3 is at least 12.

15. When 2 is subtracted from a number, the difference exceeds one-half of the original number of 1.

16. If 5 less than a number is multiplied by 2, the product is at most 24.

17. In 7 years Hector will be 3 years less than two times his current age.

CHAPTER 2

LINEAR EQUATIONS AND INEQUALITIES

2.1 SOLVING ONE-STEP EQUATIONS

KEY IDEAS

A **linear** or **first-degree equation** is an equation such as $4x - 3y = 12$ in which no variable has an exponent other than 1. The equation $2x + 1 = 7$ is an example of a linear equation in one variable. Some one-variable linear equations such as $x + 8 = 3$ can be solved in one step using the properties of equality.

Open Sentences and Solutions

The equation $2x + 1 = 7$ is an **open sentence** because its truth cannot be determined until the variable is replaced by a specific number. This equation is true when x is replaced by 3 since $2 \cdot 3 + 1$ evaluates to 7. Thus, 3 is a *root* or *solution* of the equation.

- A **root** or **solution** of a one-variable equation is a number that when substituted for that variable makes the equation a true statement.
- The **solution set** of a one-variable equation is the set of all of its solutions. The solution set is a subset of the replacement set of the variable. Unless otherwise indicated, the replacement set of a variable is assumed to be the set of real numbers or the largest possible subset of it.
- **Solving an equation** is the process of finding its solution set.

Equivalent Equations

The equations $2x + 1 = 7$, $x - 1 = 2$, and $2x = 6$ are *equivalent* since $\{3\}$ is the solution set for each equation. **Equivalent equations** are equations that have the same solution set. Solving a linear equation in one variable generally involves transforming the equation into a series of one or more equivalent equations until the variable is isolated on one side with the final equation having the form, *variable* = number.

Properties of Equality

An equation can be changed into an equivalent equation by

- adding the same quantity to, or subtracting the same quantity from, both sides.
- multiplying or dividing both sides by the same nonzero quantity.
- interchanging the two sides, as when $4 = x$ is rewritten as $x = 4$.

A one-variable linear equation is typically solved by using the properties of equality to transform the equation into a simpler one in which the variable equals a number.

Isolating a Variable in One Step

An equation in which the variable is involved in only one arithmetic operation can be solved in one step by performing the *inverse* of that arithmetic operation on *both* sides of the equation. **Inverse operations** are operations that undo each other. Addition and subtraction are inverse operations as are multiplication and division. Performing the same inverse operation on both sides of the equation insures that the equality or "balance" of the two sides is not disturbed.

- To solve $x - 7 = -5$, isolate x by *adding* 7 to both sides of the equation:

$$\begin{array}{r} x-7=-5 \\ +7=+7 \\ \hline x+0=2 \end{array} \quad \text{so} \quad x=2$$

- To solve $x + 8 = 3$, isolate x by *subtracting* 8 from both sides of the equation:

$$\begin{array}{r} x+8=3 \\ -8=-8 \\ \hline x+0=-5 \end{array} \quad \text{so} \quad x=-5$$

- To solve $\frac{p}{4} = -1.5$, isolate p by multiplying both sides of the equation by 4:

$$\frac{p}{4}=-1.5$$

$$\not{4}\times\left(\frac{p}{\not{4}}\right)=4\times(-1.5)$$

$$1\cdot p=-6.0$$

$$p=-6.0$$

- To solve $-3y = -21$, isolate y by dividing both sides of the equation by -3:

$$-3y = -21$$
$$\frac{-3y}{-3} = \frac{-21}{-3}$$
$$1 \cdot y = 7$$
$$y = 7$$

Example 1

Find the solution set for $b + 1.7 = 5.9$.

Solution: Isolate b by subtracting 1.7 from both sides of the equation:

$$\begin{aligned} b + 1.7 &= 5.9 \\ -1.7 &= -1.7 \\ \hline b + 0 &= 4.2 \\ b &= 4.2 \end{aligned}$$

The solution set is **{4.2}**.

Example 2

Solve for y: $\frac{3}{2}y = 21$.

Solution: Isolate y by multiplying both sides of the equation by $\frac{2}{3}$, the multiplicative inverse or reciprocal of the fractional coefficient of y:

$$\frac{3}{2}y = 21$$
$$\left(\frac{2}{3} \times \frac{3}{2}\right)y = \frac{2}{3} \times 21$$
$$1 \cdot y = \frac{42}{3}$$
$$\mathbf{y = 14}$$

Transposing Terms of an Equation

Solving the equation $x + 8 = 3$ requires subtracting 8 from both sides, which changes the original equation into $x = 3 - 8$. Solving the equation $x - 7 = -5$ requires adding 7 to both sides, which produces the equivalent equation, $x = -5 + 7$. In each case, the number on the left side of the original equation was moved or **transposed** to the opposite side simply by changing its sign. Here are a few more examples:

- if $b + 1.7 = 5.9$, then $b = 5.9 - 1.7$ so $b = 4.2$.
- if $y - 6 = -1$, then $y = -1 + 6$ so $y = 5$.
- if $3 = x - 1$, then $3 + 1 = x$ or, interchanging sides, $x = 4$.

Checking Possible Solutions

A number **satisfies** an equation if the equation is true when the variable is replaced by that number. To verify a number is an actual solution or root of an equation, check that it satisfies the *original* equation.

Example 3

Determine if $x = -3$ is a root of the equation $2x + 13 = 7$.

Solution: Replace x with -3 and then determine if the left side of the equation matches the right side:

Replace x with -3: $2(-3)+13 \boxed{?} 7$

Evaluate the left side: $-6+13 \boxed{?} 7$

Compare the two sides: $7 = 7$ ✓

Yes, $x = -3$ is a root of the equation.

Check Your Understanding of Section 2.1

1–21. *Solve for the variable and check.*

1. $x + 5 = -9$

2. $-3x = 12$

3. $-7 = x - 1$

4. $-0.5w = -35$

5. $-4 = x + 2$

6. $8 + n = -7$

7. $-7.8 = 1.4 + x$

8. $1.2 = m + 3.5$

9. $t + 2 = \frac{4}{3}$

10. $\frac{2x}{3} = 16$

11. $-21 = \frac{7x}{8}$

12. $\frac{x}{5} = -1.7$

13. $y - \frac{1}{3} = -2$

14. $\frac{r}{3} = -8.4$

15. $-10 = \frac{-5a}{3}$

16. $3x = 1\frac{1}{8}$

17. $\frac{x}{4} = 2\frac{1}{3}$

18. $0.7t = 4.9$

19. $\frac{a}{-3} = 1.6$

20. $b + \frac{1}{3} = 2\frac{1}{6}$

21. $7.6 = 1.9 - x$

2.2 SOLVING MULTISTEP EQUATIONS

KEY IDEAS

Isolating the variable in an equation may require performing more than one inverse operation on both sides of the equation.

Solving Equations with Two Arithmetic Operations

In an equation in which the variable is involved in two different arithmetic operations, such as $3x - 5 = 22$, isolate the variable by undoing any addition or subtraction *before* undoing any multiplication or division.

Example 1

Find the solution set of $3x - 5 = 22$.

Solution: The variable in the equation $3x - 5 = 22$ is involved in two operations: multiplication since $3x$ means 3 times x and subtraction as 5 is being subtracted from $3x$. Undo the subtraction before undoing the multiplication.

$$3x - 5 = 22$$

Transpose 5:

$$3x = 22 + 5$$

Divide both sides by 3:

$$\frac{\not{3}x}{\not{3}_1} = \frac{27}{3}$$

$$x = 9$$

The solution set is **{9}**.

Example 2

Solve for y and check: $\frac{y}{3} - 2 = 13$.

Solution: Add 2 to both sides of the equation and then multiply both sides by 3.

$$\frac{y}{3} - 2 = 13$$

$$\frac{y}{3} = 13 + 2$$

$$\not{3}\left(\frac{y}{\not{3}_1}\right) = 3(15)$$

$$\mathbf{y = 45}$$

CHECK:

$$\frac{y}{3} - 2 = 13$$

Set $y = 45$:

$$\frac{45}{3} - 2 \boxed{?} 13$$

$$15 - 2 \boxed{?} 13$$

$$13 = 13 \checkmark$$

Example 3

Solve for x: $0.03x - 0.7 = 0.8$.

Solution: When combining decimal numbers, make sure you align the decimal points. If you need to divide decimal numbers, use your calculator to help avoid making careless errors.

$$\begin{array}{rcl} 0.03x - 0.7 &=& 0.8 \\ +0.7 &=& +0.7 \\ \hline 0.03x &=& 1.5 \end{array}$$

$$\frac{0.03x}{0.03} = \frac{1.5}{0.03}$$

Use a calculator:

$$x = 1.5 \div 0.03$$

$$\mathbf{x = 50}$$

Example 4

Solve for n: $11 = 8 - n$.

Solution: Transpose n to the left side of the equation:

$$n + 11 = 8$$
$$n = 8 - 11$$
$$\mathbf{n = -3}$$

Solving Equations with Parentheses

If an equation contains parentheses, remove the parentheses using the distributive property.

Example 5

Solve for x and check: $3(1 - 2x) + 8 = -13$.

Solution 1: Remove the parentheses by multiplying each term inside the parentheses by the number in front of the parentheses.

$$3(1-2x)+8=-13$$
$$3(1)+3(-2x)+8=-13$$
$$11-6x=-13$$
$$-6x=-13-11$$
$$\frac{-6x}{-6}=\frac{-24}{-6}$$
$$\mathbf{x=4}$$

CHECK:

$$3(1-2x)+8 = -13$$

Set $x = 4$: $3(1-2\cdot 4)+8 \boxed{?} -13$

$$3(-7)+8 \quad -13$$
$$-21+8 \quad -13$$
$$-13 = -13 \checkmark$$

Solution 2: Transpose 8 and then divide both sides of the equation by 3.

$$3(1-2x)+8=-13$$
$$3(1-2x)=-13-8$$
$$\frac{\cancel{3}(1-2x)}{\cancel{3}}=\frac{-21}{3}$$
$$1-2x=-7$$
$$-2x=-7-1$$
$$\frac{-2x}{-2}=\frac{-8}{-2}$$
$$\mathbf{x=4}$$

Check Your Understanding of Section 2.2

1–21. *Solve for the variable and check.*

1. $0.4y + 5 = 1$

2. $3x - 1 = -16$

3. $7 - 2x = -13$

4. $-32 = 7 - 3w$

5. $\frac{x}{5} + 8 = 10$

6. $0.2x + 0.3 = 8.1$

7. $3(2x - 1) = 7$

8. $-17 = 2(5 + m)$

9. $19 - (1 - x) = 18$

10. $7 - 3(x - 1) = -17$

11. $-3(9 - 7p) = -12$

12. $\frac{3t}{2} + 5 = 17$

13. $0 = 18 - 2(h + 1)$

14. $1.04x + 8 = 60$

15. $2(3x - 5) + 12 = 0$

16. $-(8x - 3) = 19$

17. $0.25(3x - 5) = 2.5$

18. $6\left(\frac{x}{2} + 1\right) = -9$

19. $13 - \frac{3x}{4} = -8$

20. $2\left(7 - \frac{y}{4}\right) = 3$

21. $\frac{1}{2}(5x - 7) + 3 = 11$

22. Rewrite $\{x \mid 3x - 1 = 23, x \text{ is an integer}\}$ in roster-form.

23. Rewrite $\{x \mid 2x + 5 = 12, x \text{ is an integer}\}$ in roster-form.

24. If $A = \{x \mid 2x + 19 = 5\}$ and $B = \{x \mid 3(x + 1) = 21\}$, find the union of sets A and B where x is an integer.

25. If $-2x + 3 = 7$ and $3x + 1 = 5 + y$, what is the value of y?

26. If $2x + 5 = -25$ and $-3m - 6 = 48$, what is the product of x and m?

27. If the sum of three times a number and 5 is 32, what is the number?

28. When 2 is subtracted from one-third of a number, the result is 3. What is the number?

29. If 37 exceeds three times the sum of a number and 5 by 1, what is the number?

30. If 45 is 9 greater than the difference obtained by subtracting 7 from a number, what is the number?

2.3 SOLVING EQUATIONS WITH LIKE TERMS

KEY IDEAS

When isolating the variable in an equation, it may be necessary to first combine like terms. For example, if $5x + 3x = 24$, then $8x = 24$, so $x = \frac{24}{8} = 3$.

Equations with Like Terms on the Same Side

If the same variable appears in different terms on the same side of an equation, combine the like variable terms.

Example 1

Solve for y and check: $3(2y + 5) - 8y = 1$.

Solution: First remove the parentheses. Then combine like terms.

$$\begin{aligned} 3(2y+5)-8y&=1 \\ 3(2y)+3(5)-8y&=1 \\ 6y-8y&=1-15 \\ -2y&=-14 \\ \frac{-2y}{-2}&=\frac{-14}{-2} \\ y&=7 \end{aligned}$$

CHECK:

$$\begin{aligned} &3(2y+5)-8y = 1 \\ \text{Set } y=7:\quad &3(2\cdot 7+5)-8(7) \overset{?}{=} 1 \\ &3(14+5)-56 \quad 1 \\ &3(19)-56 \quad 1 \\ &57-56 = 1 \checkmark \end{aligned}$$

Example 2

The cost, C, of manufacturing x hammers is $C = 4x + 170$. If each hammer is sold for \$10, how many hammers must be sold to make a profit of \$100?

Solution: *Profit* is the difference between what an item costs to make and the amount it can be sold for, which is called *revenue*. Each hammer can be sold for \$10. So the revenue for selling x hammers is $10x$. If the profit for selling x hammers is \$100:

$$\underbrace{10x}_{\text{Revenue}} - \underbrace{(4x+170)}_{\text{Cost}} = \underbrace{100}_{\text{Profit}}$$

$$10x - 4x - 170 = 100$$

$$6x = 100 + 170$$

$$\frac{6x}{6} = \frac{270}{6}$$

$$x = 45$$

To make a profit of \$100, **45** hammers must be sold.

Equations with Like Terms on Both Sides

If like variable terms appear on both sides of an equation, collect the like variable terms on one side of the equation and the number terms on the opposite side. To solve $5x - 9 = 2(x + 3)$:

Remove the parentheses:	$5x - 9 = 2x + 6$
Transpose 9 and $2x$:	$5x - 2x = 6 + 9$
Divide both sides by 3:	$\frac{3x}{3} = \frac{15}{3}$
	$\boldsymbol{x = 5}$

Example 3

Solve for n and check: $2(3n - 14) - 3 = 5(2n + 1)$.

Solution: After removing the parentheses, collect like variable terms on the left side of the equation and collect number terms on the right side.

Original equation:	$2(3n - 14) - 3 = 5(2n + 1)$
Remove the parentheses:	$6n - 28 - 3 = 10n + 5$
Simplify:	$6n - 31 = 10n + 5$
Transpose -31 and $10n$:	$6n - 10n = 5 + 31$
Combine like terms:	$-4n = 36$
Divide both sides by -4:	$\frac{-4n}{-4} = \frac{36}{-4}$
	$\boldsymbol{n = -9}$

CHECK: Substitute -9 for n in the original equation:

$$\begin{array}{r|l}
2(3n-14)-3 & = \; 5(2n+1) \\
2(3(\mathbf{-9})-14)-3 & \boxed{?} \; 5(2(\mathbf{-9})+1) \\
2(-27-14)-3 & \; 5(-18+1) \\
2(-41)-3 & \; 5(-17) \\
-82-3 & \; -85 \\
-85 & = \; -85 \; \checkmark
\end{array}$$

Example 4

In 7 years Maria will be twice as old as she was 3 years ago. What is Maria's present age?

Solution: If x represents Maria's present age, then

$x + 7$ = Maria's age 7 years from now,
$x - 3$ = Maria's age 3 years ago.

$$\underbrace{x+7}_{\text{Maria's age in 7 years}} \underbrace{=}_{\text{will be}} \underbrace{2(x-3)}_{\text{twice age 3 years ago}}$$

$$x+7=2x-6$$
$$x-2x=-6-7$$
$$-x=-13$$
$$x=13$$

Maria's present age is **13 years.**

CHECK: In 7 years, Maria will be 13 + 7 = 20 years old. Three years ago, Maria was 13 – 3 = 10 years old. Because 20 = 2 × 10, the answer is correct.

Example 5

Three times the sum of a number and 7 is the same as 9 times the difference obtained when 5 is subtracted from the number. What is the number?

Solution: Translate the conditions of the problem into an algebraic equation where x is the unknown number:

$$\underbrace{3(x+7)}_{\text{3 times the sum of a number and 7}} \underbrace{=}_{\text{is the same as}} \underbrace{9(x-5)}_{\text{9 times the difference when 5 is subtracted from the number}}$$

$$3x+21=9x-45$$
$$3x-9x=-45-21$$
$$-6x=-66$$
$$\frac{-6x}{-6}=\frac{-66}{-6}$$
$$x=11$$

The number is **11**. The check is left for you.

Consecutive Integer Problems

When a list of consecutive integers is arranged in increasing order, each integer after the first is one more than the integer that comes before it, as in –2, –1, 0, 1, 2, 3, and 4. In a list of increasing consecutive *even* integers, each even integer after the first is *two* more than the integer that comes before it, as in –2, 0, 2, 4, 6, and 8. Similarly, consecutive odd integers also differ by 2. In general,

- If n is an integer, a set of consecutive integers that begins with n is

$$\{n, n+1, n+2, n+3, \ldots\}$$

- If n is an even (or odd) integer, a set of consecutive even (or odd) integers that begins with n is

$$\{n, n+2, n+4, \ldots\}$$

Example 6

Find three consecutive odd integers such that twice the sum of the second and the third is 43 more than three times the first.

Solution: If x represents an odd integer, then $x + 2$ and $x + 4$ represent the next two consecutive odd integers. Thus,

$$\begin{aligned} 2[(x+2)+(x+4)] &= 43 + 3x \\ 4x + 12 &= 43 + 3x \\ 4x - 3x &= 43 - 12 \\ x &= 31 \\ \text{so } x + 2 = 33 \quad &\text{and} \quad x + 4 = 35 \end{aligned}$$

The three consecutive odd integers are **31, 33, and 35.** The check is left for you.

Check Your Understanding of Section 2.3

A. *Multiple Choice.*

1. If $15x = 3(x + 7)$, then $x =$
(1) $\frac{4}{7}$ (2) $\frac{7}{12}$ (3) 1.75 (4) 2.75

2. If $12x - 4(2x - 3) = 10$, then $x =$
(1) 3 (2) 2 (3) $-\frac{1}{3}$ (4) $-\frac{1}{2}$

3. If $n + 4$ represents an odd integer, the next larger odd integer is represented by
(1) $n + 2$ (2) $n + 3$ (3) $n + 5$ (4) $n + 6$

4. A girl can ski down a hill five times as fast as she can climb up the same hill. If she can climb up the hill and ski down in a total of 9 minutes, how many minutes does it take her to climb up the hill?
(1) 1.8 (2) 4.5 (3) 7.2 (4) 7.5

5. If $5(2x - 7) = 15x - 10$, then $x =$
(1) 1 (2) 0.6 (3) –5 (4) –9

6. If $4(2x+1) = 22 + 3(2x-5)$, then $x =$
(1) $\frac{2}{3}$ (2) $\frac{3}{2}$ (3) $-\frac{1}{2}$ (4) 8

7. If four times a number is increased by 15, the result is three less than six times the number. What is the number? Which equation can be used to find the number?
(1) $4(x+15) = 6x - 3$ (3) $4x + 15 = 6(x-3)$
(2) $4x + 15 = 6x - 3$ (4) $4(x+15) = 3 - 6x$

8. Josh and Mae work at a concession stand. They each earn \$8 per hour. Josh worked three hours more than Mae. If Josh and Mae earned a total of \$120, how many hours did Josh work?
(1) 6 (2) 9 (3) 12 (4) 15

9. Rhona has \$1.35 in nickels and dimes in her pocket. If she has six more dimes than nickels, which equation can be used to determine x, the number of nickels she has?
(1) $0.05(x+6) + 0.10x = 1.35$ (3) $0.05 + 0.10(6x) = 1.35$
(2) $0.05x + 0.10(x+6) = 1.35$ (4) $0.15(x+6) = 1.35$

B. *Show how you arrived at your answer.*

10–21. *Solve for the variable and check.*

10. $3x + 4x = -28$

11. $2x - 5x = -27$

12. $7t = t - 42$

13. $0.54 - 0.07y = 0.2y$

14. $3s + 2(8 - s) = 3$

15. $3.3 - x = 3(x - 1.7)$

16. $9b = 2b - 3(8 - b)$

17. $z + 1.5z = 35 - z$

18. $3(5 - 2n) = 2n - 9$

19. $5(6 - q) = -3(q + 2)$

20. $3(4c - 7) = 2(3c + 11) + 5$

21. $2x = 10\left(\frac{3}{2} - x\right)$

22. Samantha, Lauren, and Jerry own shares of the same stock with a total value of \$7,650. If Samantha owns fifty shares, Lauren owns seventy shares, and Jerry owns thirty shares, what is the value in dollars of Lauren's stock?

23. Sal keeps quarters, nickels, and dimes in his change jar. He has a total of 52 coins. He has three more quarters than dimes and five fewer nickels than dimes. How many dimes does Sal have?

24. How old is David if his age 6 years from now will be twice his age 7 years ago.

25. Three years ago Jane was one-half as old as she will be 2 years from now. What is Jane's present age?

26. A geologist collected 13 rocks that have exactly the same weight. If 9 of these rocks with an additional 5-ounce weight at one end of a balance-scale can balance the remaining rocks and a 23-ounce weight at the other end of the scale, what is the number of ounces in the weight of one of these rocks?

27. Twice the sum of a number and 9 is the same as four times the difference obtained when 6 is subtracted from the number. What is the number?

28. Find four consecutive odd integers such that the sum of three times the second integer and the last integer is 104.

29. Find four consecutive integers whose sum is 15 less than 5 times the first.

30. Find three consecutive even integers such that the sum of the first integer and three times the last integer is 20 less than 5 times the second integer.

31. Find three consecutive even integers such that five times the sum of the second and third integers is 18 less than 6 times the sum of the first and second integers.

32. Find four consecutive odd integers such that five times the sum of the first and third integers exceeds four times the sum of the second and last integers by 14.

2.4 ALGEBRAIC MODELING

KEY IDEAS

A **mathematical model** is a representation of a real-world situation or process using tools such as graphs, diagrams, tables, and equations. A mathematical model typically involves some simplifying assumptions that reduces a complex situation to a form that can be studied further. An algebraic model translates real-world conditions into one or more related equations.

Algebraic Modeling

Here is a general four-step procedure for solving unfamiliar types of word problems:

- **STEP 1: Assign a variable.** After carefully reading the problem, identify the unknown quantity that you are required to find and represent it by a variable. When two or more related quantities are unknown, try to represent each quantity in terms of the same variable.
- **STEP 2: Think of a relationship.** Decide how the unknown quantity is related to the other quantities or facts in the problem. Then write the relationship as an English sentence.
- **STEP 3: Write an equation.** Express the sentence written in step 2 as an algebraic equation. Solve the equation in step-by-step fashion.
- **STEP 4: Check.** Because it is possible that a word problem can be translated into an *incorrect* equation that is then solved correctly, check the answer in the original problem statement.

Example 1

The EZ-Car Rental Agency charges \$35 for the first day of a car rental, and \$22 for each additional day. If John's car rental bill is \$189, for how many days did John rent the car?

Solution:

- ASSIGN A VARIABLE: If d represents the total number of days that John rented the car, $d - 1$ is the number of days the car was rented at \$22 per day.
- THINK OF A RELATIONSHIP:

$$\underbrace{\text{charge for first day}}_{\$35} + \underbrace{\text{charge for additional days}}_{\$22 \times \text{number of days}} = \$189$$

- WRITE AN EQUATION:

$$
\begin{aligned}
35 + 22(d-1) &= 189\\
22d+13 &= 189\\
2d &= 189-13\\
\frac{22d}{22} &= \frac{176}{22}\\
d &= 8
\end{aligned}
$$

- CHECK: Check the answer in the original problem:

$$1 \times \$35 \div 7 \times \$22 = \$35 + \$154 = \$189 \checkmark$$

total of 8 days

John rented the car for **8** days.

Problems Involving a Fixed Sum

If the sum of two unknown numbers is 20 and x represents one of these numbers, then $20 - x$ represents the other number. When a word problem involves a fixed sum of two quantities, the variable can be set equal to either of the two quantities. The remaining unknown quantity is the difference between the given sum and the variable.

Example 2

A soda machine contains 20 coins. Some of the coins are nickels, and the rest are quarters. If the value of the coins is \$4.40, find the number of coins of each kind.

Solution:

- ASSIGN VARIABLES: If x = the number of nickels, then $20 - x$ = the number of quarters.
- THINK OF A RELATIONSHIP:

$$\underbrace{\text{dollar value of nickels}}_{\$0.05\times\text{nickels}} + \underbrace{\text{dollar value of quarters}}_{\$0.25\times\text{quarters}} = \$4.40$$

- WRITE AN EQUATION:

$$
\begin{aligned}
0.05x+0.25(20-x) &= 4.40\\
0.05x+5-0.25x &= 4.40\\
-0.20x &= 4.40-5\\
\frac{-0.20x}{-0.20} &= \frac{-0.60}{-0.20}\\
x &= 3\\
\text{and}\quad 20-x &= 17
\end{aligned}
$$

- CHECK: $3 + 17 = 20$ and $(3 \times 0.05) + (17 \times 0.25) = 0.15 + 4.25 = 4.40$ ✓

The soda machine contains **3 nickels** and **17 quarters**.

Example 3

The tickets sold for a dance recital cost $5.00 for adults and $2.00 for children. If the total number of tickets sold was 295 and the total amount collected was $1,220, how many tickets were sold to adults and to children?

Solution:

- ASSIGN VARIABLES: If x is the number of adult tickets sold, then $295 - x$ is the number of tickets sold to children.
- THINK OF A RELATIONSHIP:

$$\underbrace{\text{amount collected from adults}}_{\$5.00\times\text{adult tickets sold}} + \underbrace{\text{amount collected from children}}_{\$2.00\times\text{children tickets sold}} = \$1{,}220$$

- WRITE AN EQUATION: $\$5x + \$2(295 - x) = \$1{,}220$

$$\begin{aligned} 5x + 590 - 2x &= 1{,}220 \\ 3x + 590 &= 1{,}220 \\ 3x &= 1{,}220 - 590 \\ \frac{3x}{3} &= \frac{630}{3} \\ x &= 210 \\ 295 - x = 295 - 210 &= 85 \end{aligned}$$

210 tickets were sold to adults, and **85** tickets were sold to children. The check is left for you.

Problems Involving Comparisons

When two or more unknown quantities are being compared, set the variable equal to the quantity to which the others are being compared. Then express the other quantities in terms of the same variable.

Example 4

Jim keeps quarters, nickels, and dimes in his change jar. He has three more quarters than dimes and five fewer nickels than dimes. If there are 52 coins in the change jar, how many dimes are in the jar?

Solution: The numbers of quarters and nickels in the jar are being compared to the number of dimes in the jar.

- ASSIGN VARIABLES: If x represents the number of dimes in the jar, then $x + 3$ represents the number of quarters and $x - 5$ represents the number of nickels in the jar.
- THINK OF A RELATIONSHIP: The numbers of the three different types of coins in the jar add up to 52.

- WRITE AN EQUATION:

$$
\begin{aligned}
x+(x+3)+(x-5)&=52\\
3x-2&=52\\
3x&=54\\
\frac{3x}{3}&=\frac{54}{3}\\
x&=18
\end{aligned}
$$

The jar contains **18 dimes**. The check is left for you.

Example 5

Ticket sales for a music concert totaled \$2,160. Three times as many tickets were sold for the Saturday night concert as were sold for the Sunday afternoon concert. Two times as many tickets were sold for the Friday night concert as were sold for the Sunday afternoon concert. Tickets for all three concerts sold for \$2.00 each. Find the number of tickets sold for the Saturday night concert.

Solution: The numbers of tickets sold for the concerts on Friday and Saturday nights are being compared to the number of tickets sold for the Sunday afternoon concert.

- ASSIGN VARIABLES: If x = the number of tickets sold for the Sunday afternoon concert, then

  $3x$ = the number of tickets sold for the Saturday night concert

  $2x$ = the number of tickets sold for the Friday night concert

- THINK OF A RELATIONSHIP:

$$\underbrace{\text{Friday's receipts}}_{\$2\times\text{tickets sold}} + \underbrace{\text{Saturday's receipts}}_{\$2\times\text{tickets sold}} + \underbrace{\text{Sunday's receipts}}_{\$2\times\text{tickets sold}} = \$2{,}160$$

- WRITE AN EQUATION:

$$
\begin{aligned}
\$2\cdot x+\$2\cdot 3x+\$2\cdot 2x&=\$2160\\
2x+6x+4x&=2160\\
\frac{12x}{12}&=\frac{2160}{12}\\
x&=180
\end{aligned}
$$

The number of tickets sold for the *Saturday* night concert = $3x = 3(180)$ = **540**. The check is left for you.

Check Your Understanding of Section 2.4

A. *Multiple Choice.*

1. A postal clerk sold 50 postage stamps for $16.05. Some were 39-cent stamps and the rest were 24-cent stamps. How many 39-cent stamps were sold?
   (1) 19 (2) 23 (3) 27 (4) 31

2. Mario paid $44.25 in taxi fare from the hotel to the airport. The cab charged $2.25 for the first mile plus $3.50 for each additional mile. How many miles was it from the hotel to the airport?
   (1) 10 (2) 11 (3) 12 (4) 13

B. *Show how you arrived at your answer.*

3. Mike calculates that from his day's intake of 2,156 calories, four times as many calories were from carbohydrates as from protein, and twice as many calories were from fat as from protein. How many calories were from carbohydrates?

4. There were 100 more balcony tickets than main-floor tickets sold for a concert. The balcony tickets sold for $4 and the main-floor tickets sold for $12. The total amount of sales for both types of tickets was $3,056. How many balcony tickets were sold?

5. In a town election, candidates A and B were running for mayor. Two-thirds of the people eligible to vote actually voted. Candidate A received $\frac{3}{4}$ of the number of votes that were cast for candidate B. If 5,460 people were eligible to vote in this election, how many people voted for candidate A?

6. Using only 39-cent and 24-cent stamps, Charlie put $7.14 postage on a package he mailed to his sister. If he used twice as many 39-cent stamps as 24-cent stamps, how many 24-cent stamps did he use?

7. A group of 148 people are spending five days at a summer camp. The cook ordered 12 pounds of food for each adult and 9 pounds of food for each child. A total of 1,410 pounds of food was ordered. Find the total number of adults in the group.

8. The ninth graders at a high school are raising money by selling T-shirts and baseball caps. The number of T-shirts sold was three times the number of caps. The profit they received for each T-shirt sold was $5.00, and the profit on each cap was $2.50. If the students made a total profit of $210, how many T-shirts were sold?

9. Allan has nickels, dimes, and quarters in his pocket. The number of nickels is 1 more than twice the number of quarters. The number of dimes is 1 less than the number of quarters. If the value of the change in his pocket is 85 cents, how many of each coin does Allan have?

10. Tickets to a concert that were purchased in advance of the day of a concert cost $4.50 each and tickets purchased at the box office on the day of the concert cost $8.00 each. The total amount of money collected in ticket sales was the same as if every ticket purchased had cost $6.00. If 180 tickets were purchased in advance of the day of the concert, what was the total number of tickets purchased for the concert?

11. Carlos earns $9 per hour, and on weekends he earns twice as much per hour. Last week he earned a total of $378 including the weekend. Of the 35 hours Carlos worked for that week, how many hours did he work during the weekend?

12. Tickets for a concert cost $4.00 for a balcony seat and $7.50 for an orchestra seat. If ticket sales totaled $1,585 for 300 tickets sold, how many more tickets for balcony seats were sold than tickets for orchestra seats?

13. A bank contains 30 coins, consisting of nickels, dimes, and quarters. There are twice as many nickels as quarters and the remaining coins are dimes. If the total value of the coins is $3.35, what is the number of dimes in the bank?

14. Seth has one less than twice the number of compact discs (CDs) that Jason has. Raoul has 53 more CDs than Jason. If Seth gives Jason 25 CDs, Seth and Jason will have the same number of CDs. What is the total number of CDs that the three boys had to begin with?

15. Jennifer's Bakery sold half as many apple pies as cheesecakes. The price of an apple pie is $6, and the price of a cheesecake is $8.50. If the total amount of the sales for these cakes was $391, what was the total number of each kind of cake that was sold?

16. Mr. Perez owns a sneaker store. He bought 350 pairs of basketball sneakers and 150 pairs of soccer sneakers from the manufacturers for $62,500. He sold all the sneakers and made a 25% profit. If he sold the soccer sneakers for $130 per pair, how much did he charge for one pair of basketball sneakers?

2.5 LITERAL EQUATIONS AND FORMULAS

KEY IDEAS

A **literal equation** is an equation in which the coefficients of one or more variables are letters, as in $ax + by = c$. Sometimes it is helpful to transform a literal equation so that a particular letter or variable in it is solved for in terms of the other members of the equation. A **formula** is a special type of literal equation in which the variables represent real-world quantities.

Solving a Literal Equation for a Specified Variable

To find the value of y in the equation $ax + by = c$ when $a = 3$, $x = -2$, $b = 4$, and $c = 26$, substitute the numbers in the equation and then solve for y:

$$\begin{aligned} ax+by&=c \\ (3)(-2)+4y&=26 \\ -6+4y&=26 \\ \frac{4y}{4}&=\frac{32}{4} \\ y&=8 \end{aligned}$$

If you needed to calculate the values of y repeatedly for a *series* of values of x, a, b, and c, it might be more efficient to first solve the equation for y and then do the number substitutions. To solve $ax + by = c$ for y, isolate y in the usual way by treating all of the other members of the equation as constants.

$$\begin{aligned} ax+by&=c \\ by&=c-ax \\ y&=\frac{c-ax}{b} \end{aligned}$$

The value of y can now be calculated directly:

$$y=\frac{c-ax}{b}=\frac{26-(3)(-2)}{4}=\frac{26+6}{4}=8$$

Example 1

If $P = mgh$, which expression can be used to represent g?

(1) $P - m - h$ (2) $P - mh$ (3) $\frac{P}{m} - h$ (4) $\frac{P}{mh}$

Solution: Since the given equation can be rewritten as $P = mh \cdot g$, solve for g by dividing both sides of the equation by the literal coefficient:

$$\frac{P}{mh} = \frac{\overset{1}{\cancel{mh}} \cdot g}{\cancel{mh}}$$

$$\frac{P}{mh} = g$$

The correct choice is **(4)**.

Example 2

If $3x - c = -4$, then x equals

(1) $4 - c$ (2) $\frac{4-c}{3}$ (3) $\frac{c-4}{3}$ (4) $c - 4$

Solution: Isolate x in the usual way. Since $3x - c = -4$, $3x = c - 4$. Divide both sides of the equation by 3, the coefficient of x:

$$\frac{\overset{1}{\cancel{3}}x}{\cancel{3}} = \frac{c-4}{3}$$

$$x = \frac{c-4}{3}$$

The correct choice is **(3)**.

Example 3

If $\frac{ey}{n} + k = t$, what is y in terms of e, n, k, and t?

Solution: Isolate the term involving y by subtracting k from both sides of the equation:

$$\frac{ey}{n} = t - k$$

The literal coefficient of y is $\frac{e}{n}$. Make the coefficient of y equal to 1 by multiplying both sides of the equation by the reciprocal of $\frac{e}{n}$:

$$\left(\frac{\cancel{n}}{\cancel{e}}\right)\left(\frac{\overset{1}{\cancel{e}}}{\cancel{n}}y\right) = \left(\frac{n}{e}\right)(t - k)$$

$$\boldsymbol{y = \frac{n(t-k)}{e}}$$

Example 4

Solve the equation $3ay + 1 = a(x + y)$ for y in terms of x where $a \neq 0$.

Solution: After removing the parentheses, collect the y terms on the left side of the equation and all other terms on the opposite side.

Remove the parentheses: $3ay + 1 = ax + ay$
Collect the y terms: $3ay - ay = ax - 1$
Combine the y terms: $2ay = ax - 1$
Divide each side by $2a$: $\mathbf{y = \frac{ax-1}{2a}}$

Working with Formulas

Formulas typically describe relationships based on physical laws of science or mathematical principles. A formula can also be derived from a series of experimental measurements.

Example 5

For a certain group of teenage boys, body weight and height are related by the formula $W = 6(H - 44)$, where W represents weight in pounds and H represents height in inches.

a. What is the weight in pounds of a teenage boy whose height is 5 feet 10 inches?
b. Solve for H in terms of W. Use the formula to approximate, to the *nearest inch*, the height of a teenage boy who weighs 124 pounds.

Solution: a. Because 5 feet = 5 × 12 = 60 = 60 inches, $H = 60 + 10 = 70$ inches. Hence,

$$\begin{aligned} W &= 6(H - 44) \\ &= 6(70 - 44) \\ &= 6 \times 26 \\ &= 156 \end{aligned}$$

The boy weighs **156** pounds.

b. Remove the parentheses and then isolate H:

$$\begin{aligned} W &= 6H - 264 \\ 6H &= W + 264 \\ \mathbf{H} &= \mathbf{\frac{W+264}{6}} \end{aligned}$$

If $W = 124$, then

$$H = \frac{124+264}{6} = \frac{388}{6} \approx 64.7$$

To the *nearest inch*, the height of the boy is **65** inches.

Perimeter, Area, and Volume

The **perimeter** of a figure is the distance around it. The **area** of a figure is the number of 1 by 1 unit squares that the figure can enclose. **Volume** is a measure of capacity and is the number of 1 by 1 unit cubes that can be placed inside a solid figure. Table 2.1 reviews some familiar formulas, where P represents perimeter, A represents area, and V represents volume.

TABLE 2.1 PERIMETER, AREA, AND VOLUME FORMULAS

Figure	Diagram	Formula
Rectangle	w, ℓ	$P = 2(\ell + w)$ and $A = \ell \times w$
Square	s, s, s, s	$P = 4s$ and $A = s^2$
Rectangular Box	base, h, w, ℓ	$V = \ell \times w \times h$
Cube	base, e, e, e	$V = e^3$

Example 6

What is the edge length of a cube that has the same volume as a rectangular box with dimensions 18 inches by 8 inches by 1.5 inches?

Solution: If e represents the edge length of the cube, then the volume of the cube is e^3.

- The volume of the rectangular box is 18 in. × 8 in. × 1.5 in. = 216 in.^{3}.
- Because the cube has the same volume as the rectangular box, $e^3 = 216$.
- As $6 \times 6 \times 6 = 216$, $e = 6$.

The edge length of the cube is **6 inches**.

Example 7

The dimensions of a rectangular brick, in inches, are 4 by 6 by 8. How many such bricks are needed to have a total volume of exactly 1 cubic foot?

Solution: The volume of the brick is 4 in. × 6 in. × 8 in. = 192 in.^{3}. A volume of 1 cubic foot is equivalent to a volume of

$$12 \text{ in.} \times 12 \text{ in.} \times 12 \text{ in.} = 1728 \text{ in.}^3$$

Because 1728 in.^{3} ÷ 192 in.^{3} = 9, **9** bricks are needed to have a total volume of exactly 1 cubic foot.

Check Your Understanding of Section 2.5

A. *Multiple Choice.*

1. If $a(x + b) = c$, what is x in terms of a, b, and c?

(1) $\frac{c-b}{a}$ (2) $\frac{c-ab}{a}$ (3) $\frac{b+c}{a}$ (4) $\frac{ac-b}{a}$

2. If $9x + 2a = -3a + 4x$, then x equals

(1) 1 (2) 0 (3) $-a$ (4) $-5a$

3. Assume body weight and height are related by the formula $W = 2H + 13$, where W represents weight in pounds and H represents height in inches. What is the weight in pounds of a person whose height is 5 feet 6 inches?

(1) 128 (2) 135 (3) 145 (4) 148

4. A cube whose edge length is 6 has the same volume as a rectangular box whose length is 12 and whose width is 9. The height of the rectangular box is
(1) 6 (2) 2 (3) 3 (4) 4

5. If the width of a rectangle is represented by w and the perimeter of the rectangle is represented by k, then the length of the rectangle expressed in terms of w and k is
(1) $k - 2w$ (2) $k - \frac{w}{2}$ (3) $\frac{k - 2w}{2}$ (4) $2k - w$

6. If $4x + y = H$, what is x in terms of y and H?
(1) $\frac{H}{4} - y$ (2) $\frac{H}{4} + y$ (3) $\frac{H + y}{4}$ (4) $\frac{H - y}{4}$

7. In terms of c, y, and a, what is the value of x in the equation $2ax + 2y = c$?
(1) $\frac{c - y}{a}$ (2) $\frac{c - 2y}{2a}$ (3) $c - 2y - 2a$ (4) $\frac{c + 2y}{2a}$

8. The lengths of two adjacent sides of a square are represented by $5x + 11$ and $7x - 3$. What is the perimeter of the square?
(1) 184 (2) 28 (3) 31 (4) 112

9. In terms of S, n, and a, what is the value of b in the equation $S = \frac{n}{2}(a + b)$?
(1) $\frac{2S - an}{n}$ (2) $\frac{2(S - an)}{n}$ (3) $\frac{S + an}{2n}$ (4) $\frac{2S + a}{2n}$

10. A formula used for calculating velocity is $v = \frac{1}{2}at^2$. What is a expressed in terms v and t?
(1) $a = \frac{2v}{t}$ (2) $a = \frac{2v}{t^2}$ (3) $a = \frac{v}{t}$ (4) $a = \frac{v}{2t^2}$

11. If $l = a + (n - 1)d$, then n equals:
(1) $\frac{l + d}{a}$ (2) $\frac{l + d - a}{d}$ (3) $\frac{l - d + a}{d}$ (4) $\frac{l - a + 1}{d}$

B. *Show how you arrived at your answer.*

12–15. *Solve each formula for the specified letter.*

12. $A = p + prt$ for t.

13. $ay + 2x = b + x$ for y.

14. $J = 4S + 28$ for S.

15. $ax - b = -c$ for x.

16. The formula $C=\frac{5}{9}(F-32)$ is used to convert Fahrenheit temperature, F, to Celsius temperature, C. What temperature, in degrees Fahrenheit, is equivalent to a temperature of 10° Celsius?

17. Tina's preschool has a set of cardboard building blocks, each of which measures 9 inches by 9 inches by 4 inches. How many of these blocks will Tina need to build a wall 4 inches deep, 3 feet high, and 12 feet long?

18. From each of the four corners of a rectangular piece of cardboard x inches wide and 12 inches long, a 2-inch square is cut off, as shown in the accompanying figure. The four sides of the new figure are folded straight up to form a box that is open at the top. If the volume of the box is 80 in^3, find x.

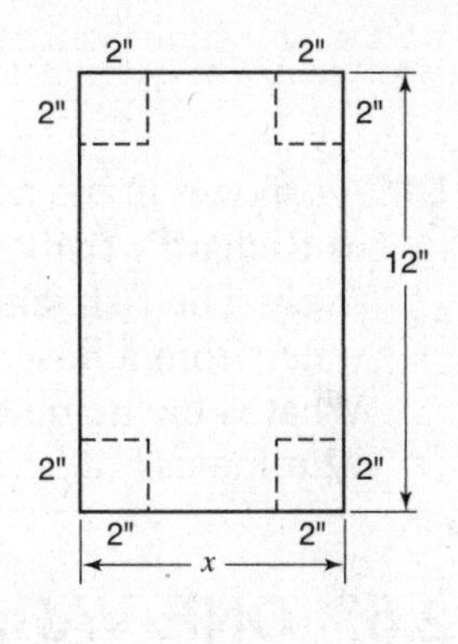

19. The lengths of the sides of a triangle are consecutive even integers. The perimeter of the triangle is the same as the perimeter of a square whose side is 5 less than the shortest side of the triangle. Find the length of the longest side of the triangle.

20. A candy store sells 8-pound bags of mixed hazelnuts and cashews. The price p of each bag can be found using the formula $p = 2.59c + 1.72(8 - c)$, where c is the number of pounds of cashews that are in the bag. If the price of a bag is \$18.11, how many pounds of cashews does it contain?

21. In a triangle in which the sides have different lengths, the length of the shortest side is 5 less than the length of the next longest side. The length of the longest side of the triangle is 1 more than three times the length of the shortest side. If the perimeter of the triangle is 21, what are the lengths of the three sides of the triangle?

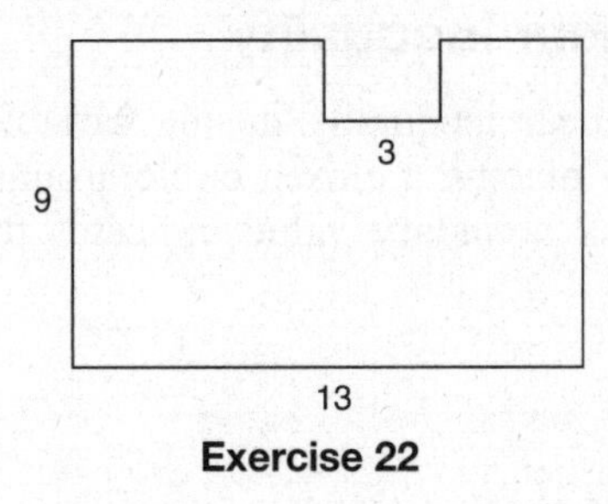

Exercise 22

22. A square is cut from a rectangular board to make the figure shown in the accompanying diagram. The rectangular board measured 9 inches by 13 inches before the cut, and the square cutout measures 3 inches on each side. Find the perimeter and area of the new figure.

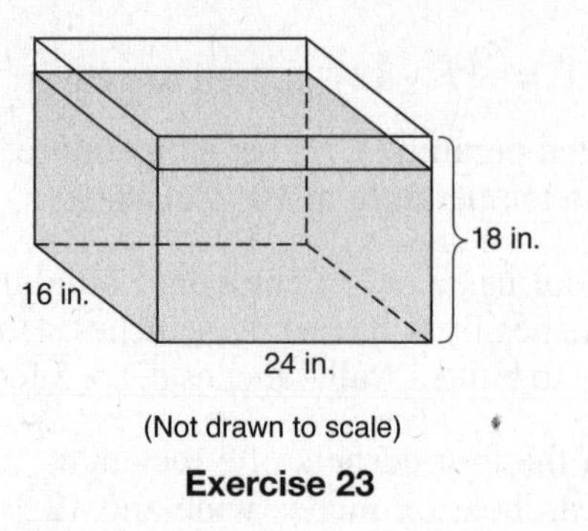

(Not drawn to scale)

Exercise 23

23. As shown in the accompanying diagram, the length, width, and height of Richard's fish tank are 24 inches, 16 inches, and 18 inches, respectively. The fish tank is empty when Richard begins to fill the tank with water from a hose at the rate of 480 cubic inches of water per minute. What is the number of inches in the depth of the water in the tank after 12 minutes?

2.6 ONE-VARIABLE LINEAR INEQUALITIES

KEY IDEAS

A linear inequality in one variable is solved in much the same way as a linear equation. The solution set may include a specific number of solutions or an infinite number of solutions depending on the replacement set for the variable. In either case, the solution set can be represented by a number line graph.

Graphing a Linear Inequality

When graphing a linear inequality in one variable, show that a boundary value is included by placing a closed circle around it, as in Figure 2.1. An open circle around a boundary value indicates that the interval does *not* include that value.

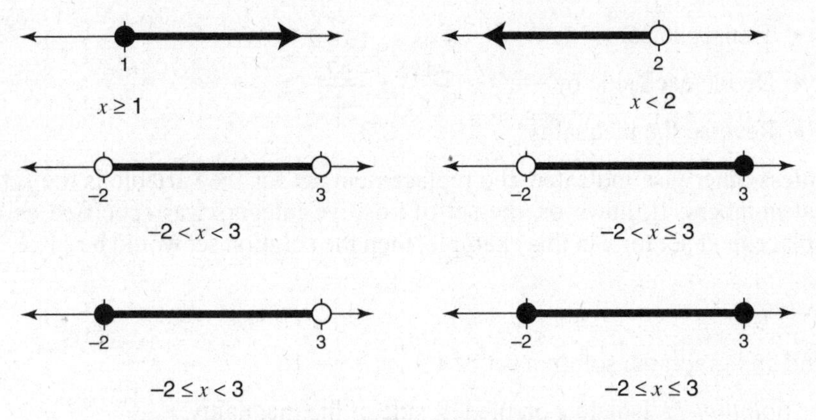

Figure 2.1 Graphing a One-Variable Linear Inequality

Example 1

In order to be admitted to a certain ride at an amusement park, a child must be greater than or equal to 36 inches tall and less than 48 inches tall. If h represents the child's height, write and graph an inequality that represents these conditions.

Solution: The inequality is $\mathbf{36 \leq h < 48}$:

24 26 28 30 32 34 36 38 40 42 44 46 48 50 52 54

Writing Equivalent Inequalities

An equivalent inequality results when

- the same quantity is added to, or subtracted from, both sides.
- both sides are multiplied or divided by the same positive quantity.
- both sides are multiplied or divided by the same **negative** quantity *and* the direction of the inequality sign is reversed. For example,

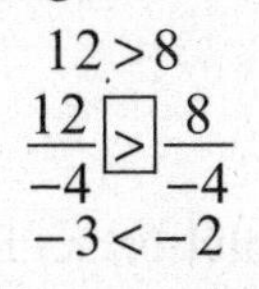

- the sides of the inequality are interchanged *and* the inequality sign is reversed. For example, $2 < 5$, so $5 > 2$.

Solving a Linear Inequality

To solve a linear inequality in one variable, isolate the variable in the usual way *except* that the direction of the inequality must be reversed when both sides of an inequality are multiplied or divided by the same *negative* number. To find the solution set of $25 - 4x \geq 13$:

- Transpose 25: $-4x \geq 13-25$
- Divide each side by -4: $\frac{-4x}{-4} \boxed{\geq} \frac{-12}{-4}$
- Reverse the inequality: $x \leq 3$

Unless otherwise indicated, the replacement set for the variable is the set of real numbers. If, however, the set of positive integers was specified as the replacement set for x in this example, then the solution set would be $\{1, 2, 3\}$.

Example 2

Find and graph the solution set of $1 - 2x \geq x + 16$.

Solution 1: Isolate x on the left side of the inequality.

Transpose 1 and x: $-2x - x \geq 16 - 1$

$-3x \geq 15$

Divide both sides by -3 *and* reverse the inequality: $\downarrow$

$\frac{-3x}{-3} \boxed{\leq} \frac{15}{-3}$

$\boldsymbol{x \leq -5}$

Solution 2: Isolate x on the right side of the inequality.

Transpose 16 and $-2x$: $1-16 \geq x+2x$

$-15 \geq 3x$

Divide both sides by 3: $\frac{-15}{3} \geq \frac{3x}{3}$

$-5 \geq x$

Interchange the sides and reverse the inequality sign: $\boldsymbol{x \leq -5}$

Graph the solution set. Since the boundary value of -5 is included, place a closed circle around it:

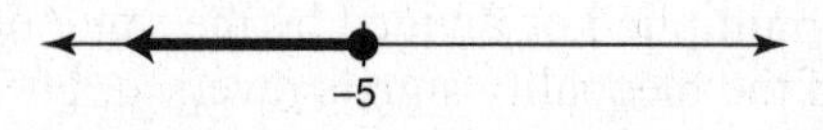

Example 3

Which graph represents the solution set for the intersection of $\{x \mid 1 - 2x \leq 9\}$ and $\{x \mid 3x + 1 < 7\}$?

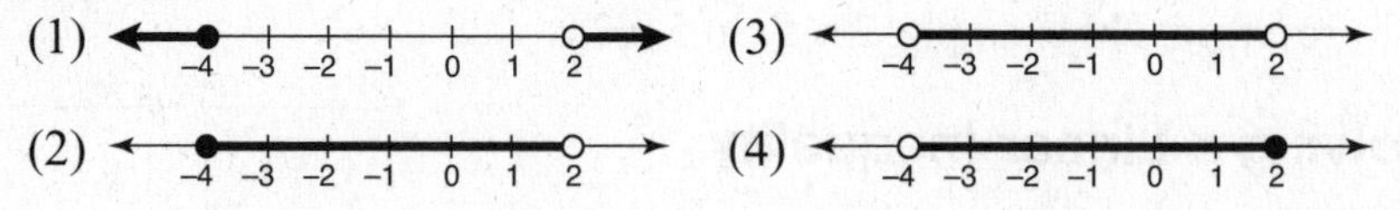

Solution: Find the intersection of the solution sets of the two given inequalities:

- If $1 - 2x \le 9$, then $-2x \le 8$ so $\frac{-2x}{-2} \ge \frac{8}{-2}$ and $x \ge -4$.
- If $3x + 1 < 7$, then $3x < 6$ so $\frac{2x}{2} < \frac{6}{2}$ and $x < 2$.
- Because the intersection of the two given sets is equivalent to the intersection of $\{x \mid x \ge -4\}$ and $\{x \mid x < 2\}$, the graph is the interval on the number line for which x is greater than or equal to -4 *and*, at the same time, less than 2. This interval is described by the inequality $-4 \le x < 2$, which is represented by the graph in choice **(2)**.

Example 4

Chris rents a booth at a flea market at a cost of $75 for one day. At the flea market Chris sells picture frames each of which costs him $6.00. If Chris sells each picture frame for $13, how many picture frames must he sell to make a profit of *at least* $200 for that day?

Solution: Use the relationship: revenue − total cost = profit. If Chris sells x picture frames, then the total cost is what the x frames cost Chris, which is $6x$, plus the $75 rental fee. Thus,

$$\overbrace{13x}^{\text{revenue}} - \overbrace{(6x+75)}^{\text{total cost}} \ge \overbrace{200}^{\text{profit}}$$

$$13x - 6x - 75 \ge 200$$

$$7x \ge 200 + 75$$

$$\frac{7x}{7} \ge \frac{275}{7}$$

$$x \ge 39.28$$

Since x must be a whole number, Chris must sell at least **40** picture frames.

Check Your Understanding of Section 2.6

A. *Multiple Choice.*

1. Which number is in the solution set of $5x + 3 > 38$?
(1) 5 (2) 6 (3) 7 (4) 8

2. When x is an integer, what is the solution set of $5 \le x < 8$?
(1) $\{5, 6, 7, 8\}$ (2) $\{5, 6, 7\}$ (3) $\{6, 7, 8\}$ (4) $\{6, 7\}$

3. What is the value of x in the inequality $14 \geq 3x + 2$?
(1) $-4 \geq x$ (2) $-4 \leq x$ (3) $4 \geq x$ (4) $4 \leq x$

4. What is the solution set of $2x - 3 < 5$ when the replacement set is the set of nonnegative integers?
(1) $\{0, 1, 2, 3\}$ (2) $\{1, 2, 3\}$ (3) $\{0, 1, 2, 3, 4\}$ (4) $\{1, 2, 3, 4\}$

5. The expression $-6 \leq x + 4$ is equivalent to
(1) $x \leq 10$ (2) $x \leq -10$ (3) $x \geq -2$ (4) $x \geq -10$

6. Which inequality is equivalent to $\frac{2}{3}x - 5 < 11$?
(1) $x < 6$ (2) $x < 9$ (3) $x < 16$ (4) $x < 24$

7. Which inequality is represented in the accompanying graph

−3 0 4

(1) $\{-3 < x \leq 4\}$
(2) $\{-3 \leq x < 4\}$
(3) $\{-3 \leq x \leq 4\}$
(4) $\{-3 < x < 4\}$

8. Which graph represents the inequality $-1 \leq x < 4$?

(1) −2 −1 0 1 2 3 4 5

(2) −2 −1 0 1 2 3 4 5

(3) −2 −1 0 1 2 3 4 5

(4) −2 −1 0 1 2 3 4 5

9. The statement "$x + 1 \geq 5$ and $2x - 4 < 6$" is true when x is equal to
(1) 1 (2) 10 (3) 5 (4) 4

10. What is the solution set for n when $3n - 2 \geq n + 6$ and n is any real number?
(1) $\{n \mid n \geq 2\}$
(2) $\{n \mid n = 2\}$
(3) $\{n \mid 3 < n < 6\}$
(4) $\{n \mid n \geq 4\}$

11. Thelma and Louis start a lawn-mowing business and buy a lawn mower for \$225. They plan to charge \$15 to mow one lawn. What is the *minimum* number of lawns they need to mow if they wish to earn a profit of at least \$750?
(1) 50 (2) 55 (3) 60 (4) 65

B. *Show how you arrived at your answer.*

12. Rewrite $\{x \mid 17 > 3x + 2$, where x is a whole number$\}$ in roster-form.

13. Find the solution set of $-3x + 1 \leq 17$, where the replacement set is the set of negative integers.

14–17. Find and graph the solution set.

14. $3(x + 1) > -6$ **15.** $1 - 5x \geq -9$ **16.** $7 \leq 3x - 2$ **17.** $3(x - 1) < 5x + 7$

18. Bart is saving to buy a flat screen computer monitor that costs \$345. He has already saved \$37. What is the least amount of money Bart must save each week so that at the end of 11 weeks he has enough money, excluding tax, to buy the monitor?

19. Forty-seven fewer boys than girls attend a school. If the school has at most 925 students in attendance, find the greatest number of boys and the greatest number of girls who attend the school.

20. Mr. Allen has \$75.00 to spend on pizzas and soda pop for a picnic. Pizzas cost \$9.00 each, and the drinks cost \$0.75 each. Five times as many drinks as pizzas are needed. What is the maximum number of pizzas that Mr. Allen can buy?

21. The cost of a telephone call from Wilson, NY, to East Meadow, NY, is \$0.60 for the first three minutes plus \$0.17 for each *additional* minute. What is the greatest number of whole minutes of a telephone call if the cost cannot exceed \$2.50?

22. The Eye Surgery Institute just purchased a new laser machine for \$500,000 to use during eye surgery. After the first 8 uses of the laser machine, the Institute must pay the inventor \$550 each additional time the machine is used. If the Institute charges \$2,000 for each laser surgery, what is the *minimum* number of laser eye surgeries that must be performed in order for the Institute to make a profit?

23. A factory packs CD cases into cartons for a music company. Each carton is designed to hold 1,152 CD cases. The Quality Control Unit in the factory expects an error of less than 5% over or under the desired packing number. Let x represent the number of CD cases that are packed in a particular carton. Write and then graph a single inequality that describes the set of all possible values of x that are acceptable to the Quality Control Unit.

24. A restaurant sells large and small submarine sandwiches. Rolls for the sandwiches are ordered from a baker. The roll for a large sandwich costs \$0.25 and the roll for a small sandwich costs \$0.15. Melissa, the manager of the restaurant, ordered 130 more large rolls than small rolls. What was the greatest number of large rolls Melissa received if she spent *less than* \$63?

CHAPTER 3

PROBLEM SOLVING AND TECHNOLOGY

3.1 PROBLEM SOLVING STRATEGIES

KEY IDEAS

Problem solving strategies do not tell you the specific steps to follow when solving a problem. Instead, they suggest things to try, which may be helpful when solving unfamiliar types of problems for which a standard algebraic solution does not seem to work.

Strategy 1: Draw a Diagram

Draw a diagram to help visualize a problem situation and organize the important facts.

Example 1

Amy goes shopping and spends one-third of her money on a new dress. She then goes to another store and spends one-half of the money she has left on shoes. If Amy has \$56 left after these two purchases, how much money did she have when she started shopping?

Solution: Use a rectangle to represent the amount of money Amy had when she started shopping. Since Amy spends $\frac{1}{3}$ of her money on a dress, divide the rectangle into 3 equal parts:

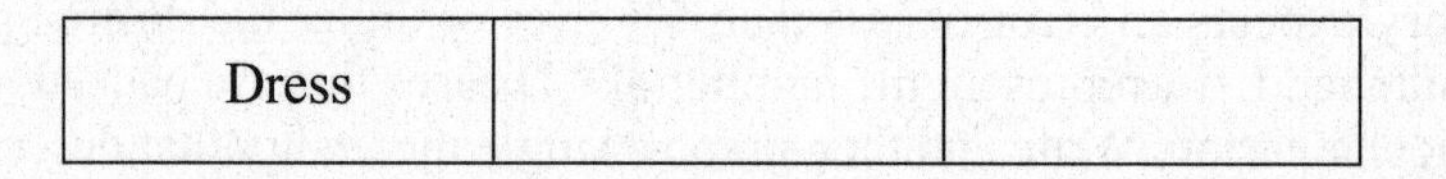

Two rectangles remain. Amy spends $\frac{1}{2}$ of the *remaining* money on shoes.

Hence, the second rectangle represents the amount of money Amy spends on shoes. Since \$56 is left, fill in the last rectangle with this amount:

Dress	Shoes	\$56

Since each of the three rectangles are equal, Amy started shopping with $3 \times \$56 = \$168$.

Strategy 2: Try Easy Numbers

If a problem does not use specific numbers, work out the problem using easy numbers.

Example 2

The perimeter of a rectangle is 10 times as great as the width of the rectangle. The length of the rectangle is how many times as great as its width?

Solution: Let ℓ represent the length of the rectangle and w its width.

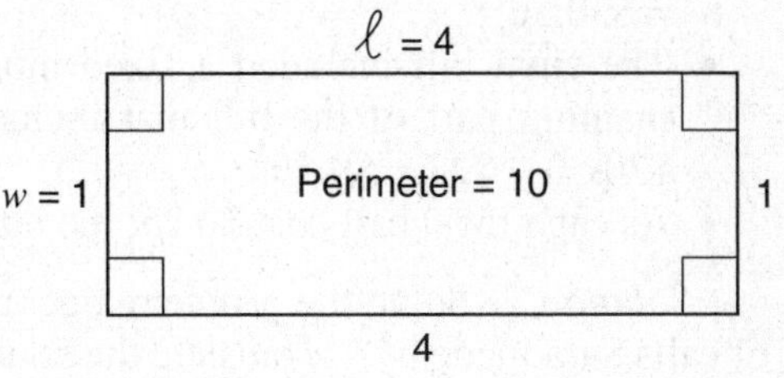

- Because the actual dimensions of the rectangle are not given, use an easy number such as 1 as the width of the rectangle. This makes the perimeter of the rectangle 10×1 or 10.
- Substituting $w = 1$ and $p = 10$ into the perimeter formula $p = 2(\ell + w)$ gives $10 = 2(\ell + 1)$ so $2\ell + 2 = 10$, $2\ell = 8$, and $\ell = \frac{8}{2} = 4$.
- Since $\ell = 4$ when $w = 1$, the length is **4 times** as great as the width.

Example 3

The length and width of a rectangular box are each doubled and the height is tripled. The volume of the new box is how many times as great as the volume of the original box?

Solution: Because the actual dimensions of the box are not given, work out the problem assuming the length, width, and height of the box are each 1 inch.

- The volume of the rectangular box is length × width × height = $1 \times 1 \times 1 = 1$ cubic inch.
- When the length and width of this box are each doubled and the height is tripled, the dimensions of the box become 2 inches by 2 inches by 3 inches. The volume of the new box is $2 \times 2 \times 3 = 12$ cubic inches.
- The volume of the new box is **12 times** the volume of the original box.

Strategy 3: Work Backward

When you only know the end result of a computation and want to find the beginning value, reverse the steps that led to that final result.

Example 4

Sara's telephone service cost \$21 per month plus \$0.25 for each local call, and long-distance calls are extra. Last month, Sara's bill was \$36.64, and it included \$6.14 in long-distance charges. How many calls did she make?

Solution: Work back from the fact that Sara's final bill was \$36.64.

- Since the final bill included \$6.14 in long-distance charges, the part of the bill that did not include any long-distance charges is \$36.64 – \$6.14 = \$30.50.
- The final bill included a fixed monthly charge of \$21. Thus, the remaining part of the bill that includes only charges for local calls is \$30.50 – \$21 = \$9.50.
- As each local call cost \$0.25, the number of local calls is $\frac{\$9.50}{\$0.25} = \mathbf{38}$.

Solution 2: Solve the problem algebraically by representing the number of calls Sara made by x. Translate the conditions of the problem into an equation:

$$\begin{aligned} 21 + 0.25x + 6.14 &= 36.64 \\ 0.25x + 27.14 &= 36.64 \\ 0.25x &= 36.64 - 27.14 \\ \frac{0.25x}{0.25} &= \frac{9.50}{0.25} \\ x &= \mathbf{38} \end{aligned}$$

Strategy 4: Find a Pattern

When the mathematical relationship between the variable quantities in a problem is not clear, try working out the problem using a series of easy numbers until you recognize a pattern that allows you to write an algebraic equation.

Example 5

A video store charges \$4.95 for a rental of two days and \$2.50 for each day the video is not returned after the second day. If John is charged \$27.45 for a rental, for how many days did John rent the video?

Solution 1: Find the cost, C, of renting the video for 1 day, 2 days, 3 days, . . . , x days, as shown in the accompanying table. When $x \geq 2$,

$$C = 4.95 + (x - 2) \times 2.50$$

To find the number of days John rented the video when he is charged \$27.45, solve the equation for x when C = \$27.45:

Number of Days Rented	C
2	4.95
3	$4.95 + 1 \times 2.50$
4	$4.95 + 2 \times 2.50$
•	•
•	•
•	•
x	$4.95 + (x - 2) \times 2.50$

$$\begin{aligned} C &= 4.95 + (x-2) \times 2.50 \\ 27.45 &= 4.95 + 2.50x - 5 \\ 27.45 &= -0.05 + 2.50x \\ 27.45 + 0.05 &= 2.50x \\ 2.50x &= 27.50 \\ \frac{2.50x}{2.50} &= \frac{27.50}{2.50} \\ x &= 11 \end{aligned}$$

John rented the video for **11** days.

Solution 2: Solve the problem numerically by working backward.

- Charge for first 2 days was \$4.95, which leaves \$27.45 − \$4.95 = \$22.50.
- At a rental fee of \$2.50 per day, the number of additional rental days was $\frac{\$22.50}{\$2.50} = 9$ days.
- Thus, John rented the video for a total of 2 days + 9 days = **11** days.

Strategy 5: Organize Data

Shopping in a supermarket for a large number of groceries becomes easier and more efficient if you first prepare an organized list of the groceries you need to buy. Similarly, organizing the facts of a problem in a convenient form such as a list or table may help you solve a problem.

Example 6

Gary and Sue are thinking of the same number. Gary says, "The number is a positive odd integer that is *at most* 25." Sue says, "The number is a prime number." If Gary's statement is true and Sue's statement is false, what are all the possible numbers?

Solution: Organize the facts of the problem by making a list of the possible numbers.

- If Gary's statement is *true*, the number may be one of the following:

  1, 3, 5, 7, 9, 11, 13, 15, 17, 19, 21, 23, 25

- A prime number is a whole number greater than 1 that is divisible only by itself and 1. If Sue's statement is *false*, the number is not a prime number, so eight numbers can be eliminated from Gary's list:

  1, ~~3~~, ~~5~~, ~~7~~, 9, ~~11~~, ~~13~~, 15, ~~17~~, ~~19~~, 21, ~~23~~, 25

The remaining numbers in Gary's list are **1, 9, 15, 21, and 25**.

Example 7

Five friends met for lunch, and they all shook hands. Each person shook the other person's right hand only once. What was the total number of handshakes?

Solution: Make a list of ordered pairs in which the five friends are referred to as A, B, C, D, and E. The ordered pair (A,B), for example, represents person A shaking hands with person B. Be careful not to include two different ordered pairs that represent the same two people shaking hands. For example, do not include both (A,B) and (B,A) since they represent the same two friends shaking hands.

1. (A,B)	**6.** (B,D)
2. (A,C)	**7.** (B,E)
3. (A,D)	**8.** (C,D)
4. (A,E)	**9.** (C,E)
5. (B,C)	**10.** (D,E)

Since 10 ordered pairs are listed, the total number of handshakes was **10**.

Strategy 6: Guess, Check, and Revise

When you can't figure out an algebraic solution, you may be able to make a reasonable numerical guess and then use that guess to arrive at a better guess. You can then repeat this process until you reach the correct answer.

Example 8

A soda machine contains 20 coins. Some of the coins are nickels, and the rest are quarters. If the value of the coins is \$4.40, find the number of coins of each kind.

Solution: Guess intelligently by reasoning as follows:

- If all 20 coins were nickels, the value of the coins in the soda machine would be $20 \times \$0.05 = \$1.00$.
- If all 20 coins were quarters, the value of the coins in the soda machine would be $20 \times \$0.25 = \$5.00$.
- Since \$5.00 is closer than \$1.00 to the desired amount of \$4.40, guess that the soda machine contains 19 quarters and 1 nickel. Then check the guess. If that guess doesn't work, reduce the number of quarters by 1 and check again. Continue this process until you reach the correct answer. To help keep track of your guesses, organize your work in a table:

Guess		
Quarters	**Nickels**	**Total Value of Coins**
19	$20 - 19 = 1$	$(19 \times \$0.25) + (1 \times \$0.05) = \$4.80 \leftarrow$ Too high
18	$20 - 18 = 2$	$(18 \times \$0.25) + (2 \times \$0.05) = \$4.60 \leftarrow$ Too high
17	$20 - 17 = 3$	$(17 \times \$0.25) + (3 \times \$0.05) = \$4.40 \leftarrow$ Correct total!

Thus, the soda machine contains **17 quarters** and **3 nickels**. When using this strategy to solve a problem on the NYS Regents exam, it is expected that you show the work for at least three guesses. If you obtain the correct answer on your first try, then demonstrate that guesses below and above the correct guess do not work.

Strategy 7: Account for All Possible Cases

Solving a problem may depend on breaking it down so that all possible cases or types of numerical values of a variable are considered.

Example 9

Ryan claims that the inequality $\frac{1}{x} \leq x^2$ is always true provided x is not equal to 0. Explain whether you agree or disagree.

Solution: Disagree. Although it is easy to see that the given inequality is true when x is greater than or equal to 1, Ryan needs to consider all possible types of number substitutions for x. Is the given inequality true for values of x between 0 and 1? Is it true for negative x values?

Test the truth of the inequality for any convenient value of x that is between 0 and 1 such as $x = \frac{1}{2}$. If $x = \frac{1}{2}$, then $\frac{1}{x} = 2$. Substituting these values in the original inequality gives $2 \leq \left(\frac{1}{2}\right)^2$ which is **false** since $\left(\frac{1}{2}\right)^2 = \frac{1}{4}$. Hence, Ryan's statement is *not* always true.

In this solution, $x = \frac{1}{2}$ served as a *counterexample*. A **counterexample** is a single, specific instance that contradicts a proposed generalization. Sometimes the easiest way to disprove a supposed fact is to find a counterexample.

Selecting a Problem Solving Strategy

A problem can often be solved in more than one way. When solving a nonroutine problem in which you are not required to use a particular method, you should use whatever strategy or combination of strategies works for you, even if the strategy is not one of those summarized in the accompanying table.

Problem Solving Strategies

- **Draw a diagram.** Visualize a problem situation by drawing a diagram that may help you organize the given information and find the clues needed to solve the problem.
- **Try easy numbers**. When no numbers are given, make up easy numbers and then work out the problem using those numbers.
- **Work backward.** When the final result but not the starting value is given, reverse the steps needed to get that value.
- **Find a pattern.** Look at a series of specific cases until you find a pattern that answers the question or that can be generalized into an algebraic equation.

- **Organize data.** Arrange the data in an organized list or table so that important facts are not overlooked while any patterns or logical relationships become easier to see.
- **Guess, check, and revise.** Make an intelligent guess, and then use that guess to obtain a better guess. Continue this process until you get the correct answer. When using this strategy, show the work for at least *three* guesses.
- **Account for all possible cases.** Consider all possible types of values of a variable including 0 and 1 as well as numbers that are between 0 and 1, greater than 1, and less than 0.

Check Your Understanding of Section 3.1

A. *Multiple Choice.*

1. If n represents an odd number, which computation results in an answer that is an even number?
(1) $2 \times n + 1$ (2) $2 \times n - 1$ (3) $3 \times n - 2$ (4) $3 \times n + 1$

2. If $\frac{b-2}{5}$ represents a positive integer, what is the remainder when b is divided by 5?
(1) 1 (2) 2 (3) 3 (4) 4

3. The length of a rectangular box is doubled, and its width is tripled. In order for the volume of the box to remain the same, the original height of the box must be
(1) multiplied by 6
(2) divided by 6
(3) divided by 5
(4) decreased by 6

4. If n pencils cost c cents, what is the cost in cents of p pencils?
(1) $\frac{pc}{n}$ (2) $\frac{nc}{p}$ (3) $\frac{c}{np}$ (4) $c - \frac{p}{n}c$

5. Parking charges at Superior Parking Garage are $5.00 for the first hour and $1.50 for each additional 30 minutes. If Margo has $12.50, what is the maximum number of hours she will be able to park her car at the garage?
(1) $2\frac{1}{2}$ (2) $3\frac{1}{2}$ (3) 6 (4) $6\frac{1}{2}$

6. The perimeter of a rectangle is six times as great as the perimeter of a square. The width of the rectangle is double the length of a side of the square. The length of the rectangle is how many times as great as its width?
(1) 10 (2) 8 (3) 5 (4) 4

7. The length of rectangle I is 10% less than the length of rectangle II. The width of rectangle I is 20% less than the width of rectangle II. The area of rectangle I is what percent of the area of rectangle II?
(1) 70% (2) 72% (3) 30% (4) 28%

B. *Show how you arrived at your answer.*

8. Sue bought a skirt on sale for half off the original price. The store charged \$3.40 tax and her final cost was \$22. What was the original price of the skirt?

9. Bob and Ray are describing the same number. Bob says, "The number is a positive even integer less than or equal to 20." Ray says, "The number is divisible by 4." If Bob's statement is true and Ray's statement is false, what are all the possible numbers?

10. Carl runs $\frac{3}{4}$ miles in 5 minutes 30 seconds and Sonya runs $\frac{2}{3}$ mile in 5 minutes 20 seconds. At the same average rates, how many seconds longer will Sonya take to run a mile than Carl will take?

11. After Frank unloaded $\frac{1}{8}$ of the cartons in a truck and Tom unloaded $\frac{2}{7}$ of the remaining cartons in the truck, there were 45 cartons left. How many cartons were originally in the truck?

12. Cindy makes a list of numbers in which each number after the first is 3 more than twice the number that comes before it. If the third number in Cindy's list is 19, what is the first number in the list?

13. Dalia and Miguel are describing the same positive number. Dalia says, "If the number is increased by 4, it is a perfect square less than 100." Miguel adds, "If 3 is subtracted from the number, the result is a prime number greater than 2." What is the number that Dalia and Miguel are describing?

14. The number of bacteria doubles every 20 minutes. The bacteria population was 16 million at 1:00 PM. How many bacteria, expressed in thousands, were present at 11:00 AM. on the same day?

15. Margarita spends one-fifth of her salary on a belt and one-fourth of the remaining money on a shirt. If she has \$72 left after making these two purchases, what is her salary?

16. Rick, Mark, Vanessa, Sandy, and Ariela each solved the same math problem in a different amount of time: 5 minutes, 6 minutes, 9 minutes, 10 minutes, and 14 minutes. Vanessa needed 4 minutes longer than Sandy. Only one person needed more time than Rick. Ariela finished before Vanessa. How many minutes did each person need to solve the problem?

17. A doughnut shop charges \$0.65 for each doughnut and \$0.35 for a carryout box. Shirley has \$5.00 to spend. At most, how many doughnuts can she buy if she also wants them in one carryout box?

18. Three boys agree to divide a package of baseball cards in the following way: Jose takes one-half of the cards, Shawn takes one-third of the remaining cards, and Yin takes the remaining cards. If Yin takes 12 cards, how many cards does Jose take?

19. In his will, a man leaves one-half of his money to his wife, one-half of what is then left to his older child, and one-half of what is then left to his younger child. His two cousins divide the remainder equally, each receiving \$2,000. What was the total amount of money in the man's will?

20. Charles claims that $x^2 > y^2$ whenever $x > y$. Explain whether you agree or disagree.

21. Ray claims that if n is a positive integer, then $n^2 + n + 41$ represents a prime number. Valerie proved that Ray was incorrect. Show how Valerie could prove her case.

22. Beth plans to swim *at least* 100 laps during a 6-day period. During this period, she will increase the number of laps completed each day by one lap. What is the *least* number of laps Beth must complete the first day?

23. The pages of a book are numbered using the digits 0, 1, 2, 3, . . . , 9. For example, page number 31 contains the two digits, 3 and 1. If 897 digits are needed to number the pages of a book, how many numbered pages does the book contain?

24. Tamara has two sisters. One of the sisters is 7 years older than Tamara. The other sister is 3 years younger than Tamara. The product of Tamara's sisters' ages is 24. How old is Tamara?

25. The accompanying figure shows a correctly worked out addition problem in which each letter represents a different digit from 1 to 9. What digit does D represent?

$$\begin{array}{r} 83A \\ +\,DBB \\ \hline CAC2 \end{array}$$

3.2 USING A GRAPHING CALCULATOR

KEY IDEAS

A graphing calculator can be used to quickly see what the graph of an equation looks like. Calculator discussions are based on the widely used Texas Instruments TI-83/84 graphing calculator. If you are using a different graphing calculator, you may have to make minor adjustments in the calculator procedures that are described throughout this book.

The Coordinate Plane

A **coordinate plane** is formed when a horizontal number line, called the ***x*-axis**, intersects a vertical number line, called the ***y*-axis**, at a point called the **origin**. As shown in Figure 3.1, the coordinate axes divide the coordinate plane into four quadrants numbered counterclockwise. The location of a point in the coordinate plane is given by an ordered pair of numbers (x, y) which are the **coordinates** of the point. The order in which numbers appear within parentheses matters. If the coordinates of point A are (4, 3), then the x-coordinate is 4 and the y-coordinate is 3. As illustrated in Figure 3.1,

- the x-coordinate of a point corresponds to its horizontal distance from the origin. The sign of x indicates whether the point is to the right (+) or to the left (–) of the origin.
- the y-coordinate of a point indicates its vertical distance from the origin. The sign of y tells whether the point is above (+) or below (–) the origin.

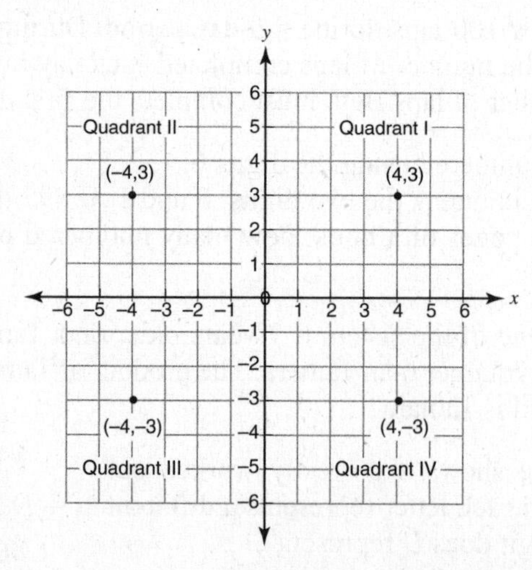

Quadrant	(x, y)
I	(+, +)
II	(–, +)
III	(–, –)
IV	(+, –)

Figure 3.1 The Coordinate Plane

The Viewing Window

The rectangular viewing window of a graphing calculator shows only a small region of the coordinate plane. In a **standard window** the positive and negative coordinate axes each have 10 tick marks, as shown in Figure 3.2.

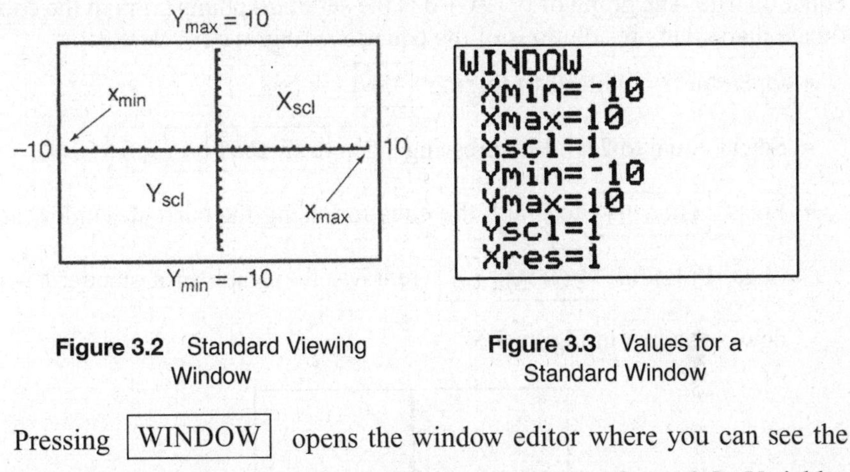

Figure 3.2 Standard Viewing Window

Figure 3.3 Values for a Standard Window

Pressing WINDOW opens the window editor where you can see the current values of the window variables, as shown in Figure 3.3. Variables Xscl and Yscl determine the scales of the coordinate axes. For example, Xscl = 1 scales the *x*-axis so that consecutive tick marks along this axis are 1 unit apart. Setting Xscl = 5 scales the *x*-axis so that consecutive tick marks are 5 units apart. The variable X_{res} refers to the screen resolution.

Changing the Viewing Window

When the values of the window screen variables are changed, different sections of the coordinate plane can be brought into view. The value of a window variable can be changed in the window editor either by overwriting the current value or by entering an arithmetic expression that evaluates to the desired value. For example, the value of Xmax in Figure 3.3 can be changed from 10 to 20 by positioning the blinking cursor after Xmax = 10 and multiplying that value by 2, as shown in Figure 3.4. Pressing ENTER , or moving the cursor to another line, sets Xmax = 20.

```
WINDOW
  Xmin=-10
  Xmax=10*2
  Xscl=1
  Ymin=-10
  Ymax=10
  Yscl=1
  Xres=1
```

Figure 3.4 Changing Xmax to 20

Graphing a Linear Equation

A first degree equation in two variables such as $y = 2x + 3$ is also called a **linear equation** since its graph is a straight line. The ordered number pair (1, 5) is a **solution** of $y = 2x + 3$ since substituting 1 for x and 5 for y makes the equation true. The graph of $y = 2x + 3$ is the set of *all* points (x, y) in the coordinate plane that are solutions of the equation. To graph $y = 2x + 3$:

- Open the $Y=$ editor by pressing [$Y=$].
- Set Y_1 equal to $2x + 3$ by pressing [2] [x, T, θ, n] [+] [3].
- Press [GRAPH] to graph the equation using the current window settings. Pressing [ZOOM] [6] redraws the graph in a standard window, as shown in Figure 3.5.

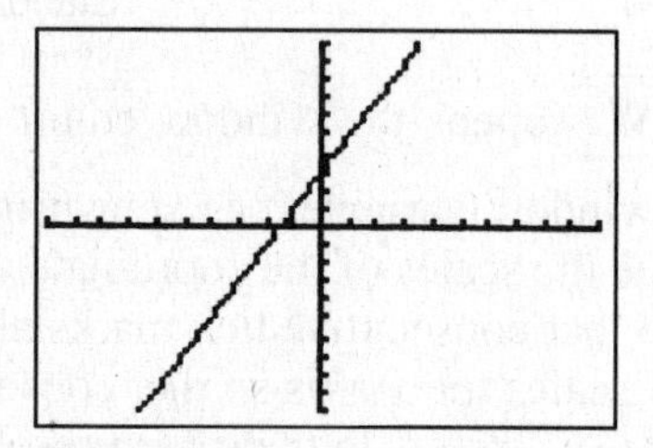

Figure 3.5 Graph of $y = 2x + 3$

MATH FACTS

- Before an equation can be entered in the $Y=$ editor, it must be in a form in which y is solved for in terms of x. To graph $4y - 3x = 8$, first solve for y: $4y = 3x + 8$ so $Y_1 = \frac{3}{4}x + 2$. If Y_1 has already been used, you can set $Y_2 = \frac{3}{4}x + 2$. Up to 10 different equations can be stored in the $Y=$ editor.
- If the coefficient of x is negative, as in $Y_1 = -2x$, the negative sign is entered by pressing [(–)] rather than [–] which indicates subtraction.
- When storing an equation such as $Y_1 = 6(x - 1) - x$, it is *not* necessary to simplify the right side of the equation before entering it.

The Trace Feature

By pressing TRACE you can move the cursor from one plotted point on a graph to the next. The equation of the graph appears in the upper corner of Quadrant II while the coordinates of the cursor's current location on the graph are displayed at the bottom of the screen, as shown in Figure 3.6.

- Pressing the right or left cursor arrow keys moves the cursor up or down with the coordinates of its current position appearing at the bottom of the screen.

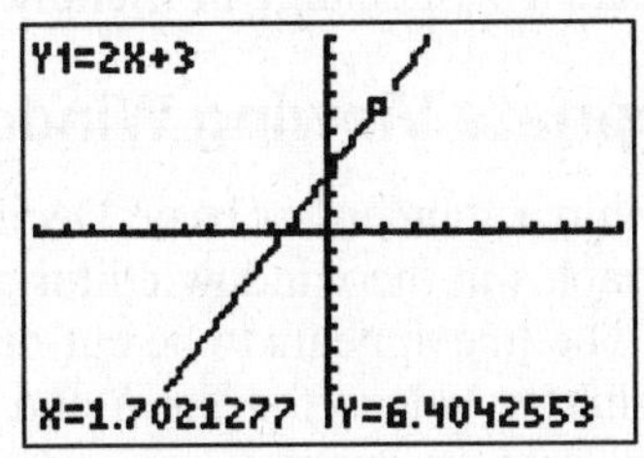

Figure 3.6 Trace of $y = 2x + 3$

- Because the line in Figure 3.6 is graphed in a standard window, the trace coordinates are not "friendly" numbers. To remedy this situation, use the ZDecimal command by pressing ZOOM 4 . The screen dimensions are instantly changed to the preset values in Figure 3.7, called a **Decimal window**, and the graph is redrawn using these values. The x-coordinate of the trace cursor now changes in friendly steps of 0.1, as shown in Figure 3.8.

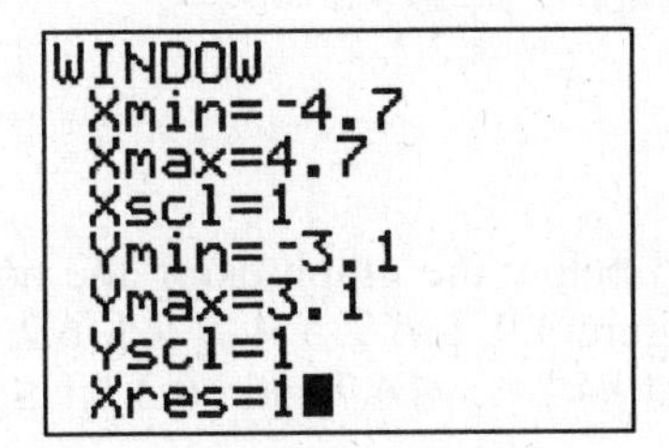

WINDOW
Xmin=-4.7
Xmax=4.7
Xscl=1
Ymin=-3.1
Ymax=3.1
Yscl=1
Xres=1

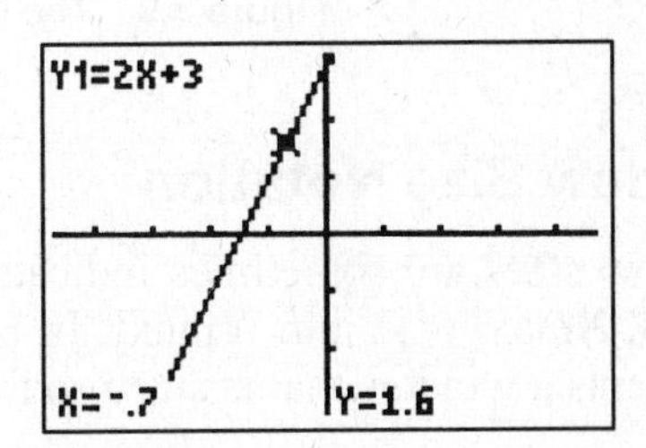

Figure 3.7 Decimal Window Values **Figure 3.8** Trace of $y = 2x + 3$

Decimal Windows

Graphs viewed in a Decimal window maintain their true size proportions so that the graph of a circle, for example, has a circular rather than an oval shape.

- Different Decimal windows can be created by multiplying the values of the four screen variables in Figure 3.7 by the same positive integer.
- When only the values of Xmin and Xmax in Figure 3.7 are multiplied by the same positive integer, a new *friendly* window is created. If the width of the viewing rectangle is adjusted so that Xmin = -4.7×2 and Xmax = 4.7×2, then the x-coordinate of the cursor will change in friendly steps of 0.2.
- A friendly window is created whenever the difference between Xmax and Xmin is a multiple of 9.4. If Xmin = −0 and Xmax = 9.4, then the x-coordinate of the cursor will change in friendly steps of 0.1.

Finding an Appropriate Viewing Window

If a graph doesn't fit within a standard or basic Decimal window, change the values of the screen variables in the window editor as needed. Look back at the graph in Figure 3.8. The line appears to be cut-off where it crosses the y-axis. Multiplying the window values for Ymin and Ymax by 2 produces a "complete graph" in which its key features are now visible, as shown in Figure 3.9.

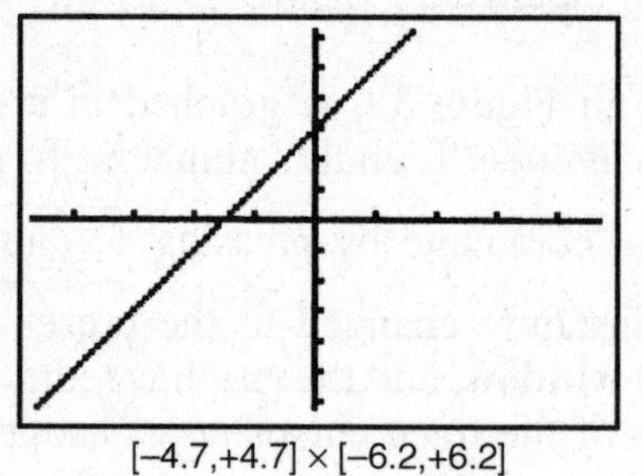

[−4.7,+4.7] × [−6.2,+6.2]

Figure 3.9 The Graph of $y = 2x + 3$

Window Size Notation

Window sizes are sometimes indicated below the graph using the notation [Xmin, Xmax] × [Ymin, Ymax]. In Figure 3.9, [−4.7, + 4.7] × [−6.2, +6.2] represents a window that is sized so that $-4.7 \leq x \leq 4.7$ and $-6.2 \leq y \leq 6.2$.

Using Zoom Commands

The ZStandard and ZDecimal commands are useful when first trying to find an appropriate viewing window. To further refine the viewing window, selecting other Zoom commands from the ZOOM menu may be helpful:

- ZoomFit: Pressing [ZOOM] [0] adjusts the Ymin and Ymax values so that the graph fits between the current values of Xmin and Xmax. Figure 3.10 shows the result of applying the ZoomFit command to the graph of $y = 2x + 3$ in Figure 3.8. The ZoomFit command is especially useful when you already know the Xmin and Xmax values that work and you want the calculator to figure out the corresponding Ymin and Ymax values that produces a complete graph.

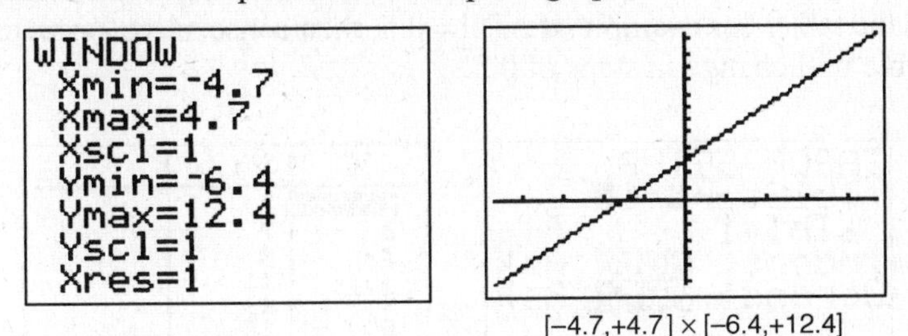

$[-4.7,+4.7] \times [-6.4,+12.4]$

Figure 3.10 Using the ZoomFit Command for $y = 2x + 3$

- ZInteger: Pressing ZOOM 8 ENTER redraws the graph centered at the current cursor location with the x- and y-axes automatically scaled to integer values. The x-coordinate of the trace cursor will now have an integer value. Before pressing ENTER, move the cursor to the point on the screen that you want to become the center of the new screen.

Second Function Keys and Bracket Notation

Most keys on your graphing calculator have two functions. The second function is printed on top of the key. To access the second function, press 2nd and then the key that has the desired function printed above it. For example, the ON key has the label OFF above it. To turn the calculator off, press 2nd [OFF] in succession, where the brackets refer to the calculator key that has OFF printed *above* it.

Using the Table-Building Feature

To list the coordinates of a series of points on the graph of $y = 2x + 3$, set $Y_1 = 2x + 3$ and continue as follows:

- Enter the TABLE SETUP mode by pressing 2nd [TBLSET], where TBLSET is the label above the WINDOW key.

- Set TblStart = 1, as shown in Figure 3.11. This makes 1 the beginning value of X when the table is first displayed. When ΔTbl = 1, consecutive values of X in the table will increase by 1. By adjusting the value of ΔTbl, you can change the amount by which X changes from line to line in the table. For example, if ΔTbl = 0.5, then consecutive X-values in the table will change in steps of 0.5.

Figure 3.11 Table Setup

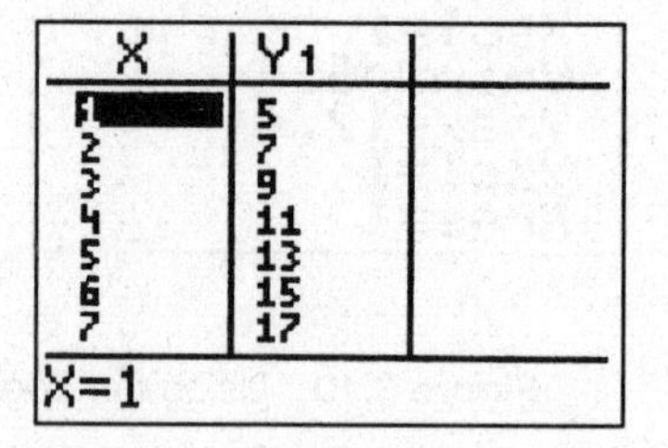

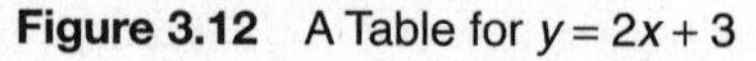

Figure 3.12 A Table for $y = 2x + 3$

- Press [2nd] [TABLE], where TABLE is the label printed above the [GRAPH] key. The table in Figure 3.12 should now appear in the viewing window. To see table values that are not currently in view, press the up or down cursor arrow keys.

Check Your Understanding of Section 3.2

A. *Multiple Choice.*

1. If $x = -3$ and $y = 2$, which point on the accompanying graph represents $(-x, -y)$?

(1) P (2) Q (3) R (4) S

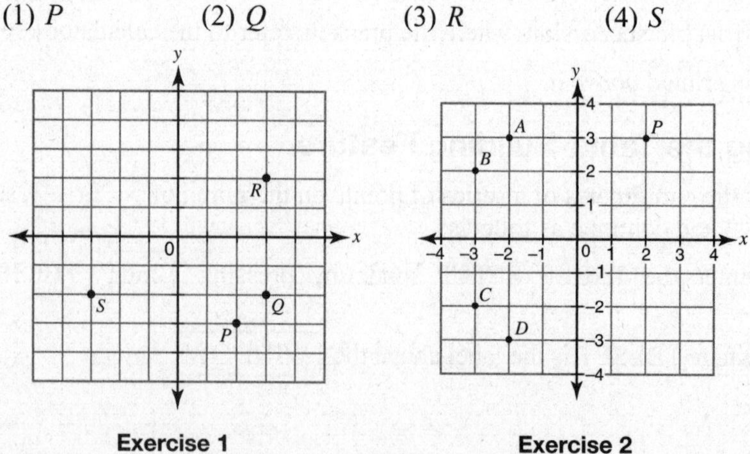

2. In the accompanying graph, if point P has coordinates (a, b), which point has coordinates $(-b, a)$?
(1) A (2) B (3) C (4) D

3. The coordinates of three of the vertices of rectangle $RECT$ are $R(-1, 1)$, $E(3, 1)$, and $C(3, 5)$. What are the coordinates of vertex T?
(1) $(-5, 3)$ (2) $(1, 3)$ (3) $(-1, 5)$ (4) $(3, -5)$

4. If $x = (-2) - (-3)$ and $y = \frac{-9+5}{2}$, in which quadrant is point (x, y) located?
(1) I (2) II (3) III (4) IV

B. *Show how you arrived at your answer.*

5-8. *Graph each equation in an appropriate viewing window.*

5. $y = 2x - 1$ **6.** $y + 3x = 5$

7. $2y - x = 6$ **8.** $y = 13x + 52$

9. Create table in which $Y_1 = x + 3$, $Y_2 = 3x - 5$, and ΔTbl = 1. Using the table, find the ordered pair that satisfies both equations.

10. Create a table in which Y_1 is equal to one side of the equation $2(x + 5) = 1 - x$, and Y_2 is equal to the other side. Use the table to solve the equation.

3.3 COMPARING MATHEMATICAL MODELS

KEY IDEAS

The information contained in a problem situation can be represented or "modeled" in more than one way: algebraically by an equation, numerically by a table or chart, or visually by a graph.

The Cable TV Problem

Suppose you are signing up for a cable television service plan. Plan A costs \$11 per month plus \$7 for each premium channel. Plan B costs \$27 per month plus \$3 for each premium channel. For what number of premium channels will the two plans cost the same? This problem can be solved in at least three different ways.

Solve Algebraically

To solve the cable TV problem algebraically, represent the cost of each plan when x premium channels are ordered. Then form an equation by setting the costs of the two plans equal to each other.

- If you have difficulty representing the cost of each plan algebraically, find the cost for 1 premium channel, 2 premium channels, and so forth until you see a pattern:

Number of Premium Channels	Cost of Plan A	Cost of Plan B
1	$11 + 7 \times \underline{1}$	$27 + 3 \times \underline{1}$
2	$11 + 7 \times \underline{2}$	$27 + 3 \times \underline{2}$
...	...	...
Generalize: n	$11 + 7 \times \underline{n}$	$27 + 3 \times \underline{n}$

- Write and solve an equation.

$$\begin{aligned} \text{cost of plan } A &= \text{cost of plan } B \\ 11+7n &= 27+3n \\ 7n - 3n &= 27 - 11 \\ 4n &= 16 \\ n &= \frac{16}{4} = \mathbf{4} \text{ premium channels} \end{aligned}$$

- Verify that the two plans have the same cost when $n = 4$:

$$\begin{aligned} \text{cost of plan } A &= \$11+\$7n \\ &= \$11+\$7(4) \\ &= \$39 \end{aligned} \qquad \begin{aligned} \text{cost of plan } B &= \$27+\$3n \\ &= \$27+\$3(4) \\ &= \$39 \end{aligned}$$

Solve Numerically

The same problem can be solved by making a chart or table. Calculate the costs of the plans for successive numbers of premium channels until the two costs are the same:

Number of Premium Channels	Cost of Plan A	Cost of Plan B
1	$\$11 + \$7 \times 1 = \$18$	$\$27 + \$3 \times 1 = \$30$
2	$\$11 + \$7 \times 2 = \$25$	$\$27 + \$3 \times 2 = \$33$
3	$\$11 + \$7 \times 3 = \$32$	$\$27 + \$3 \times 3 = \$36$
4	$\$11 + \$7 \times 4 = \underline{\$39}$	$\$27 + \$3 \times 4 = \underline{\$39}$ ✓

A graphing calculator can be used to build a table of values that compares the costs of the two plans for x channels where $Y_1 = 11 + 7x$ represents the cost of plan A and $Y_2 = 27 + 3x$ gives the cost of plan B .

- Press $\boxed{Y=}$ to open the $Y=$ editor. Set $Y_1 = 11 + 7x$ and $Y_2 = 27 + 3x$.
- Enter the TABLE SETUP mode and set TblStart = 1, as shown in Figure 3.13.

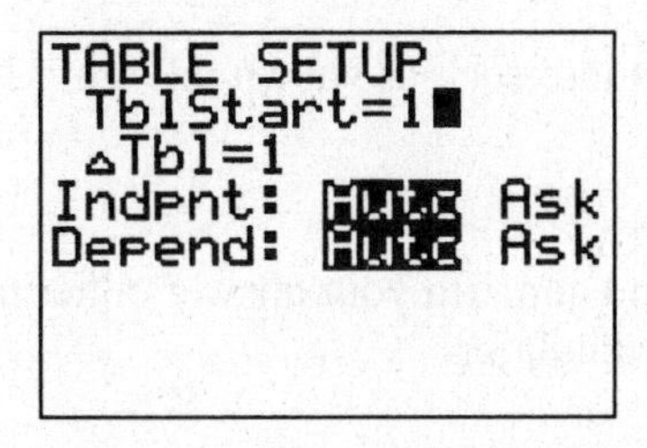

X	Y1	Y2
1	18	30
2	25	33
3	32	36
4	39	39
5	46	42
6	53	45
7	60	48

X=4

Figure 3.13 Table Setup **Figure 3.14** Comparing plans A and B

- Display the table by pressing $\boxed{\text{2nd}}$ [TABLE]. The plans will have the same cost for the value of X that makes $Y_1 = Y_2$. According to the table in Figure 3.14, when $X = 4$ premium channels, $Y_1 = Y_2 = \$39$.

Solve Graphically

The plans will have the same cost at the point at which their graphs intersect. After entering the two cost equations, continue as follows:

- Press $\boxed{\text{WINDOW}}$ to adjust the values of the screen variables. Because x and y cannot be negative, set $X\text{min} = 0$ and $Y\text{min} = 0$. Set $Y\text{max}$ to an appropriately large value such as $Y\text{max} = 62$. Create a friendly window by choosing a value for $X\text{max}$ such as $X\text{max} = 9.4$ that makes the width of the screen a multiple of 4.7.
- Press $\boxed{\text{GRAPH}}$ to display the graphs in Figure 3.15.

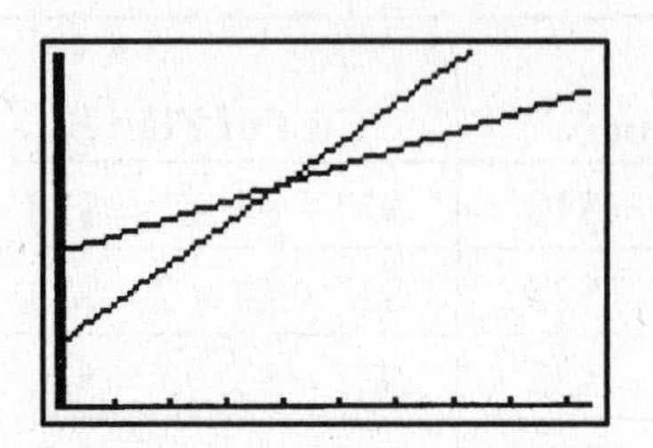

Figure 3.15 Graphs of Plans A and B

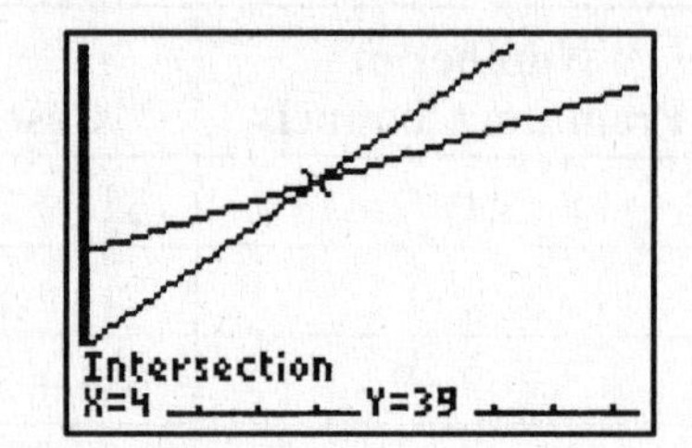

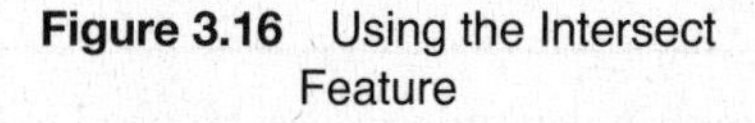

Figure 3.16 Using the Intersect Feature

- Press 2nd [CALC] to open the CALCULATE menu, where the label CALC is above the TRACE key. Press 5 to select the **intersect** feature. Press ENTER three times to locate the x-coordinate of the point at which the two lines intersect, as shown in Figure 3.16.

Example 1

Solve $5x + 7 = 2(x - 1)$ algebraically and confirm your answer either numerically or graphically using a graphing calculator.

Solution:

$$
\begin{aligned}
5x+7&=2(x-1)\\
5x+7&=2x-2\\
5x-2x&=-2-7\\
\frac{3x}{3}&=\frac{-9}{3}\\
\boldsymbol{x}&=\mathbf{-3}
\end{aligned}
$$

CHECK: Set one side of $5x + 7 = 2(x - 1)$ equal to Y_1 and the other side equal to Y_2. You do not need to simplify $2(x - 1)$ before entering it as Y_2.

- Solve the equation graphically by graphing Y_1 and Y_2 in a friendly window in which the graph fits such as $[-4.7,+4.7] \times [-12,+3.1]$. If you have difficulty determining appropriate window values for *Y*min and *Y*max, use the ZoomFit command. The root of the equation corresponds to the x-coordinate of the point at which the graphs intersect. Use the intersect feature to find that the root is $x = -3$, as shown in the accompanying figure.

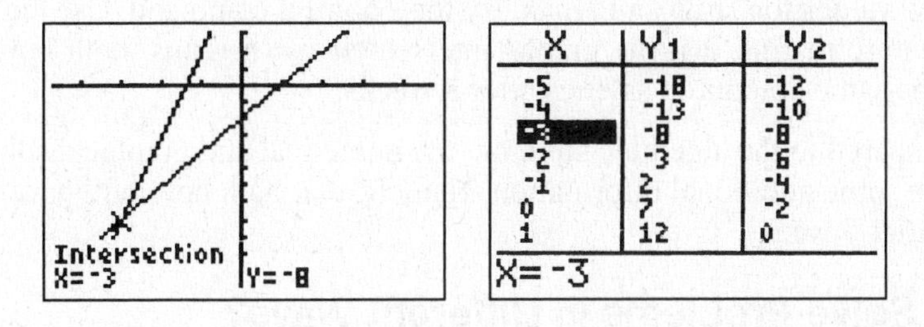

X	Y1	Y2
-5	-18	-12
-4	-13	-10
-3	-8	-8
-2	-3	-6
-1	2	-4
0	7	-2
1	12	0

X=-3

- Solve the equation numerically by pressing 2nd [TABLE] to create a table. Scroll up or down until you find the row on which $Y_1 = Y_2$, as shown in the accompanying figure. Be aware that if the solution is not an integer, you will need to adjust the value of ΔTb1 in the table setup or use a different approach.

Example 2

Carlos has \$365 in savings and saves \$20 each week. His brother has \$590 in savings and spends \$25 of his savings each week. After how many weeks will Carlos and his brother have the same amount in savings?

Solution 1: Solve algebraically. If x represents the number of weeks it takes for Carlos and his brother to have the same amount in savings, then

$$\overbrace{365+20x}^{\text{amount Carlos saved}} = \overbrace{590-25x}^{\text{amount brother saved}}$$
$$20x+25x=590-365$$
$$45x=225$$
$$\frac{45x}{45}=\frac{225}{45}$$
$$\mathbf{x=5}$$

Carlos and his brother have the same amount in savings after **5** weeks.

Solution 2: Solve numerically. Use a graphing calculator to build a table of values for $Y_1 = 365 + 20x$ and $Y_2 = 590 - 25x$, as shown at the right. Because $Y_1 = Y_2$ when $x = 5$, the solution is **5 weeks**.

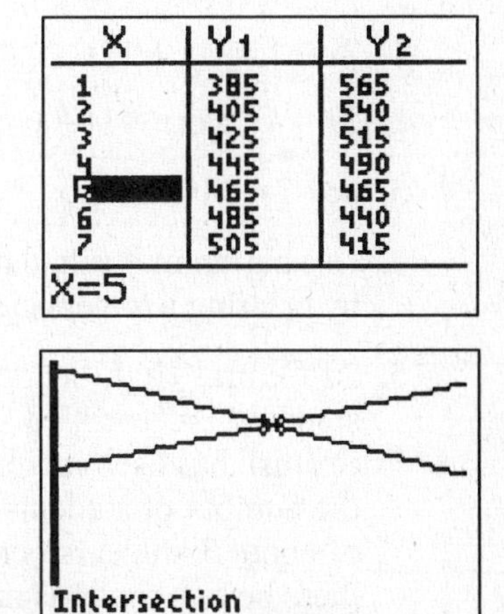

X	Y1	Y2
1	385	565
2	405	540
3	425	515
4	445	490
5	465	465
6	485	440
7	505	415

X=5

Solution 3: Solve graphically. Graph $Y_1 = 365 + 20x$ and $Y_2 = 590 - 25x$ in an appropriate window such as $[0,+9.4] \times [0,+600]$, as shown in the accompanying graph. If you have difficulty determining appropriate window values for *Y*min and *Y*max, try the ZoomFit command. Use the intersect feature to find that the graphs intersect at $x = 5$. Thus, both boys will have the same amount in savings after **5 weeks**.

Compared to the algebraic solution, the numerical and graphical solutions provide some additional information. Namely, that both boys will have saved **$465** after 5 weeks.

Why Solve Problems in Different Ways?

The Cable TV problem was solved in three different ways: algebraically, numerically, and graphically. As you become more comfortable in representing and solving problems in different ways, you will become better at solving new and unfamiliar problems. There are some additional advantages.

- One way of solving a particular problem may be easier than another. Algebraic solutions for some problems may prove difficult or impractical, while numerical or graphical approaches may be easier and less prone to error.
- A table or a graph may provide additional information or insights that may not be apparent or easily obtained from an algebraic solution.
- When a problem is solved in two different ways, one method can be used as a check on the other method.

Check Your Understanding of Section 3.3

Show how you arrived at your answer.

1-3. *Solve each equation algebraically. Use a graphing calculator to confirm your answer graphically or numerically by creating a table.*

1. $2x - 1 = 5(x + 1)$

2. $21 - 4x = 6(x - 1) - x$

3. $(x + 2) - 4 = 2(3 - x) + 1$

4. Use a graphing calculator to solve $7(x - 1) = 5(2x + 3) - 13$ numerically by creating a table. Check your answer algebraically.

5. At Central High School, 434 students are enrolled in Spanish, and 271 students are enrolled in French. The number of students enrolling in Spanish has been increasing at a rate of about 21 students per year, while the number of students enrolling in French has been decreasing at a rate of about 3 students per year. Using these rates, in how many years will there be two times as many students taking Spanish as taking French?

6. Juan has a cellular phone that costs $12.95 per month plus 25 cents per minute for each call. Tiffany has a cellular phone that costs $14.95 per month plus 15 cents per minute for each call. For what number of minutes do the two plans cost the same?

7. At Ron's Rental, a person can rent a big-screen television for $10 a month plus a one-time "wear-and-tear" fee of $100 and no delivery charge. At Joe's Rental, the charge is $20 a month and an additional charge of $20 for delivery with no "wear-and-tear" fee. After how many months will Joe's cost equal Ron's cost?

8. A hotel charges $150 for the use of its dining room and $4.50 for each breakfast plate. An association gives a breakfast and charges $7 a plate but invites five nonpaying guests. If each person has one breakfast plate, how many paying persons must attend for the association to collect the exact amount needed to pay the hotel?

9. The Excel Cable Company has a monthly fee of $32.00 and an additional charge of $8.00 for each premium channel. The Best Cable Company has a monthly fee of $26.00 and an additional charge of $10.00 for each premium channel.

   a. For what number of premium channels will the total monthly subscription fee for the two cable companies be the same?
   b. If a family subscribes to 2 premium channels for a period of one year with one of these two companies, how much money will the family save by choosing the less expensive cable company?

10. Mr. Day and Ms. Knight gave the same math quiz consisting of 16 questions. Mr. Day gave each student 5 points for each correct answer and then added 20 points to the total. Ms. Knight gave each student 7 points for each correct answer and then subtracted 6 points from the total. A student in Mr. Day's class answered the same number of questions correctly as a student in Ms. Knight's class and received the same grade. What grade did each student receive on the test?

11. An empty tub has a 40 gallon capacity. Kristin turns on the hot-water faucet, which releases water into the tub at a constant rate of 0.9 gallons per minute. Exactly three minutes later, Kristin turns on the cold-water faucet, which releases water into the tub at a constant rate of 1.35 gallons per minute. Kristin turns both faucets off when the tub has received the same number of gallons of hot water as cold water.

    a. How many minutes after Kristin turned on the hot-water faucet does she turn both faucets off?
    b. After Kristin turns off both faucets, what percent of the tub is filled with water?

CHAPTER 4

RATIOS, RATES, AND PROPORTIONS

4.1 RATIOS AND RATES

If a drama club has 12 girl members and 9 boy members, then the *ratio* of girl to boy members is $\frac{12}{9}$ which simplifies to $\frac{4}{3}$. The ratio $\frac{4}{3}$ states that for every 4 girl members in the drama club there are 3 boy members, but it does not indicate the actual number of boys and girls in the club.

Ratio

If Bill is 24 years old and Glen is 16 years old, then

$$\frac{\text{Bill's age}}{\text{Glen's age}} = \frac{24 \cancel{\text{ years}}}{16 \cancel{\text{ years}}} = \frac{24}{16} = \frac{3}{2}$$

The *ratio* of Bill's age to Glen's age is $\frac{3}{2}$ which is read "3 to 2." This ratio may also be written as 3 : 2. Because $\frac{3}{2} = \frac{1.5}{1}$, Bill is 1.5 times as old as Glen. When the second part of a ratio is 1, the ratio tells how many times as great the first part is compared to the second part.

MATH FACTS

A **ratio** is a comparison by division of two quantities measured in the same units. The ratio of a to b can be written as $\frac{a}{b}$, $a : b$, or "a to b," provided $b \neq 0$.

Rate

If quantities a and b are measured in *different* units, then $\frac{a}{b}$ is the **rate of *a* per *b***. Suppose Jack was paid $304 for working 16 hours and Jill earned $246 for working 12 hours. To determine which person earns more money per hour, compare their *unit rates* :

- Jack's unit rate is

$$\frac{304 \text{ dollars}}{16 \text{ hours}} = \frac{304 \text{ dollars} \div 16}{16 \text{ hours} \div 16} = \frac{19 \text{ dollars}}{1 \text{ hour}} = 19 \text{ dollars per hour}$$

- Jill's unit rate is

$$\frac{246 \text{ dollars}}{12 \text{ hour}} = \frac{246 \text{ dollars} \div 12}{12 \text{ hour} \div 12} = \frac{20.5 \text{ dollars}}{1 \text{ hour}} = 20.5 \text{ dollars per hour}$$

Jill will earn more money than Jack when they work the same number of hours.

MATH FACTS

A **unit rate** is a rate in which the second part is 1. To change the rate $\frac{a}{b}$ into a unit rate, divide both parts of the rate by b.

Example 1

A car travels 110 miles in 2 hours. At the same rate of speed, how far will the car travel in h hours?

Solution: To find how far the car will travel in h hours, multiply the unit rate by h hours.

- The rate of the car is $\frac{110 \text{ miles}}{2 \text{ hours}}$ so the unit rate is

  $\frac{110 \text{ miles} \div 2}{2 \text{ hours} \div 2} = \frac{55 \text{ miles}}{1 \text{ hour}}$ or 55 miles per hour.
- Because the car travels 55 miles in 1 hour, in h hours the car will travel $55 \times h = \mathbf{55h}$ miles.

Example 2

A machine that prints logos on T-shirts can print 600 T-shirts in 2.5 hours. What is the hourly rate at which the machine prints T-shirts? At this rate, how many T-shirts can the machine print in 4 hours?

Solution: The given rate is $\frac{600 \text{ T-shirts}}{2.5 \text{ hours}}$. Change to a unit rate by dividing both the numerator and the denominator by 2.5:

$$\begin{aligned} \text{rate} &= \frac{600 \text{ T-shirts}}{2.5 \text{ hours}} \\ &= \frac{600 \text{ T-shirts} \div 2.5}{2.5 \text{ hours} \div 2.5} \\ &= 240 \frac{\text{T-shirts}}{\text{hours}} \end{aligned}$$

To find the number of T-shirts that can be printed in 4 hours, multiply the unit rate by 4:

$$\begin{aligned} \text{number of T-shirts} &= 240 \frac{\text{T-shirts}}{\cancel{\text{hour}}} \times 4 \cancel{\text{hours}} \\ &= \mathbf{960} \text{ T-shirts} \end{aligned}$$

Example 3

Nicole's aerobics class exercises to fast-paced music. If the rate of the music is 120 beats per minute, how many beats would there be in a class that is 0.75 hour long?

(1) 90 (2) 160 (3) 5,400 (4) 7,200

Solution: The unit rate is given as 120 $\frac{\text{beats}}{\text{minute}}$. To find the number of beats in a given amount of time, multiply the unit rate by the amount of time expressed in *minutes*.

- Since there are 60 minutes in 1 hour:

$$0.75 \text{ hours} = 0.75 \cancel{\text{hours}} \times 60 \frac{\text{minutes}}{\cancel{\text{hour}}} = 45 \text{ minutes}$$

- To find the number of beats in 0.75 hours, multiply the unit rate by 45 minutes:

$$120 \frac{\text{beats}}{\cancel{\text{minute}}} \times 45 \cancel{\text{minutes}} = 5{,}400 \text{ beats}$$

The correct choice is **(3)**.

Ratios and Variables

The ratio of two quantities does not indicate the actual amounts of each quantity. If you are told that the ratio of Bill's age to Glen's age is 3 : 2, then their actual ages could be 3 and 2, 6 and 4, 9 and 6, and so forth. In each case, the possible ages are multiples of the base values of 3 and 2.

MATH FACTS

If two quantities are in the ratio $a : b$, then the two quantities can be represented by ax and bx, respectively. If the ratio of Bill's age to Glen's age is 3 : 2, then their ages can be expressed as $3x$ and $2x$, respectively.

Example 4

The perimeter of a rectangle is 70 cm. If the ratio of the length of the rectangle to its width is 4 to 3, what are the dimensions of the rectangle?

Solution: Represent the width of the rectangle by $3x$ and the length by $4x$. Because the sum of the lengths of the four sides is 70 cm:

$$\begin{aligned} 3x+4x+3x+4x &= 70 \\ 14x &= 70 \\ \frac{14x}{14} &= \frac{70}{14} \\ x &= 5 \end{aligned}$$

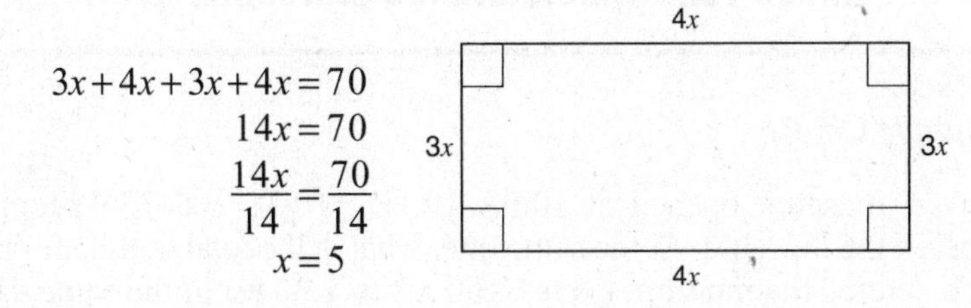

Hence, $3x = 15$ and $4x = 20$.

The width of the rectangle is **15 cm** and the length is **20 cm**. The check is left for you.

Example 5

The lengths of an opposite pair of sides of a square are each multiplied by 5 while the lengths of the other two sides remain the same. What is the ratio of the perimeter of the rectangle thus formed to the perimeter of the original square?

Solution:

- Since the actual dimensions of the original square are not given, assume each side measures 1 unit so the perimeter of the square is 4 units.
- The length of the rectangle formed is 5 units and its width is 1 unit so its perimeter is $2(5 + 1) = 12$ units.

- Hence, the ratio of the perimeter of the rectangle to the perimeter of the original square is $\frac{12}{4}$ or **3 to 1**.

Example 6

The ratio of the number of girls to the number of boys who attended a yoga class was 5 : 2. If 12 more girls than boys attended the class, how many girls attended the class?

Solution: Represent the number of girls by $5x$ and the number of boys by $2x$. Since the number of girls = number of boys + 12:

$$
\begin{aligned}
5x &= 2x + 12 \\
5x - 2x &= 12 \\
\frac{3x}{3} &= \frac{12}{3} \\
x &= 4
\end{aligned}
$$

Hence, $5x = 5 \cdot 4 =$ **20 girls** attended the yoga class. The check is left for you.

Check Your Understanding of Section 4.1

A. *Multiple Choice.*

1. An art museum opened at 10:00 AM. By 10:30 AM, 356 people had entered the museum. At the same rate, what is the total number of people who entered the museum from 10:00 AM to 1:30 PM of the same day?
(1) 1246 (2) 2492 (3) 2679 (4) 2848

2. The perimeter of a square is four times as great as the perimeter of a smaller square. What is the ratio of the area of the smaller square to the area of the larger square?
(1) 1 to 2 (2) 1 to 4 (3) 1 to 8 (4) 1 to 16

3. The length of each side of a cube is multiplied by 3. What is the ratio of the volume of the larger cube to the volume of the smaller cube?
(1) 3 to 1 (2) 9 to 1 (3) 27 to 1 (4) 81 to 1

4. If the ratio of p to q is 3 to 2, what is the ratio of $2p$ to $2q$?
(1) 1 : 3 (2) 2 : 3 (3) 3 : 1 (4) 3 : 4

5. The population of a bacteria culture doubles in number every 12 minutes. The ratio of the number of bacteria at the end of 1 hour to the number of bacteria at the beginning of that hour is
(1) 60 : 1 (2) 32 : 1 (3) 16 : 1 (4) 8 : 1

6. Rain is falling at the rate of 2 inches per hour. At this rate, how many inches of rain will fall in x minutes?

(1) $\frac{x}{30}$ (2) $\frac{30}{x}$ (3) $\frac{1}{2x}$ (4) $2x$

7. Sterling silver is made of an alloy of silver and copper in the ratio of 37 : 3, respectively. How many grams of silver does a 600 gram mass of sterling silver contain?

(1) 48 (2) 200 (3) 450 (4) 555

8. A bicyclist travels 6 miles in 20 minutes. Which expression does not represent the rate of speed of the bicyclist?

(1) 18 miles per hour
(2) $3\frac{1}{3}$ minutes per mile
(3) 20 miles per hour
(4) $\frac{3}{10}$ mile per minute

9. The ratio of Tariq's telephone bill to Pria's telephone bill was 7 : 5. If Tariq's bill was $14 more than Pria's bill, what was Tariq's bill?

(1) $21 (2) $28 (3) $35 (4) $49

***B.** Show how you arrived at your answer.*

10. Seth bought a used car that had been driven 20,000 miles. After he owned the car for 2 years, the total mileage of the car was 49,400. Find the average number of miles he drove per month during those 2 years.

11. A paper copy machine needs 4 minutes and 20 seconds to duplicate 520 pages. What is the rate at which the machine duplicates, in pages per second?

12. At the Phoenix Surfboard Company, $306,000 in profits was made last year. This profit was shared by the four partners in the ratio 3 : 3 : 5 : 7. How much *more* money did the partner with the largest share make than one of the partners with the smallest share?

13. While bicyling at a constant speed, Mike covers 20 miles in $2\frac{1}{3}$ hours. If he maintains this speed, how many *minutes* will he take to bicycle the next 6 miles?

14. At a certain college, the ratio of the number of students who are liberal arts majors to the number of students who are science majors is 8 to 3. If 480 students are science majors, how many students are liberal arts majors?

15. From January through October, ACE Autos sold 387 cars. The number of cars sold in November and the number of cars sold in December were in the ratio of 3 : 4. If the total number of cars sold in that year was 709, find the number of cars sold in December.

16. Three numbers are in the ratio of 2 : 3 : 5. If the smallest number is multiplied by 8, the result is 32 more than the sum of the second and third numbers. What is the smallest of the three numbers?

17. In the accompanying diagram, *ADFE* is a square and *ABCD* is a rectangle. The perimeter of *ABCD* is 132 and

$EB : AB = 3 : 7$.

a. Find the length of $\overline{AE}$.
b. Find the ratio of the area of rectangle *EBCF* to the area of rectangle *ABCD*, in simplest form.

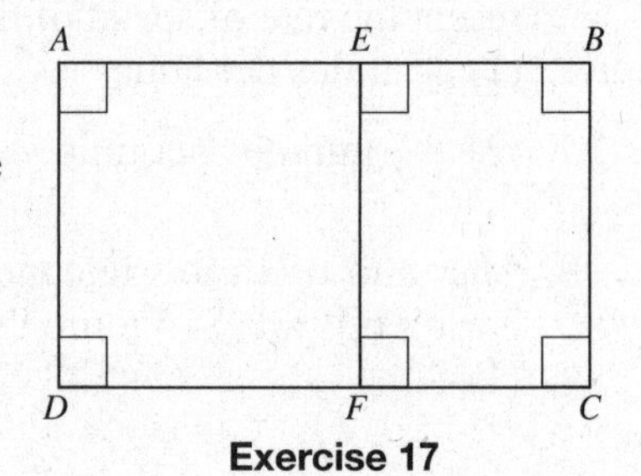

Exercise 17

4.2 PROPORTIONS

KEY IDEAS

A **proportion** is an equation that states that two ratios are equal. In a true proportion such as $\frac{12}{9} = \frac{4}{3}$, the products of opposite pairs of terms are equal: $9 \times 4 = 12 \times 3$.

Terms of a Proportion

The proportion $\frac{a}{b} = \frac{c}{d}$ is read, "*a* is to *b* as *c* is to *d*." The same proportion may also be written in the form $a : b = c : d$. The two inside terms, *b* and *c*, are the **means**. The two outside terms, *a* and *d*, are the **extremes**. In the proportion $\frac{12}{9} = \frac{4}{3}$ or, equivalently, $12 : 9 = 4 : 3$, 9 and 4 are the

extremes

$a : b \quad = \quad c : d$

means

means, and 12 and 3 are the *extremes*. Because the proportion is true, $9 \times 4 = 12 \times 3$. In any true proportion, **the product of the means is equal to the product of the extremes** or, more simply, the cross-products are equal.

Solving a Proportion

To solve a proportion that contains a variable, set the cross-products equal. Then solve for the variable. If $\frac{3}{4} = \frac{x}{7}$, then $4x = 21$ and $x = \frac{21}{4}$.

Example 1

On a certain map, $\frac{5}{8}$ of an inch represents 100 miles. If two towns on this map are $1\frac{3}{4}$ inches apart, what is the number of miles between the two towns?

Solution: Rewrite $\frac{5}{8}$ as 0.625 and $1\frac{3}{4}$ as 1.75. Use x to represent the number of miles that corresponds to 1.75 map inches. Because the map scale ratio is $\frac{0.625}{100}$,

$$\frac{\text{number of map inches}}{\text{actual number of miles}} = \frac{0.625}{100}$$

$$\frac{1.75}{x} = \frac{0.625}{100}$$

$$0.625x = 1.75(100)$$

$$x = \frac{175}{0.625}$$

$$= 280$$

The two towns are **280 miles** apart.

Example 2

A boy takes 3 minutes to read an article of 312 words. How many minutes will it take him to read an article of 884 words at the same rate?

Solution: Let x represent the number of minutes it takes the boy to read an article of 884 words. Because the rate of words read per minute is the same for each article,

$$\begin{aligned}\frac{312}{3} &= \frac{884}{x}\\ 312x &= 3(884)\\ x &= \frac{2{,}652}{312}\\ &= 8.5\end{aligned}$$

It will take the boy **8.5 minutes** to read the article.

Example 3

Steve ran a distance of 135 meters in $1\frac{1}{2}$ minutes. What is his average speed in meters per hour?

Solution: Solve using either unit rates or by using a proportion.

METHOD 1 Use unit rates.

- Find the unit rate in meters per minute:

$$\begin{aligned}\text{rate} &= \frac{135 \text{ meters}}{1\frac{1}{2} \text{ minutes}}\\ &= \frac{135 \text{ meters} \div 1.5}{1.5 \text{ minutes} \div 1.5}\\ &= 90\frac{\text{meters}}{\text{minutes}}\end{aligned}$$

- Change the unit rate to meters per hour:

$$\begin{aligned}90\frac{\text{meters}}{\text{minutes}} &= 90\frac{\text{meters}}{\text{minutes}} \times \frac{60}{60}\\ &= 5{,}400\frac{\text{meters}}{60 \text{ minutes}}\\ &= \mathbf{5{,}400\frac{meters}{hour}}\end{aligned}$$

METHOD 2 Use a proportion.

If x represents the number of meters Steve can run in 1 hour or, equivalently, 60 minutes, then his hourly rate can be expressed as $\frac{x}{60 \text{ minutes}}$. Since the minute and hourly rates must be equal:

$$\frac{135 \text{ meters}}{1\frac{1}{2} \text{ minutes}} = \frac{x}{60 \text{ minutes}}$$

$$1.5x = 60 \times 135 \text{ meters}$$

$$\frac{1.5x}{1.5} = \frac{8{,}100}{1.5}$$

$$x = 5{,}400 \text{ meters}$$

Steve's average speed is **5,400** meters per hour.

Example 4

Tammy takes 45 minutes to ride her bike 5 miles. At this rate, how long will she take to ride 8 miles?

(1) 0.89 hour
(2) 1.125 hours
(3) 0.8 hours
(4) 1.2 hours

Solution: Use a proportion where x represents the number of *minutes* Tammy takes to ride 8 miles:

$$\frac{5 \text{ miles}}{45 \text{ minutes}} = \frac{8 \text{ miles}}{x}$$

$$5x = 8 \cdot 45 \text{ minutes}$$

$$\frac{5x}{5} = \frac{360}{5} \text{ minutes}$$

$$x = 72 \text{ minutes}$$

Change from minutes to hours to match the units given in the answer choices:

$$x = 72 \text{ \sout{minutes}} \times \frac{1 \text{ hour}}{60 \text{ \sout{minutes}}}$$

$$= \frac{72}{60} \text{ hours}$$

$$= 1.2 \text{ hours}$$

The correct choice is **(4)**.

Example 5

Solve the proportion $\frac{n-3}{4}=\frac{5n+19}{3}$. Check your answer.

Solution: Set the cross-products equal:

$$\begin{aligned}4(5n+19)&=3(n-3)\\20n+76&=3n-9\\20n-3n&=-76-9\\17n&=-85\\\frac{17n}{17}&=\frac{-85}{17}\\n&=-5\end{aligned}$$

CHECK: $\frac{n-3}{4} = \frac{5n+19}{3}$

Set $n=-5$: $\frac{-5-3}{4} \overset{?}{=} \frac{5(-5)+19}{3}$

$\frac{-8}{4} \overset{?}{=} \frac{-6}{3}$

$-2 = -2$ ✓

Converting Between Units of Measurement

To convert between measurement units within the same measurement system, write a proportion in which one side represents the conversion factor. For example, because 1 mile is equivalent to 5,280 feet, the conversion factor when converting between miles and feet is $\frac{1\text{ mile}}{5{,}280\text{ feet}}$. To convert 14,520 feet into an equivalent number of miles, write the proportion

$$\frac{x\text{ miles}}{14{,}520\text{ feet}}=\frac{1\text{ mile}}{5{,}280\text{ feet}}$$

where x represents the number of miles equivalent to 14,520 feet. Solve the proportion:

$$\begin{aligned}5280x&=14{,}520\\x&=\frac{14{,}520}{5{,}280}\\&=2.75\end{aligned}$$

Thus, 14,520 feet is equivalent to 2.75 miles.

Check Your Understanding of Section 4.2

A. Multiple Choice.

1. If $\frac{x}{4} - \frac{a}{b} = 0, b \neq 0$, then x is equal to

(1) $\frac{4a}{b}$ (2) $\frac{a}{4b}$ (3) $-\frac{4a}{b}$ (4) $-\frac{a}{4b}$

2. A car is traveling at an average rate of 60 miles per hour. How many miles per minute is the car traveling?

(1) 1 (2) $\frac{1}{60}$ (3) $\frac{1}{360}$ (4) 3600

3. A cake recipe calls for 1.5 cups of milk and 3 cups of flour. Seth made a mistake and used 5 cups of flour. How many cups of milk should he use to keep the recipe proportions correct?
(1) 1.75 (2) 2.0 (3) 2.25 (4) 2.5

B. Show how you arrived at your answer.

4–9. Solve each proportion and check your answer.

4. $\frac{1}{x+1} = \frac{10}{5}$

5. $\frac{2}{3} = \frac{2-k}{12}$

6. $\frac{7y-5}{3} = \frac{9y}{4}$

7. $\frac{p-4}{p} = \frac{5}{7}$

8. $\frac{4}{11} = \frac{x+6}{2x}$

9. $\frac{10-h}{5} = \frac{7-h}{2}$

10. An astronaut weighs 174 pounds on Earth and 29 pounds on the Moon. If his daughter weighs 108 pounds on Earth, what is the daughter's weight on the Moon, in pounds?

11. If one meter is approximately 3.25 feet, how many meters are equivalent to 26 feet?

12. If 8 ounces of a sports drink contains 110 milligrams of sodium, how many milligrams of sodium are contained in 20 ounces of the same sports drink?

13. If one ounce is approximately 28.4 grams, what is the weight, in grams, of a quarter-pound hamburger?

14. During an hour of prime time television programming, the ratio of the number of minutes of television programs to the number of minutes of commercial interruptions is about 11 to 2. What is the approximate number of minutes of commercial interruptions in $1\frac{1}{2}$ hours of prime time television programming?

15. A blueprint of a house is drawn to scale so that 1 inch on the blueprint represents to 2.5 feet. If a bedroom on the blueprint is 4.8 inches in width and 7.2 inches in length, what is the number of square feet in the area of the actual bedroom?

16. A rocket car on the Bonneville Salt Flats is traveling at a rate of 640 miles per hour. How many *minutes* would it take for the rocket car to travel 384 miles at this rate?

17. The denominator of a fraction is 4 less than twice the numerator. If 3 is added to both the numerator and the denominator, the new fraction is equal to $\frac{2}{3}$. What is the original fraction?

18. One knot is one nautical mile per hour, and one nautical mile is 6,080 feet. If a cruiser ship has an average speed of 3.5 knots, what is the average speed of the ship in miles per hour?

19. Running at a constant speed, Andrea covers 15 miles in $2\frac{1}{2}$ hours. At this speed, how many *minutes* will it take her to run 2 miles?

20. Roger bought a generator that will run for 1.5 hours on one liter of gas. If the gas tank has the shape of a rectangular box that is 25 cm by 20 cm by 16 cm, how long will the generator run on a full tank of gas? [1 liter = 1,000 cc]

4.3 SOLVING MOTION PROBLEMS

KEY IDEAS

Motion problems depend on the formula, rate × time = distance. A train traveling at an average rate of 40 miles per hour for 3 hours travels a distance of

$$40\,\frac{\text{miles}}{\cancel{\text{hour}}} \times 3\ \cancel{\text{hours}} = 120\text{ miles}$$

Using Rate × Time = Distance

The formula "rate (R) times time (T) equals distance (D)" can be expressed in equivalent ways:

$$R \times T = D \qquad R = \frac{D}{T} \qquad T = \frac{D}{R}$$

Example 1

Kimberly rides her bicycle from her home to school at an average rate of 12 miles per hour. If she takes 20 minutes to get to school, how many miles is her home from her school?

Solution: Before you can use the formula $R \times T = D$, where $R = 12$ miles per hour and $T = 20$ minutes, you must make sure the units are consistent. Since 20 minutes is equivalent to $\frac{20 \text{ minutes}}{60 \text{ minutes/hour}} = \frac{1}{3}$ hour:

$$D = 12\frac{\text{miles}}{\cancel{\text{hour}}} \times \frac{1}{3}\cancel{\text{hour}}$$

$$= \frac{12}{3} \text{ miles}$$

$$= \mathbf{4 \text{ miles}}$$

Example 2

Hannah took a trip to visit her cousin. She drove 120 miles to reach her cousin's house and the same distance back home.

a. She took 1.2 hours to get halfway to her cousin's house. What was Hannah's average speed, in miles per hour, for the first 1.2 hours of the trip?
b. Hannah's average speed for the remainder of the trip to her cousin's house was 40 miles per hour. How long, in hours, did she take to drive the remaining distance?
c. While traveling home along the same route, Hannah drove at an average rate of 55 miles per hour. After 2 hours, her car broke down. How many miles was she from home?

Solution:

a. Halfway to Hannah's cousin's house represents a distance of $\frac{1}{2} \times 120$ miles = 60 miles. Use the formula $R = \frac{D}{T}$, where $D = 60$ miles and $T = 1.2$ hours:

$$R = \frac{60 \text{ miles}}{1.2 \text{ hours}} = \mathbf{50\frac{miles}{hour}}$$

b. To find how long Hannah took to drive the remaining distance of 60 miles to her cousin's house at a speed of 40 miles per hour, use the formula $T = \frac{D}{R}$, where $D = 60$ miles and $R = 40\frac{\text{miles}}{\text{hour}}$:

$$T = \frac{60 \text{ miles}}{40 \text{ miles/hour}} = \mathbf{1.5\ hours}$$

c. To find the distance Hannah drove in 2 hours at a rate of 55 miles per hour, multiply the rate by the time traveled:

$$D = 55\frac{\text{miles}}{\cancel{\text{hour}}} \times 2\,\cancel{\text{hours}} = \mathbf{110\ miles}$$

Since Hannah lives 120 miles from her home, her car broke down $120 - 110 =$ 10 miles from her home.

Sometimes first drawing a diagram or making a table helps when solving a motion problem.

Example 3

Two trains leave the same station at the same time and travel in opposite directions. One train travels at 80 kilometers per hour and the other at 100 kilometers per hour. In how many hours will the trains be 900 kilometers apart?

Solution 1: Solve algebraically.

- Draw a diagram to show the distances of each train in h hours:

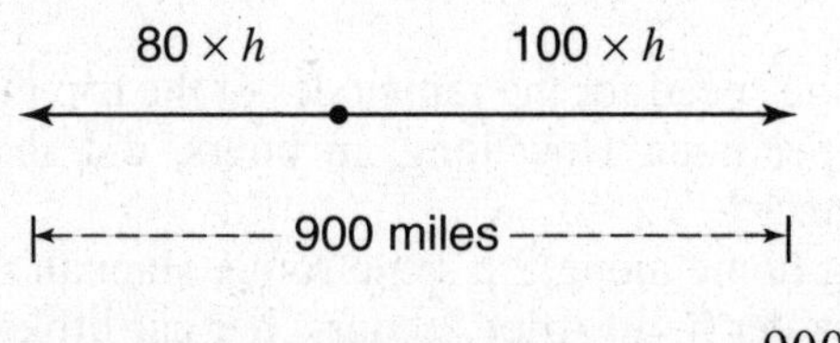

- Hence, $80h + 100h = 900$ so $180h = 900$ and $h = \frac{900}{180} = 5$.

The two trains will be 900 miles apart in **5 hours**.

Solution 2: Solve numerically by making a chart in which the distance between the two trains increases at a rate of $80 + 100 = 180$ miles each hour:

Time Elapsed	Distance Apart
2 hours	180 miles × 2 = 360 miles
3 hours	180 miles × 3 = 540 miles
4 hours	180 miles × 4 = 720 miles
5 hours	180 miles × <u>5</u> = 900 miles ✓

Example 4

A truck traveling at a constant rate of 45 miles per hour leaves Albany at 9:00 AM. One hour later a car traveling at a constant rate of 60 miles per hour leaves from the same place traveling in the same direction on the same highway. At what time will the car overtake the truck, if both vehicles continue in the same direction on the highway?

Solution: Assume x represents the number of hours the car has traveled when it overtakes the truck.

- Since the truck started 1 hour before the car, the truck has traveled $x + 1$ hours when it is overtaken by the car.
- Organize the facts in a table.

	Rate	× Time	= Distance
Truck	45 mph	$x + 1$	$45(x + 1)$
Car	60 mph	x	$60x$

- Write an equation. When the car overtakes the truck, the two vehicles have traveled the same distance. Thus,

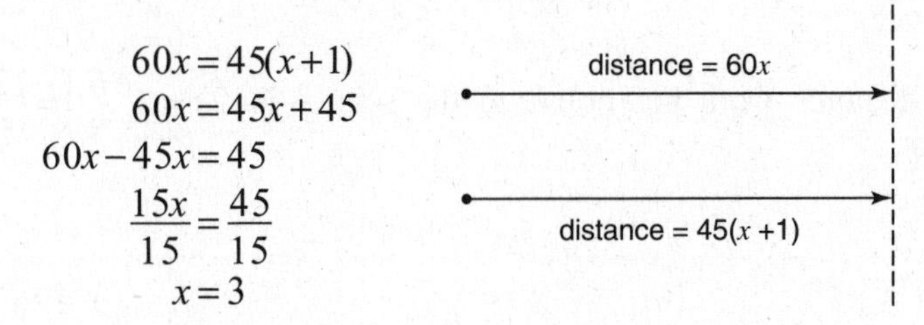

The car starts at 10:00 AM since it leaves one hour after the truck. Hence, the car overtakes the truck 3 hours after it starts which is at **1:00 PM**.

Example 5

Amy walked from her house to her school at the rate of 2 miles per hour and returned on a bicycle riding at a rate of 8 miles per hour. If the round trip took a total of $1\frac{1}{2}$ hours, how many miles is the school from her house?

Solution: If x represents the time it took Amy to walk to school, then $1\frac{1}{2}-x$ or, equivalently, $\frac{3}{2}-x$, is the time it took her to bicycle back to her house.

- Organize the given information in a table.

	Rate	× Time	= Distance
Walking to school	2 mph	x	$2x$
Bicycling from school	8 mph	$\frac{3}{2}-x$	$8\left(\frac{3}{2}-x\right)$

- Since the distances each way must be equal,

$$2x=8\left(\frac{3}{2}-x\right)$$
$$2x=12-8x$$
$$2x+8x=12$$
$$\frac{10x}{10}=\frac{12}{10}$$
$$x=\frac{6}{5}$$

- The distance from her house to the school is $2x=2\left(\frac{6}{5}\right)=\frac{12}{5}=$ **2.4 miles**.

Check Your Understanding of Section 4.3

A. Multiple Choice.

1. If Jamar can run $\frac{3}{5}$ of a mile in 2 minutes 30 seconds, what is his average rate in miles per minute?

(1) $\frac{4}{5}$ (2) $\frac{6}{25}$ (3) $3\frac{1}{10}$ (4) $4\frac{1}{6}$

2. Two cars leave from the same location and travel in opposite directions. One car travels at a rate of *m* miles per hour. The rate of the other car exceeds the rate of the first car by 15 miles per hour. Which expression represents the number of miles that the two cars will be apart after 4 hours?

(1) $4m + 15$ (2) $8m^2 + 15m$ (3) $8m + 60$ (4) $4m^2 + 60$

3. A bicyclist leaves Bay Shore traveling at an average rate of 12 miles per hour. Three hours later, a car leaves Bay Shore, on the same route, traveling at an average speed of 30 miles per hour. How many hours after the car leaves Bay Shore will the car catch up to the cyclist?

(1) 8 (2) 2 (3) 5 (4) 4

4. A truck travels 40 miles from point *A* to point *B* in exactly 1 hour. When the truck is halfway between point *A* and point *B*, a car starts from point *A* and travels at 50 miles per hour. How many miles has the car traveled when the truck reaches point *B*?

(1) 25 (2) 40 (3) 50 (4) 60

5. Two objects are 2.4×10^{20} centimeters apart. A signal from one object travels to the other at a rate of 1.2×10^5 centimeters per second. How many seconds does it take the signal to travel from one object to the other?

(1) 1.2×10^{15} (2) 2.0×10^4 (3) 2.0×10^{15} (4) 2.88×10^{25}

6. The distance from Earth to the imaginary planet Zota is 1.74×10^7 miles. If a spaceship is capable of traveling at an average rate of 1,450 miles per hour, what is the approximate number of *days* it will take the spaceship to reach Zota?

(1) 50 (2) 1200 (3) 500 (4) 8333

B. Show how you arrived at your answer.

7. A train leaves a station at 9:00 AM traveling at a constant rate of speed. Two hours later a second train traveling at an average rate of 25 miles per hour more than the first train leaves the same station traveling in the same direction on a parallel track. If the faster train overtakes the first train at 3:00 PM on the same day, what is the average rate of speed of the first train?

8. A freight train and a passenger train start toward each other at the same time from two towns that are 500 miles apart. After three hours the trains are 80 miles apart. If the rate of the passenger train is 20 miles per hour faster than the rate of the freight train, find the rate of the freight train.

9. A truck left a rest stop on the highway at 12 noon going north at a rate of 50 miles per hour. One hour later a car left the same rest stop going south at the rate of 60 miles per hour. At what time were the two vehicles 325 miles apart?

10. Ms. Ruiz drove from her home to her job at an average speed of 60 miles per hour and returned home along the same route at an average speed of 40 miles per hour. If her total driving time for the trip was two hours, how many *minutes* did it take her to drive from her job to her home?

11. It takes a man 20 minutes longer to travel from his home to his place of business if he drives at a rate of 30 mph than if he drives at a rate of 40 mph. How many miles is his home from his place of business?

12. John rode his bicycle from his house to Vincent's house at the rate of 15 miles per hour. He left the bicycle with Vincent, and walked home along the same route at the rate of 3 miles per hour. If the entire trip took 3 hours, how many hours did it take John to walk back to his house?

13. Gregory hiked from the base of a mountain to its top at an average rate of 2 miles per hour and hiked down the mountain following the same path at an average rate of 3 miles per hour. If Gregory hiked, excluding rest stops, for a total of 8 hours, how many miles was it from the base to the top of the mountain?

14. A train travels 30 miles in $\frac{5}{6}$ the amount of time it takes a man to drive a car the same distance. If the rate of the train exceeds the rate of the car by 8 miles per hour, find the rate at which the car is traveling.

15. Brian is walking along a trail at an average rate of 3 miles per hour. Carol is bicycling along the same trail in the same direction at an average rate of 9 miles per hour. At 12:00 noon Brian is 5 miles ahead of Carol. If Brian and Carol maintain these rates, at what time will Carol overtake Brian?

16. A boy riding a bicycle at the rate of 10 miles per hour and a car traveling 40 miles per hour are traveling in the same direction on the same road. If the car passes a road sign 30 minutes after the bicycler does, in how many minutes will the car overtake the bicycler?

4.4 SOLVING PERCENT PROBLEMS

KEY IDEAS

The familiar percent notation of $p\%$ means $\frac{p}{100}$. For example, $35\% = \frac{35}{100} = 0.35$. Problems involving percent can be solved using linear equations or proportions.

Basic Types of Percent Problems

There are three basic types of percent problems that can be solved by translating the problem into an algebraic equation:

- Finding a percent of a given number.

EXAMPLE: What is 15% of 80?

$$n = 0.15 \times 80 = 12$$

- Finding a number when a percent of it is given.

EXAMPLE: 30% of what number is 12?

$$0.30 \times n = 12$$

$$n = \frac{12}{0.30} = 40$$

- Finding what percent a number is of another.

EXAMPLE: What percent of 45 is 9?

$$\underbrace{\text{What percent}}_{\downarrow}\ \underbrace{\text{of}}_{\downarrow}\ \underbrace{45}_{\downarrow}\ \underbrace{\text{is}}_{\downarrow}\ 9?$$

$$\frac{n}{100} \times 45 = 9$$

$$45n = 900$$

$$n = \frac{900}{45} = 20\%$$

Example 1

The amount of sales tax on the purchase of a new computer system is $68.25. If the sales tax rate is 7%, what is the cost of the computer system?

Solution: If c is the cost of the computer system, then

$$\underbrace{7\%}_{\downarrow\ 0.07}\ \underbrace{\text{of}}_{\downarrow\ \times}\ \underbrace{\text{cost of computer}}_{\downarrow\ c}\ \underbrace{\text{is } \$68.25}_{\downarrow\ = \$68.25}$$

$$\frac{0.07c}{0.07} = \frac{\$68.25}{0.07}$$

$$c = \$975$$

The cost of the computer system is **$975.**

Example 2

Part of $12,000 was invested in a money market account that yielded an annual rate of return of $8\frac{1}{2}\%$, and the other part was invested in a stock market account that yielded an annual rate of return of 18%. If the total amount of annual income from both investments is $1210, find the number of dollars invested in each account.

Solution: If x represents the amount of money invested in the money market account, then $12{,}000 - x$ is the amount invested in the stock market account. To find the annual income from an investment, multiply the amount of money invested at that rate by the annual rate of return.

$$\overbrace{8.5\% \text{ of } x}^{\text{income from money market}} + \overbrace{18\% \text{ of } (12{,}000 - x)}^{\text{income from stock market}} = 1{,}210$$

$$0.085x + 0.18(12{,}000 - x) = 1{,}210$$

$$0.085x + 2160 - 0.18x = 1{,}210$$

$$0.085x - 0.18x = 1{,}210 - 2{,}160$$

$$-0.095x = -950$$

$$\frac{-0.095x}{-0.095} = \frac{-950}{-0.095}$$

$$x = 10{,}000$$

Thus, **$10,000** was invested in the money market account and $12,000 – $10,000 = **$2,000** was invested in the stock market account.

CHECK: $8.5\% \times \$10{,}000 = 0.0850 \times \$10{,}000 = \$850$
$+18\% \times \$2{,}000 = 0.18 \times \$2{,}000 = \$360$
$= \$1{,}210$

Circle Graphs

A circle graph or "pie chart" shows visually how the parts that make up a whole are related to the whole and to each other.

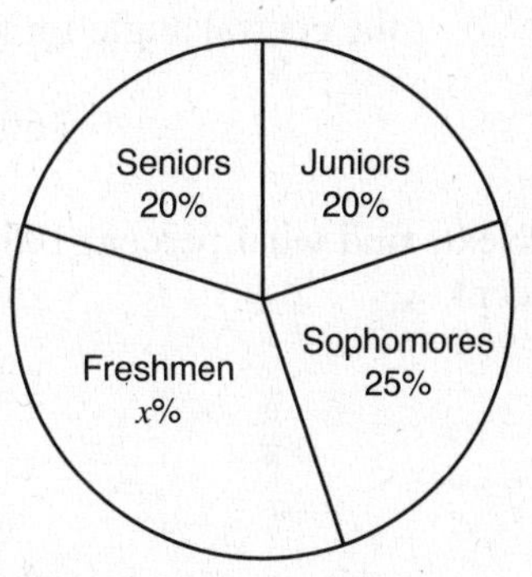

Example 3

The circle graph in the accompanying figure shows the enrollment by grade level at Central High School. If 1,200 students attend Central High School, how many students are freshmen?

Solution: Since all of the sectors of the circle graph must add up to100%,

$$x\% + 20\% + 25\% + 20\% = 100\% \text{ so } x\% = 100\% - 65\% = 35\%$$

If 1,200 students attend Central High School, then the number of freshmen is

$$35\% \text{ } of \text{ } 1{,}200 = 0.35 \times 1{,}200 = \mathbf{420}$$

Example 4

The circle graph in the accompanying figure summarizes the results of a poll in which seniors at a New York State high school were asked whether they would be attending a college in New York or a different state. The central angle for the sector that represents the number of students who will attend college in NYS measures 140° and the central angle for the sector that represents the number of students who will attend college out of NYS measures 112°.

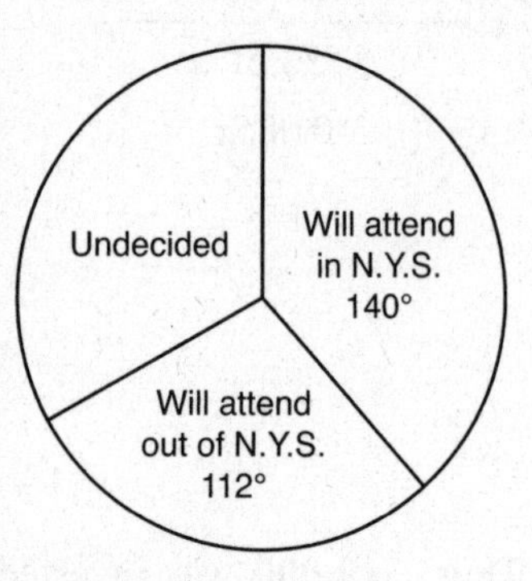

a. What percent of the students polled were undecided?
b. If 210 students were polled, how many students were undecided?

Solution:

a. Because all the central angles of a circle graph must add up to 360°, the central angle for the undecided group measures

$$360° - (140° + 112°) = 108°$$

Next, find what percent 108 is of 360, where n stands for the unknown percent:

$$\frac{n}{100} \times 360 = 108$$
$$36n = 1{,}080$$
$$\frac{36n}{36} = \frac{1{,}080}{36}$$
$$n = 30\%$$

Thus, **30%** of the students polled were undecided.

b. Since 30% of 210 = 0.30 × 210 = 63, **63** students were undecided.

Solving Percent Problems with Proportions

The fact that 4 is 50% of 8 can be expressed as a percent proportion:

$$\frac{4 \text{ (number that is being compared to base)}}{8 \text{ (base)}} = \frac{50 \text{ (amount of percent)}}{100}$$

In comparing 4 to 8, 8 is the *base* and 4 is the number that is being compared to that base.

MATH FACTS

The three basic types of percent problems can be solved using the percent proportion:

$$\frac{\text{number being compared to the base}}{\text{base}} = \frac{\text{amount of \%}}{100}$$

Example 5

Sara conducted a poll of 45 students and found that 27 of these students are members of a club. What percent of the students polled are members of a club?

Solution: Write a percent proportion in which 27 is compared to a base of 45 and p is the amount of percent:

$$\frac{27}{45} = \frac{p}{100}$$

Write $\frac{27}{45}$ in lowest terms:

$$\frac{3}{5} = \frac{p}{100}$$

$$5p = 300$$

$$\frac{5p}{5} = \frac{300}{5}$$

$$p = 60\%$$

Thus, **60%** of the students polled are members of a club.

Example 6

If a multiple-choice test contains 60 questions, how many questions would need to be answered correctly in order to get 85% of the questions correct?

Solution: Assume x represents the number of questions that must be answered correctly. Write the percent proportion where x is being compared to a base of 60:

$$\frac{x}{60}=\frac{85}{100}$$
$$100x=5{,}100$$
$$\frac{100x}{100}=\frac{5{,}100}{100}$$
$$x=51$$

Thus, **51** questions need to be answered correctly.

Percents of Increase and Decrease

In a six-month period, the price of a stock increased from \$25 a share to \$29 a share. To find the percent of increase in the share price, compare the change in price to the original price:

$$\begin{aligned}\text{percent increase} &= \frac{\text{change in price}}{\text{original price}}\times 100\% \\ &= \frac{\$29-\$25}{\$25}\times 100\% \\ &= \frac{4}{\cancel{25}}\times \overset{4}{\cancel{100}}\% \\ &= \mathbf{16\%}\end{aligned}$$

MATH FACTS

When a quantity x increases or decreases, the

$$\text{percent of change in } x = \frac{\text{change in } x}{\text{starting value of } x}\times 100\%$$

Example 7

A pair of sneakers that regularly sells for \$72 is on sale at a discounted price of \$45. What is the percent of the discount?

Solution: To find the percent of discount, find the percent of decrease in the price of the sneakers.

$$\begin{aligned}\text{Percent of discount} &= \frac{\text{change in price}}{\text{original price}} \times 100\% \\ &= \frac{\$72 - \$45}{\$72} \times 100\% \\ &= \frac{27}{72} \times 100\% \\ &= 0.375 \times 100\% \\ &= \mathbf{37.5\%}\end{aligned}$$

Example 8

Craig weighs exactly 149.0 pounds. When Craig weighs himself on a defective scale, he weighs 156.0 pounds. What is the percent of error in measurement of the scale to the *nearest tenth*?

Solution: To find the percent of error in measurement of the scale, find the percent of increase in Craig's weight.

$$\begin{aligned}\text{percent of error in measurement} &= \frac{\text{amount of error}}{\text{original weight}} \times 100\% \\ &= \frac{156.0 - 149.0}{149.0} \times 100\% \\ &= \frac{7.0}{149.0} \times 100\% \\ &\approx \mathbf{4.7\%}\end{aligned}$$

Check Your Understanding of Section 4.4

A. *Multiple Choice.*

1. John, Gary, and Melissa played a computer game for a total of 3 hours. If John played the game for 28% of the total time, and Gary played the game for 52% of the total time, how many minutes did Melissa play the game?

(1) 30 (2) 36 (3) 42 (4) 48

2. Marcia paid $36 for a dress that was on sale for 25% of the original price. What was the original price of the dress?
(1) $48 (2) $60 (3) $108 (4) $144

3. Carla paid $24 for slacks that were on sale. If the original price of the slacks was $32, what was the percent of discount?
(1) 20 (2) 25 (3) 75 (4) 80

4. Vincent paid $48 for a computer game that was on sale for 60% off of the original price. What was the original price of the computer game?
(1) $76.80 (2) $80 (3) $120 (4) $128

5. After a 20% increase, the new price of a share of a computer stock is $78.00. What was the original price of a share of the computer stock?
(1) $60 (2) $62.40 (3) $65 (4) $97.50

6. One year ago the average price of a home computer system was $1,500. Today, the average price of a home computer system is $1,200. By what percent has the average cost of a home computer system decreased?
(1) 20% (2) 25% (3) 75% (4) 80%

7. In a factory that manufactures light bulbs, 0.04% of the bulbs manufactured are defective. What is the minimum number of bulbs manufactured for which you would expect to find one defective light bulb?
(1) 250 (2) 1,250 (3) 2,500 (4) 4,000

8. Rashawn bought a CD that cost $18.99 and paid $20.51, including sales tax. What was the rate of the sales tax?
(1) 5% (2) 2% (3) 7% (4) 8%

9. On January 1, Mr. Warner and his son jointly invest a sum of money at an annual rate of $6\frac{3}{4}\%$. On January 1 of the following year, their total investment is worth $5,978. If Mr. Warner invested $4,500 of his own money, how much money did his son invest?
(1) $1,478 (2) $1,100 (3) $900 (4) $750

10. A planned building was going to be 100 feet long, 75 feet deep, and 30 feet high. The owner decides to increase the volume of the building by 10% without changing the dimensions of the depth and the height. What will be the new length of this building?
(1) 106 ft (2) 108 ft (3) 110 ft (4) 112 ft

11. If the length of each side of a square is increased by 10%, by what percent does the area of the square increase?
(1) 21% (2) 20% (3) 40% (4) 100%

***B.** Show how you arrived at your answer.*

12. A recent survey shows that the average man will spend 141,288 hours sleeping, 85,725 hours working, 81,681 hours watching television, 9,945 hours commuting, 1,662 hours kissing, and 363,447 hours on other tasks during his lifetime. What percent of his life, to the *nearest tenth of a percent*, does he spend sleeping?

13. Nine hundred students were asked whether they thought their school should have a dress code. A circle graph was constructed to show the results. The central angles for two of the three sectors are shown in the accompanying diagram. What is the number of students who felt that the school should have no dress code?

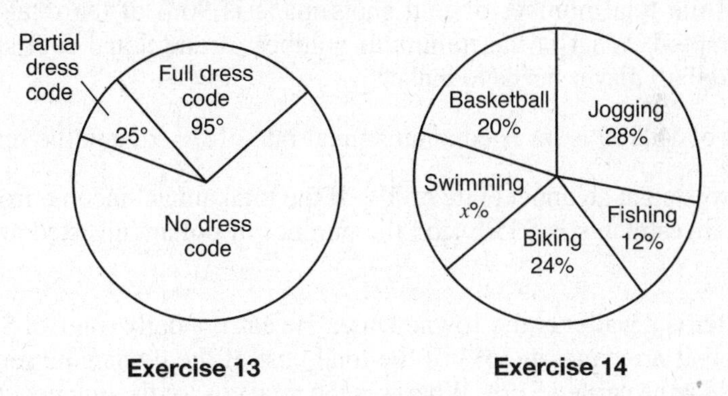

Exercise 13 **Exercise 14**

14. The accompanying circle graph summarizes the results of a survey of 1,400 students who named their favorite sports activity. If each student in the survey named exactly one activity, what is the total number of students who named either swimming or biking as their favorite sports activity?

15. In bowling leagues, some players are awarded extra points called their "handicap." The handicap in Anthony's league is 80% of the difference between 200 and the bowler's average. If Anthony's handicap is 44, what is Anthony's bowling average?

16. During a week when the 7% charge for sales tax on clothes was waived, Mary tells her friend that she saved $13.65 in sales tax on the purchase of a new coat. What was the cost of the new coat?

17. In a town election, candidates A and B were running for mayor. There were 30,500 people eligible to vote, and 80% of them actually voted. Candidate A received 6,100 votes. What *percent* of the votes cast did candidate B receive?

18. The larger of two numbers exceeds three times the smaller by 2. If the smaller number is 30% of the larger, what is the smaller number?

19. A class consists of 14 boys and 19 girls. On a certain day, all the boys were present, and some girls were absent. If the girls made up only 30% of the class attendance on that day, how many girls were absent?

20. Richard's favorite breakfast cereal has a 20% sugar content. If Richard adds 5 grams of sugar to a 45-gram serving of this cereal, what is the percent concentration of sugar in the resulting mixture?

21. A basketball player made 9 foul shots, which is 75% of the number he attempted. The coach tells the same player to keep making foul shots until the total number of foul shots made is 90% of the total number attempted. What is the minimum number of additional foul shots the basketball player needs to make?

22. Part of $5,200 is invested at an annual rate of $7\frac{1}{2}$%, and the remainder is invested at an annual rate of 5%. If the total annual income from these two investments is $350, find the number of dollars invested at the 5% rate.

23. Walter is a waiter at the Towne Diner. He earns a daily wage of $50, plus tips that are equal to 15% of the total cost of the dinners he serves. On Tuesday he earned $170. What was the total cost of the dinners he served on Tuesday?

24. A high school tennis team wins 60% of the first 15 matches it plays. What is the least number of additional matches the team must win in order to finish the season winning at least 75% of the matches it plays?

25. Sue bought a beach chair on sale for 50% off the original price. The store charged her 10% tax, and her final cost was $22.00. What was the original price of the beach chair?

26. A tank contains 50 liters of a solution that is 30% acid. How many liters of water must be evaporated from the solution in order to produce a solution that is 40% acid.

27. When Raymond was on a diet, his weight w in pounds after x days on the diet could be approximated by the formula $w = 240 - 0.6x$.

a. After how many days on the diet did Raymond's weight drop to 210 pounds?
b. What percent of his original weight did Raymond lose after 30 days on the diet?

28. Brittany ordered a belt from an internet store at 60% off the original price. The store charges 8% sales tax and a $6 delivery charge. If her final cost was $25.44, what was the original price of the belt?

29. A student answered 50% of the questions on a test correctly. If he answered 10 questions correctly out of the first 12, and 25% of the remaining questions correctly, how many questions were on the test?

30. Mr. Perez owns a sneaker store. He bought 350 pairs of basketball sneakers, and 150 pairs of soccer sneakers from the manufacturers for $62,500. He sold all the sneakers and made a 25% profit. If he sold the soccer sneakers for $130 per pair, how much did he charge for one pair of basketball sneakers?

31. A painting that regularly sells for $55 is on sale for 20% off. The sales tax on the painting is 7%. What is the final cost? Does the final total cost of the painting depend on whether the salesperson deducts the discount before adding the sales tax or takes the discount after computing the sum of the original price and the sales tax on $55? Justify your answer.

32. A scarf is on sale at 20% off the list price. The following week the price of the scarf is reduced 25% of the sale price. Carol called her friend and told her that the price of the scarf is now 45% off the original price. Do you agree with Carol? If not, what is the actual percent of the discount off of the original price?

33. Keesha wants to tile the floor shown in the accompanying diagram. If each tile measures 1 foot by 1 foot and costs $2.99, what will be the total cost, including an 8% sales tax, for tiling the floor?

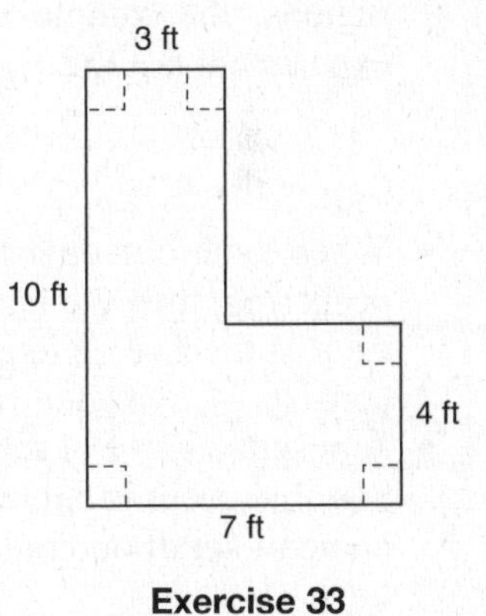

Exercise 33

4.5 PROBABILITY RATIOS

KEY IDEAS

The *probability* that an event will happen is represented by a number from 0 to 1. The more likely an event will occur, the closer its probability of happening is to 1. If an event E can happen in r out of n equally likely ways, then the **probability** that event E will occur is the ratio of r to n which is written as, $P(E)=\frac{r}{n}$.

Probability Terms

The spinner shown in Figure 4.1 consists of five equal regions, numbered from 1 to 5.

- An activity whose outcome is uncertain, such as predicting the region in which the spinner will stop, is referred to as a **probability experiment**.
- The set of all possible outcomes of a probability experiment is called the **sample space**. Since the spinner must land in one of the five numbered regions, the sample space for this probability experiment is $\{1, 2, 3, 4, 5\}$.

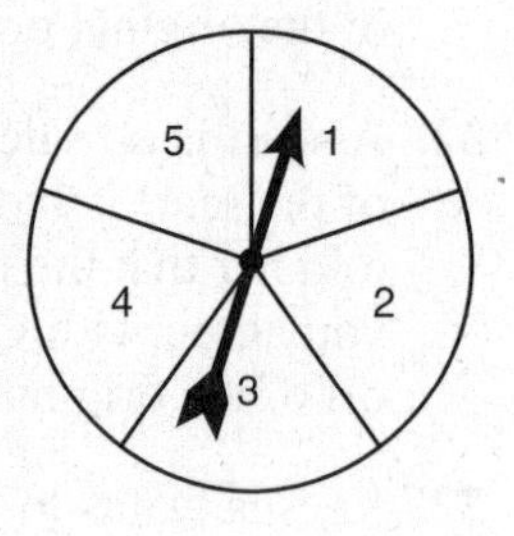

Figure 4.1 Spinner

- When each outcome in the sample space has the same probability of occurring, then the outcomes are each **equally likely** to occur. Because each of the five regions of the spinner is the same size, the sample space consists of five equally likely outcomes.
- A possible *event* of this probability experiment is spinning a 2. Another possible event is spinning an odd number. An **event** is a particular outcome or set of outcomes that are contained in the sample space.

Definition of Probability

The outcomes that make an event happen are the **favorable outcomes** or **successes** for that event. Since three of the five regions have odd numbers, the probability of spinning an odd number is $\frac{3}{5}$, where 3 is the number of favorable outcomes and 5 is the total number of possible outcomes.

MATH FACTS

- The notation $P(E)$ represents the probability that an event E will happen, where

$$P(E)=\frac{\text{number of favorable outcomes for event } E}{\text{total number of possible equally likely outcomes}}$$

- The probability of an impossibility is 0, and the probability of a certainty is 1.

Example 1

The party registration of the voters in Jonesville is shown in the accompanying table. If one of the voters is selected at random, what is the probability that the person selected will *not* be a Democrat?

Voter Party Registration	Number of Voters Registered
Democrat	8,000
Republican	5,800
Independent	4,200

Solution:

- The number of favorable outcomes is the sum of the number of voters *not* registered as Democrats: 5,800 + 4,200 = 10,000.
- The total number of possible outcomes is the total number of registered voters: 8,000 + 5,800 + 4,200 = 18,000.
- The probability that the person selected will *not* be a Democrat is $\frac{10{,}000}{18{,}000}=\frac{5}{9}$.

When Probabilities Add Up to 1

The sum of probability values is 1 when:

- the probabilities of *all* possible outcomes in the same sample space are added together. In Figure 4.1,

$$P(1)+P(2)+P(3)+P(4)+P(5)=\frac{1}{5}+\frac{1}{5}+\frac{1}{5}+\frac{1}{5}+\frac{1}{5}=1$$

- the probabilities of an event E and the event *not E* are added together. That is, $P(E) + P(not\ E) = 1$. The event *not E* is the **complement** of event E. Referring again to Figure 4.1, the probability of spinning an odd number is $\frac{3}{5}$. Hence, the probability of *not* spinning an odd number (or spinning an even number) is $1-\frac{3}{5}=\frac{2}{5}$. If there is a 30% chance of rain tomorrow, then the probability that it will *not* rain tomorrow is 1 – 0.30 = 0.70 or 70%.

Example 2

A bucket contains only red, blue, and green blocks. The probability that a red block is selected from the bucket is $\frac{3}{8}$, and the probability that a blue block is selected is $\frac{2}{8}$. If the bucket contains 9 green blocks, what is the total number of blocks in the bucket?

Solution: Because the sum of the probabilities of each of the possible outcomes is 1,

$$\begin{aligned} P(\text{red})+P(\text{blue})+P(\text{green})&=1 \\ \tfrac{3}{8}+\tfrac{2}{8} \qquad +P(\text{green})&=1 \\ P(\text{green})&=1-\tfrac{5}{8}=\tfrac{3}{8} \end{aligned}$$

If x stands for the total number of blocks in the bucket, then $P(\text{green})=\frac{9}{x}$.

Since $\frac{9}{x}$ and $\frac{3}{8}$ each represent the probability of the same event occurring,

$$\begin{aligned} \frac{9}{x}&=\frac{3}{8} \\ 3x&=72 \\ \frac{3x}{3}&=\frac{72}{3} \\ x&=24 \end{aligned}$$

The bucket contains a total of **24** blocks.

Example 3

A jar contains 110 marbles of which 50 are yellow and 60 are white. How many yellow marbles must be added to the jar so that the probability of picking a yellow marble will be $\frac{2}{3}$?

Solution 1: If x represents the number of yellow marbles that must be added to the jar,

$$P(\text{yellow}) = \frac{\text{number of yellow marbles}}{\text{total number of marbles in the jar}}$$

$$\frac{2}{3} = \frac{50+x}{110+x}$$

$$3(50+x) = 2(110+x)$$

$$150 + 3x = 220 + 2x$$

$$3x - 2x = 220 - 150$$

$$x = 70$$

Thus, **70** yellow marbles must be added to the jar.

Solution 2: Change your point of view by considering the probability of picking a *white* rather than a yellow marble:

- If the probability of picking a yellow marble is $\frac{2}{3}$, the probability of picking a *white* marble is $1 - \frac{2}{3} = \frac{1}{3}$.
- The number of white marbles is fixed at 60. The probability of picking a white marble is $\frac{1}{3}$ only when the total number of marbles in the jar is three times 60 or 180 because $\frac{60}{180} = \frac{1}{3}$.
- Since 180 – 110 = 70, **70** yellow marbles must be added to the jar.

Theoretical vs. Empirical Probability

Until now, only *theoretical probability* values have been considered. The **theoretical probability** of an event is the probability value arrived at by using mathematical reasoning and logical analysis to determine the number of favorable outcomes for an event compared with the total number of possible outcomes. Unlike empirical probability, theoretical probability does not involve performing an actual probability experiment and recording the results.

A theoretical probability value cannot always be calculated. This can happen when finding probabilities of real-world events such as the probability that a manufactured part will be defective. In such cases, *empirical probability* values can be calculated using collected data. The **empirical probability** of an event is an *estimate* that the event will occur based on sample data or the results of performing repeated trials of a probability experiment. The greater the number of sample items collected or trials performed, the more closely empirical and theoretical probability values will agree.

- Suppose in a production run of 10,000 bulbs, a sample of 250 bulbs are selected at random and tested. If 8 of these 250 bulbs fail, then the *empirical* probability that a bulb selected at random from the entire production run will fail is $\frac{8}{250}$ or 3.2%.
- When a cone-shaped cup is tossed, it lands either point-up or on its side. The theoretical probability of the cup landing point-up cannot be calculated since the two possible outcomes are *not* equally likely to occur. The probability that the cup lands point-up can be *estimated* by repeatedly dropping the cup and comparing the number of times it lands point-up to the total number of times the cup is dropped. If it lands point-up 19 out of 100 times it is dropped, the *empirical probability* of the cone landing point-up is $\frac{19}{100} = 0.19$.

Check Your Understanding of Section 4.5

A. Multiple Choice.

1. During a half hour of television programming, eight minutes is used for commercials. If a television set is turned on at a random time during the half hour, what is the probability that a commercial is *not* being shown?

(1) 1 (2) $\frac{11}{15}$ (3) $\frac{4}{15}$ (4) 0

2. A bag has five green marbles and four blue marbles. If one marble is drawn at random, what is the probability that it is *not* green?

(1) $\frac{1}{9}$ (2) $\frac{4}{9}$ (3) $\frac{5}{9}$ (4) $\frac{5}{20}$

3. An urn contains four yellow marbles and three blue marbles. How many blue marbles must be added to the urn so that the probability of picking a yellow marble is $\frac{1}{3}$?

(1) 5 (2) 2 (3) 3 (4) 4

4. A jar contains only red, white, and blue marbles. If the ratio of red to white to blue marbles is 1 to 3 to 8, what is the probability that a marble drawn at random will *not* be white?

(1) $\frac{1}{4}$ (2) $\frac{1}{3}$ (3) $\frac{2}{3}$ (4) $\frac{3}{4}$

5. The probability of drawing a red marble from a sack of marbles is $\frac{2}{5}$. Which set of marbles could the sack contain?

(1) 2 red marbles and 5 green marbles
(2) 4 red marbles and 6 green marbles
(3) 6 red marbles and 15 green marbles
(4) 2 red marbles, 1 blue marble, and 4 white marbles

6. Each face of a cube has a different whole number from 1 to 6 written on it. When the cube is rolled, which event has the greatest probability of occurring?

(1) rolling an even number
(2) rolling a prime number
(3) rolling a number greater than 3
(4) *not* rolling either 1 or 6

7. A jar contains 12 identical cards. Each card has a different letter of the word MATHEMATICAL written on it. If a card is picked at random, what is the probability that the card will *not* have the letter *H* or the letter *A* written on it?

(1) $\frac{1}{4}$ (2) $\frac{1}{3}$ (3) $\frac{2}{3}$ (4) $\frac{3}{4}$

8. The probability of guessing the correct answer to a certain test question is $\frac{x}{12}$. If the probability of *not* guessing the correct answer to the same question is $\frac{2}{3}$, what is the value of x?

(1) 9 (2) 3 (3) 4 (4) 8

B. *Show how you arrived at your answer.*

9–12. The numbers from 1 to 20, inclusive, are written on individual slips of paper and placed in a hat. A slip of paper is picked out of the hat without looking.

9. Find the probability of picking a number that is divisible by 5.

10. Find the probability of picking a prime number.

11. Find the probability of picking a number that is *at most* 13.

12. Find the probability of picking a number that is *at least* 7.

13. In a production run of 10,000 light bulbs, a random sample of 100 bulbs is selected and tested. If 2 bulbs from the sample are found to be defective, how many defective bulbs would we expect to find in the entire production run?

14. If the replacement set for x is $\{2, 3, 4, 5, 6\}$, what is the probability that a number chosen at random from the replacement set will make the expression $3x - 1$ an odd number?

15. Three students were selected to toss a cone-shaped cup and record the results which are summarized in the accompanying table.

Name	Lands on Side	Lands Point-up
Henry	28	3
Jose	24	6
Linda	22	7

a. Based on these results, express the empirical probability of the cup landing point-up to the *nearest tenth of a percent*.
b. Using your answer from part a, estimate the number of times you would expect the cup to land point-up when it is tossed 500 times.

16. From a jar of red jellybeans and white jellybeans, the probability of picking a white jellybean at random from the jar is $\frac{2}{3}$. If the jar contains 24 jellybeans, how many red jellybeans are in the jar?

17. The probability of selecting a green marble at random from a jar that contains only green, white, and yellow marbles is $\frac{1}{4}$. The probability of selecting a white marble at random from the same jar is $\frac{1}{3}$. If this jar contains 10 yellow marbles, what is the number of marbles in the jar?

18. An urn contains only green marbles and blue marbles. Eight of the marbles in the urn are green. The probability of picking a green marble is $\frac{1}{5}$. How many blue marbles must be removed from the urn so that the probability of picking a green marble is $\frac{1}{3}$?

19. One marble is drawn at random from a jar that contains x red marbles, $2x - 1$ blue marbles, and $2x + 1$ white marbles.

a. If the probability of drawing a blue marble is $\frac{1}{3}$, how many red marbles are in the jar.
b. What is the probability of *not* drawing a white marble?

CHAPTER 5

POLYNOMIAL ARITHMETIC AND FACTORING

5.1 COMBINING POLYNOMIALS

Algebraic terms like $7x$, $-3xy$, and $\frac{1}{2}ab^2$ are *monomials*. A **monomial** is a single number, a variable, or the *product* of one or more numbers and variables with positive integer exponents. The prefix "poly" means many. A **polynomial** is a monomial or the sum of monomials, as in $3x^2 - 5x + 1$. Since polynomials represent real numbers, arithmetic operations can be performed with polynomials using the properties of real numbers.

Like Monomials

Monomials such as $3xy^2$ and $4xy^2$ that differ only in their numerical coefficients are called **like monomials**. Like monomials can be added or subtracted by combining their numerical coefficients, as in

$$3xy^2 + 4xy^2 = (3+4)xy^2 = 7xy^2$$

and

$$ab - 3ab = (1-3)ab = -2ab$$

Standard Form of a Polynomial

A polynomial in one variable is in **standard form** when its terms are arranged so that the exponents decrease when the polynomial is read from left to right. To express $2x - 5x^2 + 3$ in standard form, write it as $-5x^2 + 2x + 3$.

Degree of a Polynomial

The **degree of a monomial** is the sum of the exponents of its variables. The degree of $-x^2y$ is 3 because x has an exponent of 2, y has an exponent of 1, and $2 + 1 = 3$. The **degree of a polynomial** in one variable is the greatest exponent of its terms. Thus, the degree of $-4x^2 + 8x + 3$ is 2. The polynomial $y^3 + 5y^2 - 4y + 9$ is a third-degree polynomial.

Naming Polynomials

A polynomial can be referred to by the number of unlike terms it contains.

Number of Unlike Terms	Name of Polynomial	Example
One	Monomial	$4x^3$
Two	Binomial	$7x - 3y$
Three	Trinomial	$5n^2 - n - 2$

Adding Polynomials

To add two polynomials such as $2x^2 - 5x - 1$ and $x^2 + 3x - 7$, write one polynomial underneath the other, aligning like terms in the same column. Then combine like terms in each column:

$$\begin{array}{rr} & 2x^2 - 5x - 1 \\ + & \underline{x^2 + 3x - 7} \\ \text{Sum} = & 3x^2 - 2x - 8 \end{array}$$

Example 1

The lengths of the sides of a triangle are represented by $3x + 5y$, $2x - 9y$, and $5x + y$. What is the perimeter of the triangle in terms of x and y?

Solution: The perimeter of a triangle is the sum of the lengths of its three sides, as shown at the right.

$$\begin{array}{rr} & 3x + 5y \\ & 2x - 9y \\ + & \underline{5x + y} \\ \text{Perimeter} = & \mathbf{10x - 3y} \end{array}$$

Subtracting Polynomials

To subtract a polynomial from another polynomial, change to an addition example by adding the *opposite* of each term of the polynomial that is being subtracted.

Example 2

Subtract $3a - 2b - 9c$ from $5a - 7b - 3c$.

Solution 1: Write the polynomials on the same line and change to an addition example:

$$\begin{aligned} (5a - 7b - 3c) - (3a - 2b - 9c) &= (5a - 7b - 3c) + (-3a + 2b + 9c) \\ &= (5a - 3a) + (-7b + 2b) + (-3c + 9c) \\ &= \quad 2a \quad - \quad 5b \quad + \quad 6c \end{aligned}$$

The difference is $\mathbf{2a - 5b + 6c}$.

Solution 2: Write the polynomial being subtracted on the line underneath the other polynomial and then change to an equivalent addition example:

change to addition

$$\begin{array}{r} 5a - 7b - 3c \\ -\ \underline{3a - 2b - 9c} \end{array} \quad \longrightarrow \quad \begin{array}{r} 5a - 7b - 3c \\ +\ \underline{-3a + 2b + 9c} \\ \mathbf{2a - 5b + 6c} \end{array}$$

Example 3

Find the difference when $x^2 - 2x + 1$ is subtracted from $4x^2 - 3x - 5$.

Solution: Write the polynomial being subtracted on the line underneath the other polynomial and then change to an equivalent addition example:

$$\begin{array}{r} 4x^2 - 3x - 5 \\ -\ \underline{x^2 - 2x + 1} \end{array} \quad \longrightarrow \quad \begin{array}{r} 4x^2 - 3x - 5 \\ +\ \underline{-x^2 + 2x - 1} \\ 3x^2 - \ \ x - 6 \end{array}$$

The difference is $\mathbf{3x^2 - x - 6}$.

Check Your Understanding of Section 5.1

A. *Multiple Choice.*

1. When $4a^2 - 7a - 5$ and $-6a^2 - 2a + 7$ are added, the sum is
(1) $-2a^2 - 9a + 2$
(2) $2a^5 - 5a + 2$
(3) $-10a^2 - 5a + 12$
(4) $-2a^4 - 9a^2 + 2$

2. If $2y^2 - 7y + 6$ is subtracted from $3y^2 - 2y + 5$, the difference is
(1) $5y^2 - 9y + 11$
(2) $-y^2 - 5y + 1$
(3) $y^2 + 5y - 1$
(4) $y^2 - 9y + 11$

3. If $2x^2 - 4x + 6$ is subtracted from $5x^2 + 8x - 2$, the difference is
(1) $3x^2 + 12x - 8$
(2) $-3x^2 - 12x + 8$
(3) $3x^2 + 4x + 4$
(4) $-3x^2 + 4x + 4$

4. When $a^2 + a - 3$ is subtracted from $3a^2 - 5$, the difference is
(1) $2a^2 - a - 2$
(2) $-2a^2 + a + 2$
(3) $2a^2 - a + 2$
(4) $4a^2 + a - 8$

5. The expression $-6x - 7(4 + 3x)$ is equivalent to
(1) $3x - 28$ (2) $-9x - 28$ (3) $-21x - 4$ (4) $-27x - 28$

B. *Show how you arrived at your answer.*

6–14. *Perform the indicated operation and simplify the result.*

6. $x-(3x-4)$

7. $(5y-8)+(-2+3y)$

8. $(4n^2-11)-(7n^2-6)$

9. $(-x^3+7x^2-9)+(3x^3+x^2-6x)$

10. $(2x-5y+4z)-(-3x+2y-3z)$

11. $(3a^2-11a)-(4a^2-7.5a+2)$

12. $\left(1.8b-2.4c+\frac{2}{3}e\right)-\left(\frac{3}{10}b+1.7c-\frac{1}{6}e\right)$

13. $\left(\frac{1}{2}m-3n+\frac{2}{3}p\right)+\left(\frac{1}{3}m+2n-\frac{5}{4}p\right)$

14. What is the additive inverse of $4x^3 - 5x^2 + 13$?

15. What is the difference when $5x^2 - 2x + 3$ is subtracted from $3x^2 - 4x - 3$?

16. From the sum of $(2x^3+6x^2-3)$ and (x^3-9x+7), subtract $-x^3 + 5x - 2$.

17. The value of $ax^2 + bx$ is 18 when $x = 2$ and is equal to 0 when $x = 1$. What are the values of a and b? Explain how you arrived at your answer.

5.2 MULTIPLYING AND DIVIDING POLYNOMIALS

KEY IDEAS

Monomials with a common base can be multiplied and divided using the laws of exponents. Polynomials can be multiplied and divided by a monomial in a similar way. Two binomials are multiplied together in much the same way that a two-digit number is multiplied by another two-digit number.

Multiplying a Monomial by a Monomial

To multiply one monomial by another monomial, multiply their numerical coefficients together and multiply powers of the same base:

$$\begin{aligned}(6x^3y)\left(\frac{1}{3}xy^2\right) &= \left(6\cdot\frac{1}{3}\right)(x^3\cdot x)(y\cdot y^2)\\ &= 2\cdot x^{3+1}\cdot y^{1+2}\\ &= 2x^4y^3\end{aligned}$$

Dividing a Monomial by a Monomial

To divide one monomial by another monomial, divide the numerical coefficients and divide variable factors with the same base:

$$\begin{aligned}\frac{-12a^6bc}{3a^4b^2} &= \left(\frac{-12}{3}\right)\left(\frac{a^6}{a^4}\right)\left(\frac{b}{b^2}\right)c \\ &= -4 \cdot a^{6-4} \cdot b^{1-2} \cdot c \\ &= -4a^2b^{-1}c \\ &= \frac{-4a^2c}{b}\end{aligned}$$

Simplifying Exponential Expressions

An exponential expression is in simplest form when it does not contain any negative exponents, products, or quotients of the same base, or powers raised to powers.

Example 1

Express $\left(3wz^{-2}\right)\left(-2w^3z\right)^5$ in simplest form.

Solution: Raise the second factor to the indicated power. Then multiply the two monomials.

$$\begin{aligned}\left(3wz^{-2}\right)\left(-2w^3z\right)^5 &= \left(\frac{3w}{z^2}\right)\left[(-2)^5\left(w^3\right)^5(z)^5\right] \\ &= \left(\frac{3w}{z^2}\right)\left(-32w^{15}z^5\right) \\ &= -96\left(w \cdot w^{15}\right)\left(\frac{z^5}{z^2}\right) \\ &= \mathbf{-96w^{16}z^3}\end{aligned}$$

Multiplying a Polynomial by a Monomial

To multiply a polynomial by a monomial, multiply each term of the polynomial by the monomial:

$$\begin{aligned}3a^2\left(a^2-4a+5\right) &= 3a^2\left(a^2\right)+3a^2(-4a)+3a^2(5) \\ &= 3a^4 \quad -12a^3 \quad +15a^2\end{aligned}$$

Dividing a Polynomial by a Monomial

To divide a polynomial by a monomial, divide each term of the polynomial by the monomial:

$$\begin{aligned}\frac{72x^3-32x^2+8x}{8x} &= \frac{72x^3}{8x}+\frac{-32x^2}{8x}+\frac{8x}{8x} \\ &= 9x^2-4x+1\end{aligned}$$

Multiplying a Polynomial by a Polynomial

To multiply a polynomial by another polynomial, write the second polynomial underneath the first polynomial. Then multiply each term of the second polynomial by the polynomial above it in much the same way two multi-digit numbers are multiplied:

$$\begin{array}{rrr} & & 2x + 7 \\ & & \underline{x \ + 3} \\ 3(2x+7) = & & 6x + 21 \\ x(2x+7) = & \underline{2x^2 + 7x} & \\ \text{Add like terms in each column:} & 2x^2 + 13x & + 21 \quad \leftarrow \text{final product} \end{array}$$

Multiplying Binomials Horizontally

To multiply two binomials together such as $(2x + 7)$ and $(x + 3)$, add the products of the *F*irst, *O*uter, *I*nner, and *L*ast terms of the two binomials:

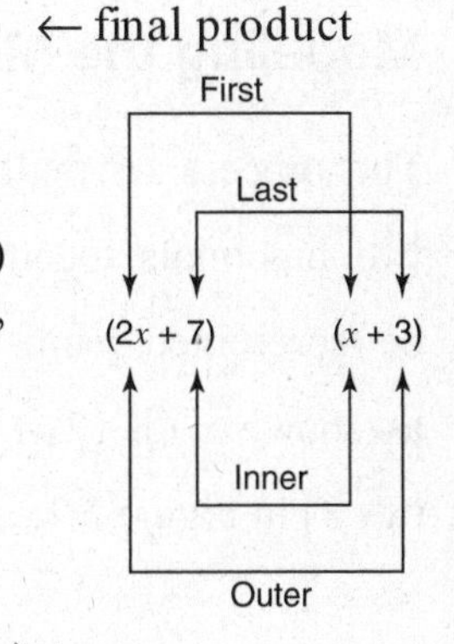

- Multiply ***F***irst terms: $(\underline{\mathbf{2x}} + 7)(\underline{\mathbf{x}} + 3) = (2x)(x) +$
- Multiply ***O***uter terms: $(\underline{\mathbf{2x}} + 7)(x + \underline{\mathbf{3}}) = 2x^2 + (2x)(3) +$
- Multiply ***I***nner terms: $(2x + \underline{\mathbf{7}})(\underline{\mathbf{x}} + 3) = 2x^2 + 6x + (7)(x) +$
- Multiply ***L***ast terms: $(2x + \underline{\mathbf{7}})(x + \underline{\mathbf{3}}) = 2x^2 + 6x + 7x + (7)(3)$
  $= 2x^2 + 13x + 21$

You can remember the four products that are needed by referring to the horizontal multiplication process as FOIL.

Example 2

Multiply $(x-6)$ by $(x+2)$.

Solution: Multiply horizontally using FOIL:

$$\begin{aligned}(x-6)(x+2) &= \overbrace{x \cdot x}^{F} + \overbrace{2 \cdot x}^{O} + \overbrace{(-6)(x)}^{I} + \overbrace{(-6)(+2)}^{L} \\ &= x^2 + \ [2x - 6x] \quad - 12 \\ &= \boldsymbol{x^2 - 4x - 12}\end{aligned}$$

Example 3

Multiply $(2a+b)$ by $(a-b)$.

Solution: Multiply horizontally using *FOIL* :

$$\begin{aligned}(2a+b)(a-b) &= \overbrace{2a\cdot a}^{F} + \overbrace{2a\cdot(-b)}^{O} + \overbrace{(a)(b)}^{I} + \overbrace{(b)(-b)}^{L}\\ &= 2a^2 + [-2ab + ab] - b^2\\ &= \mathbf{2a^2 - ab - b^2}\end{aligned}$$

Modeling the Multiplication of Binomials

The process of multiplying two binomials together can be represented geometrically as shown with $(2x+7)$ and $(x+3)$ in Figure 5.1.

	$2x$	+ 7	
x	$2x^2$	$7x$	x
3	$6x$	21	3

Figure 5.1 Representing the Product of Two Binomials Geometrically

Since the area of the large rectangle in Figure 5.1 must be equal to the sum of the areas of the four small rectangles,

$$\begin{aligned}\overbrace{(2x+7)(x+3)}^{\text{area of big rectangle}} &= \overbrace{2x^2 + 6x + 7x + 21}^{\text{sum of areas of four small rectangles}}\\ &= 2x^2 + 13x + 21\end{aligned}$$

Check Your Understanding of Section 5.2

A. Multiple Choice.

1. The product of $\left(\dfrac{2a^3}{5b}\right)\left(\dfrac{3a^2}{7b}\right)$ is

(1) $\dfrac{5a^5}{12b^2}$ (2) $\dfrac{5a^6}{12b}$ (3) $\dfrac{6a^5}{35b^2}$ (4) $\dfrac{6a^6}{35b}$

2. What is the quotient of $\frac{26x^4y^2}{13xy}$?

(1) $2x^4y^2$ (2) $13x^5y^3$ (3) $2x^3y$ (4) $13x^3y$

3. The product of $-3xy^2$ and $5x^2y^3$ is

(1) $-8x^3y^5$ (2) $-15x^3y^5$ (3) $-15x^2y^5$ (4) $-15x^3y^6$

4. What is the product of $1.45(xy)^3$ and $2.6xy^3$?

(1) $3.77x^4y^9$ (2) $4.05x^4y^9$ (3) $3.77x^3y^{18}$ (4) $4.05x^3y^9$

5. If $14x^3 - 35x^2 + 7x$ is divided by $7x$, where $x \neq 0$, the quotient is

(1) $2x^2 - 5x$
(2) $2x^2 - 5x + 1$
(3) $2x^3 - 5x^2 + x$
(4) $2x^2 - 5x + x$

6. If the length of a rectangle is represented by $4n + 5$ and the width is represented by $3n - 1$, which algebraic expression represents the perimeter of the rectangle?

(1) $14n + 8$ (2) $7n + 4$ (3) $7n^2 + 11n - 5$ (4) $\frac{1}{2}(7n+4)$

7. The fraction $\frac{-12x^3+4x^2-8x}{2x}$, $x \neq 0$, is equivalent to

(1) $-6x^2 + 2x - 4$
(2) $6x^2 + 2x - 4$
(3) $6x^2 - 2x + 4$
(4) $-6x^2 + 2x$

8. If the length of a side of a square is represented by $2x + 3$, which expression represents the area of the square?

(1) $4x^2 + 9$
(2) $8x + 12$
(3) $4x^2 + 6x + 9$
(4) $4x^2 + 12x + 9$

9. The product $(x-7)(x+3)$ is equivalent to

(1) $x^2 - 4x - 21$
(2) $x^2 - 4x + 21$
(3) $x^2 + 4x - 21$
(4) $x^2 - 21$

10. The product $(2x-3)(3x+5)$ is equivalent to

(1) $5x^2 - x - 15$
(2) $6x^2 + x + 15$
(3) $5x + 2$
(4) $6x^2 + x - 15$

B. *Show how you arrived at your answer.*

11–28. *Find each product or quotient.*

11. $\left(-\frac{1}{2}a^2b\right)\left(-8ab^3\right)$

12. $\left(-2x\right)\left(-3x^2\right)\left(-4x^3\right)$

13. $\left(0.4y^3\right)\left(-0.15xy^2\right)$

14. $\dfrac{xy^2}{x^3y}$

15. $\dfrac{-8a^2b^3}{12a^2b^5}$

16. $1.05x^5y^3 \div 0.35x^2y^4$

17. $\left(\frac{1}{2xy}\right)\div\left(\frac{1}{6x^3y^2}\right)$

18. $5y\left(y^3-8y-4\right)$

19. $\left(n-5\right)\left(n+8\right)$

20. $\left(3x+7\right)\left(x-9\right)$

21. $\left(4-y\right)\left(5+y\right)$

22. $\left(3r^2+5\right)\left(3r^2-5\right)$

23. $\left(0.4m-10\right)\left(0.6m+5\right)$

24. $\left(\frac{4}{5}k+10\right)\left(\frac{1}{2}k-20\right)$

25. $\left(m+2n\right)\left(3m-n\right)$

26. $\left(\frac{3}{4}a-\frac{2}{3}b\right)\left(a+\frac{3}{2}b\right)$

27. $\left(2x-3\right)^2$

28. $\left(4c-3\right)\left(2c+1\right)$

29–31. *Simplify.*

29. $\left(-\frac{1}{2}b^3c\right)\left(-4bc^2\right)^3$

30. $\left(56p^7q^5\right)\left(-2p^2q^2\right)^{-3}$

31. $\left(\dfrac{21x^3y^4}{14x^{-2}y}\right)^2$

32–34. *Divide.*

32. $\dfrac{18r^4-27r^3s^2}{9rs}$

33. $\dfrac{21c^3-12c^2+3c}{-3c}$

34. $\dfrac{0.14a^3-1.05a^2b}{0.7a}$

35. Roberto claims that the sum of any five consecutive integers is always evenly divisible by 5. Prove or disprove Roberto's claim.

36. In the accompanying diagram, the width of the inner rectangle is represented by x and its length by $2x - 1$. The width of the outer rectangle is represented by $x + 3$ and its length by $x + 5$.

a. Express the area of the shaded region as a trinomial in terms of x.

b. If the perimeter of the outer rectangle is 24, what is the value of x?

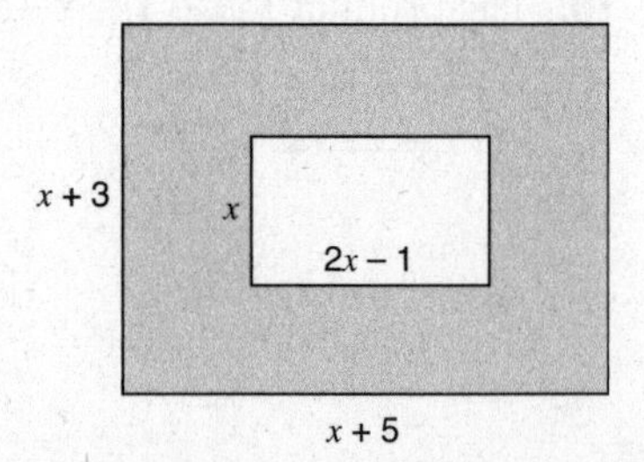

37. The length of a rectangle exceeds the width by 3. If the length is increased by 2 and the width is decreased by 1, the area remains the same. What are the dimensions of the original rectangle?

38. The square of the larger of two consecutive integers exceeds the product of these two integers by 10. Find the smaller of the two integers.

5.3 FACTORING POLYNOMIALS

KEY IDEAS

Sometimes it is useful to know what quantities were multiplied together to produce a given product. Undoing multiplication is called *factoring*:

- Multiplication: $3x(x+2) = 3x^2 + 6x$ ← use the distributive property
- Factoring: $3x^2 + 6x = 3x(x+2)$ ← reverse the distributive property

It is usually understood that when a polynomial with integral coefficients is factored, its factors will also have only integral coefficients. For example, $3x^2 + 6x$ would *not* be factored as $6x\left(\frac{1}{2}x+1\right)$.

Factoring a Polynomial by Removing the GCF

Factoring a polynomial is the process of rewriting the polynomial as the product of two or more lower degree polynomials. Each of the lower degree polynomials is a **factor** of the original polynomial. Because $3x^2 + 6x = 3x(x+2)$, $3x$ and $x + 2$ are *factors* of $3x^2 + 6x$.

For the polynomial $3x^2 + 6x$, the *greatest common factor* (GCF) of its terms is $3x$. The **GCF** of a polynomial is the greatest monomial that exactly divides *each* term of the polynomial. If you know the GCF of a polynomial, the other factor can be obtained by dividing the polynomial by the GCF, as illustrated in the next set of examples.

Example 1

Factor $21a^5 + 14a^3$.

Solution: The GCF of $21a^5 + 14a^3$ is $7a^3$ since 7 is the greatest integer that exactly divides both 21 *and* 14, and a^3 is the greatest power of a that is contained in a^5 *and* a^3.

- Find the corresponding factor by dividing the original polynomial by $7a^3$:

$$\frac{21a^5}{7a^3}+\frac{14a^3}{7a^3}=3a^2+2$$

- Write $21a^5 + 14a^3$ in factored form:

$$\mathbf{21a^5+14a^3=7a^3\left(3a^2+2\right)}$$

Check that you have factored a polynomial correctly by multiplying the factors together and verifying that the product is the original polynomial:

$$7a^3\left(3a^2+2\right)=7a^3\cdot 3a^2+7a^3\cdot 2=21a^5+14a^3\ ✓$$

Example 2

Factor $9x^4 - 3x^3 + 12x$.

Solution: The GCF of $9x^4 - 3x^3 + 12x$ is $3x$. Find the corresponding factor:

$$\frac{9x^4}{3x}-\frac{3x^3}{3x}+\frac{12x}{3x}=3x^3-x^2+4$$

Hence, $\mathbf{9x^4-3x^3+12x=3x\left(3x^3-x^2+4\right)}$. The check is left for you.

Example 3

Factor $40y^2z^5 - 16xy^3z^4$.

Solution: The GCF of $40y^2z^5 - 16xy^3z^4$ is $8y^2z^4$. Find the corresponding factor:

$$\frac{40y^2z^5}{8y^2z^4}-\frac{16xy^3z^4}{8y^2z^4}=5z-2xy$$

Hence, $\mathbf{40y^2z^5-16xy^3z^4=8y^2z^4\left(5z-2xy\right)}$. The check is left for you.

Prime Polynomials

Not all polynomials can be factored. A **prime polynomial** is a polynomial such as $5x^2 + 8$ that cannot be factored except by writing it as the product of itself and 1 (or –1).

Solving Literal Equations by Factoring

Factoring may be needed when solving a literal equation for a particular letter in it.

Example 4

If $xz = y - x$, what is x in terms of y and z?

(1) $\frac{z}{y+1}$ (2) $\frac{y}{z+1}$ (3) $\frac{y}{z-1}$ (4) $\frac{z}{y-1}$

Solution: Isolate the x terms on the left side of the equation:

$$xz + x = y$$

Factor out x: $x(z+1) = y$

Divide by the coefficient of x: $\frac{x(z+1)}{z+1} = \frac{y}{z+1}$

$$x = \frac{y}{z+1}, \text{ provided } z \neq -1$$

The correct choice is **(2)**.

Check Your Understanding of Section 5.3

A. *Multiple Choice.*

1. If $A = p + prt$, then $p =$

(1) $\frac{A}{rt}$ (2) $\frac{A}{1+rt}$ (3) $A - rt$ (4) $\frac{A}{1-rt}$

2. Expressed in factored form, the binomial $2x^2y - 4xy^3$ is equivalent to

(1) $2xy(x-2y)$ (3) $2xy(xy-4y)$

(2) $2xy(x-2y^2)$ (4) $2x^2y^3(y-2)$

3. If $ay - c = d + by$, then $y =$

(1) $\frac{c+d}{a+b}$ (2) $\frac{a-b}{c+d}$ (3) $\frac{c+d}{a-b}$ (4) $\frac{a+c}{d-b}$

4. If $a(x-y) = y(a-x)$, then $x =$

(1) $\frac{2ay}{a+y}$ (2) $\frac{ay}{2(a-y)}$ (3) $\frac{a}{y-2}$ (4) $\frac{2a}{y}$

B. *Show how you arrived at your answer.*

5–16. *Factor.*

5. $15x^2 - 6x$

6. $7p^2 + 7q^2$

7. $x^3 + x^2 - x$

8. $3y^7 - 6y^5 + 12y^3$

9. $-4a - 4b$

10. $8b^7 - 20a^4$

11. $10x^3y - 15xy^2$

12. $0.24x^2 - 0.36xy$

13. $18u^5w^2 - 30u^3w^7$

14. $\left(2x^3y\right)^4 - 12x^8y^3$

15. $\frac{1}{4}a^2b - \frac{3}{4}ab^3$

16. $6h^2k + \frac{1}{2}\left(h^2k\right)^2$

17. Solve for p: $x\left(2p-5\right) = x - 4p$.

18. Solve for y: $\frac{y-1}{b} = \frac{x+y}{a}$, where $a, b \neq 0$.

5.4 MULTIPLYING AND FACTORING SPECIAL BINOMIAL PAIRS

KEY IDEAS

The binomial factors of the product $\left(a+b\right)\left(a-b\right)$ are the sum and difference of the same two terms. Using FOIL, the product is $a^2 - b^2$. Thus, the equation

$$\left(a+b\right)\left(a-b\right) = a^2 - b^2$$

can be used as a formula to multiply any pair of binomials that have this form and their product can be factored by reversing this rule.

Multiplying the Sum and Difference of Two Terms

The product of two binomials is not always a trinomial. For example,

$$\begin{aligned}\left(y+5\right)\left(y-5\right) &= \overbrace{y \cdot y}^{F} + \overbrace{y \cdot \left(-5\right)}^{O} + \overbrace{\left(5\right)\left(y\right)}^{I} + \overbrace{\left(5\right)\left(-5\right)}^{L} \\ &= y^2 + \left[-10y + 10y\right] - 25 \\ &= y^2 - 25\end{aligned}$$

The binomials $y + 5$ and $y - 5$ represent the sum and difference of the same two terms, y and 5. Rather than using FOIL, the product $(y+5)(y-5)$ can be expanded using the rule $(a+b)(a-b)=a^2-b^2$ with $a = y$ and $b = 5$:

$$\overbrace{(y+5)}^{(a+b)}\overbrace{(y-5)}^{(a-b)} = \overbrace{(y)^2}^{a^2} - \overbrace{(5)^2}^{b^2} = y^2 - 25$$

Here are a few more examples:

- $(3x+4y)(3x-4y)=(3x)^2-(4y)^2$
  $=9x^2-16y^2$
- $(0.8+m^2)(0.8-m^2)=(0.8)^2-(m^2)^2$
  $=0.64-m^4$
- $(a-2b^3)(a+2b^3)=a^2-(2b^3)^2$
  $=a^2-4b^6$
- $\left(\frac{1}{2}x-3y\right)\left(\frac{1}{2}x+3y\right)=\left(\frac{1}{2}x\right)^2-(3y)^2$
  $=\frac{1}{4}x^2-9y^2$

Factoring the Difference of Two Squares

The **difference of two squares** is a binomial of the form $a^2 - b^2$. Factoring the difference of two squares is the reverse of the process of finding the product of the sum and difference of two terms. To factor $16x^2 - 49$, use the rule $a^2-b^2=(a+b)(a-b)$, where a^2 is $16x^2$ so a is $4x$ and b^2 is 49 so b is 7:

$$\overbrace{16x^2-49}^{a^2-b^2}=\overbrace{(4x)^2}^{a^2}-\overbrace{(7)^2}^{b^2}=\overbrace{(4x+7)}^{a+b}\overbrace{(4x-7)}^{a-b}$$

To illustrate further:

- $1-9y^4=1^2-(3y^2)^2$
  $=(1+3y^2)(1-3y^2)$
- $81a^2-25b^2=(9a)^2-(5b)^2$
  $=(9a+5b)(9a-5b)$

- $n^2 - \frac{49}{100} = (n)^2 - \left(\frac{7}{10}\right)^2$
  $= \left(n + \frac{7}{10}\right)\left(n - \frac{7}{10}\right)$

- $0.16y^4 - 0.09 = (0.4y^2)^2 - (0.3)^2$
  $= (0.4y^2 + 0.3)(0.4y^2 - 0.3)$

Check Your Understanding of Section 5.4

Show how you arrived at your answer.

1–6. *Write each product as a binomial.*

1. $\left(x+\frac{1}{2}\right)\left(x-\frac{1}{2}\right)$ **3.** $\left(\frac{x}{2}-4y\right)\left(\frac{x}{2}+4y\right)$ **5.** $(3x+2y^2)(3x-2y^2)$

2. $(5w+8)(5w-8)$ **4.** $(0.3y^2+1)(0.3y^2-1)$ **6.** $(3h^3-2k^2)(3h^3+2k^2)$

7–21. *Factor each expression as the product of two binomials.*

7. $y^2 - 144$

8. $81 - x^2$

9. $p^2 - \frac{1}{9}$

10. $1 - \frac{16x^2}{25}$

11. $9e^2 - 0.04$

12. $49e^2 - 121f^2$

13. $100a^6 - 49b^2$

14. $\frac{1}{4}r^2 - \frac{1}{9}s^2$

15. $\frac{4}{9}c^2 - 0.01$

16. $4a^2b^2 - 9c^6$

17. $16y^4 - 49x^2$

18. $0.09w^2 - 0.64z^6$

19. $x^{2n} - 1$

20. $a^{2n} - b^{2y}$

21. $9x^{-2} - 25y^{-4}$

5.5 FACTORING QUADRATIC TRINOMIALS

KEY IDEAS

By reversing the FOIL process, it may be possible to factor a quadratic trinomial as the product of two binomials. Sometimes more than one factoring technique is needed to factor a polynomial *completely* so that none of its factors can be factored further.

Factoring $x^2 + bx + c$

To factor $x^2 + 7x + 10$ as the product of two binomials:

- Write the general form of the binomial factors:

$$x^2+7x+10=\left(x+\boxed{?}\right)\left(x+\boxed{?}\right)$$

- Find the pair of integers whose product is the last number term of the quadratic trinomial $\left(x^2+7x+\boxed{10}\right)$ *and* whose sum equals the numerical coefficient of the x term $\left(x^2+\boxed{7}x+10\right)$. Since $2 \times 5 = 10$ and $2 + 5 = 7$, the missing integers are 2 and 5:

$$x^2+7x+10=(x+2)(x+5)$$

- Check by multiplying the two binomial factors together:

$$\begin{aligned}(x+2)(x+5)&=x^2+2x+5x+10\\&=x^2+7x+10 \ \checkmark\end{aligned}$$

Example 1

Factor $y^2 - 7y + 12$ as the product of two binomials.

Solution: The general form of the binomial factors of $y^2 - 7y + 12$ is $\left(y+\boxed{?}\right)\left(y+\boxed{?}\right)$.

- Find the pair of integers whose product is +12 and whose sum is −7. Since $(-3)\times(-4)=+12$ and $(-3)+(-4)=-7$, the correct factors of +12 are −3 and − 4.
- Factor: $\mathbf{y^2-7y+12=(y-3)(y-4)}$

- Check: $(y-3)(y-4) = y \cdot y - 3y - 4y + (-3)(-4)$
  $= y^2 - 7y + 12$ ✓

Example 2

Factor $n^2 - 5n - 14$ as the product of two binomials.

Solution: The general form of the binomial factors of $n^2 - 5n - 14$ is $(n+\boxed{?})(n+\boxed{?})$.

- Find the pair of integers whose product is -14 and whose sum is -5. Since $(+2)\times(-7) = -14$ and $(+2)+(-7) = -5$, the correct factors of -14 are $+2$ and -7.

- Factor: $\mathbf{n^2 - 5n - 14 = (n+2)(n-7)}$

- Check: $(n+2)(n-7) = n \cdot n + 2n - 7n + (+2)(-7)$
  $= n^2 - 5n - 14$ ✓

Factoring Completely

After a polynomial has been factored by removing the GCF of its terms, it may be possible to factor again. A polynomial is **factored completely** when none of its factors can be factored further.

Example 3

Factor $2x^3 - 50x$ completely.

Solution: The GCF of the terms of $2x^3 - 50x$ is $2x$.

Factor out $2x$: $2x^3 - 50x = 2x(x^2 - 25)$

Factor the binomial factor: $= \mathbf{2x(x+5)(x-5)}$

Example 4

Factor $3t^3 + 18t^2 - 48t$ completely.

Solution: The GCF of the terms of $3t^3 + 18t^2 - 48t$ is $3t$.

Factor out the GCF of $3t$: $3t^3 + 18t^2 - 48t = 3t(t^2 + 6t - 16)$

Factor the quadratic trinomial: $= \mathbf{3t(t+8)(t-2)}$

Factoring $ax^2 + bx + c$ when $a > 1$

Factoring a quadratic trinomial becomes more complicated when the leading numerical coefficient is greater than 1. To factor $3x^2 + 10x + 8$:

- Factor the x^2 term and set up the binomial factors:

$$3x^2 + 10x + 8 = \left(3x + \boxed{?}\right)\left(x + \boxed{?}\right)$$

- Identify possibilities for the unknown pair of integers in the binomial factors. The product of these integers must be +8, the last term of $3x^2 + 10x + 8$. The possibilities are 1 and 8, −1 and −8, 2 and 4, and −2 and − 4.
- Use trial and elimination to determine the correct pair of factors of 8 and their proper placement in the binomial factors. The factors of +8 must be chosen and placed so that the sum of the outer and inner products of the terms of the binomial factors is equal to $+10x$, the middle term of $3x^2 + 10x + 8$:

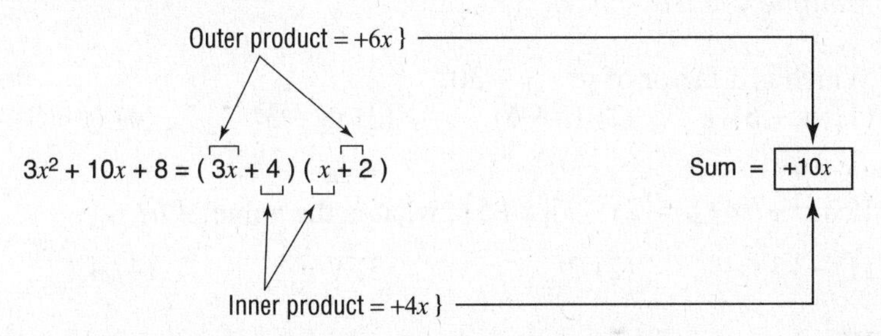

- Check that the factors work. The placement of the factors of 8 matters. Although $(3x+2)(x+4)$ contains the correct factors of 8, they are not placed correctly since the sum of the outer and inner products is $12x + 2x = 14x$ rather than $10x$.

Factoring by Decomposing the Middle Term

To eliminate some of the guess and check work when factoring a quadratic trinomial such as $3x^2 + 10x + 8$, break down the linear term, $10x$, into the sum of two x terms. Choose the two x terms so that their numerical coefficients have the same product as the product of the leading coefficient and the constant term of the original quadratic trinomial. Since

$$\overbrace{\boxed{3}x^2 + 10x + \boxed{8}}^{3 \times 8 = 24}$$

decompose $10x$ into $6x + 4x$ because 6 times 4 is also 24:

$$3x^2+10x+8=3x^2+\overbrace{6x+4x}^{10x}+8$$

- Group the first and last pairs of terms: $=\left(3x^2+6x\right)+\left(4x+8\right)$
- Factor out the GCF from each pair of terms: $=3x\left(x+2\right)+4\left(x+2\right)$
- Factor out the common *binomial* factor of $\left(x+2\right)$: $=\left(3x+4\right)\left(x+2\right)$

Check Your Understanding of Section 5.5

A. *Multiple Choice.*

1. Which is a factor of $y^2 + y - 30$?
(1) $(y - 6)$ (2) $(y + 6)$ (3) $(y - 3)$ (4) $(y + 3)$

2. If $ax^2+bx+c=\left(2x-3\right)\left(x+5\right)$, what is the value of b?
(1) -15 (2) 2 (3) 7 (4) 4

3. Which expression is a factored form of $0.08y^2 - 18$?
(1) $20\left(0.02y+3\right)\left(0.02y-3\right)$ (3) $2\left(0.2y+3\right)\left(0.2y-3\right)$
(2) $\left(0.4y+18\right)\left(0.2y-1\right)$ (4) $\left(4y+6\right)\left(0.02y-3\right)$

4. If $\left(x-4\right)$ is a factor of $x^2 - x + w$, then the value of w is
(1) 12 (2) -12 (3) 3 (4) -3

5. What is a common factor of $x^2 - 9$ and $x^2 - 5x + 6$?
(1) $x + 3$ (2) $x - 3$ (3) $x - 2$ (4) x^2

6. If $x^2+2x+k=\left(x+5\right)\left(x+p\right)$, then
(1) $p = 3$ and $k = -5$ (3) $p = -5$ and $k = -3$
(2) $p = -3$ and $k = 15$ (4) $p = -3$ and $k = -15$

7. Which product is a factored form of $4n - n^3$?

(1) $-n(n-2)(n+2)$
(3) $n(n-4)(n+4)$
(2) $(n^2+1)(4-n)$
(4) $n(2-n)(2-n)$

8. Which product is a factored form of $2x^2 + 2x - 12$?

(1) $2(x+2)(x-2)$
(3) $2(x+6)(x-1)$
(2) $2(x+3)(x-2)$
(4) $2(x+1)(x-6)$

9. Which is *not* a factor of $2x^2 - 32$?

(1) $x - 4$ (2) 2 (3) $x^2 - 32$ (4) $x^2 - 16$

10. Which expression is the complete factorization of $32 - 18y^2$?

(1) $2y(16-9y)$
(3) $2(3y-4)(3y+4)$
(2) $2(4-3y)^2$
(4) $2(4-3y)(4+3y)$

B. *Show how you arrived at your answer.*

11. Solve for x: $b^2y + ax = bx + a^2y$.

12–20. *Factor each trinomial as the product of two binomials.*

12. $x^2+8x+15$
13. $x^2-10x+21$
14. y^2+6y+9
15. $p^2-8p+16$
16. $b^2+3b-40$
17. $w^2-13w+42$
18. $a^2+4a-45$
19. $n^2-17n-60$
20. x^4-5x^2+6

21–32. *Factor completely.*

21. $2y^3-50y$
22. $-5t^2+5$
23. $4m^2-4n^2$
24. $4y^2+20y+16$
25. $8xy^3-72xy$
26. $-2y^2-14y-20$
27. $2x^3+2x^2-112x$
28. $4y^5-12y^4-160y^3$
29. $18x^3-50xy^2$
30. m^4-256
31. $\frac{1}{2}x^3-18x$
32. y^8-81x^4

33–35. *Factor each trinomial as the product of two binomials.*

33. $3x^2+2x-21$ **34.** $4n^2+11n-3$ **35.** $5s^2-14s-3$

5.6 SOLVING QUADRATIC EQUATIONS BY FACTORING

MATH FACTS

Equations such as $x^2 - 16 = 0$, $x^2 - 3x = 0$, and $x^2 + 4x = 5$ are **quadratic equations** because all the exponents are positive integers and the greatest exponent of the variable is 2. If the left side of a quadratic equation of the form $ax^2 + bx + c = 0$ can be factored, then the equation can be solved by setting each of these factors equal to 0.

The Multiplication Property of Zero

According to the Zero Product rule, if the product of two expressions is 0, then at least one of these is 0. If $x^2 - 3x = 0$, then $x(x-3)=0$ and, using the multiplication property of zero, $x = 0$ or $x - 3 = 0$. Hence, the two solutions are $x = 0$ *or* $x = 3$.

Solving Factorable Quadratic Equations

Before solving a quadratic equation, make sure the quadratic equation has the **standard form** $ax^2 + bx + c = 0$ with all the nonzero terms collected on one side of the equation and 0 isolated on the other side. To find the solution set of $x^2 + 4x = 5$, rewrite it in standard form as $x^2 + 4x - 5 = 0$. Then continue as follows:

- Factor the left side: $(x+5)(x-1)=0$
- Set each factor equal to 0: $x+5=0$ *or* $x-1=0$
  so $x=-5$ *or* $x=1$
- Write the solution set: **{–5, 1}**
- Check each root in the original equation:

Check $x = -5$:	Check $x = 1$
$x^2+4x = 5$	$x^2+4x = 5$
$(-5)^2+4(-5) \boxed{?} 5$	$(1)^2+4(1) \boxed{?} 5$
$25 - 20 = 5$ ✓	$1 + 4 = 5$ ✓

Example 1

Find the solution set and check: $6y^2 + 18y + 12 = 0$.

Solution: First simplify the equation by dividing each term by 6. Then factor.

$$\frac{6y^2}{6}+\frac{18y}{6}+\frac{12}{6}=0$$
$$y^2+3y+2=0$$
$$(y+2)(y+1)=0$$
$$y+2=0 \quad \text{or} \quad y+1=0$$
$$y=-2 \quad \Big| \quad y=-1$$

The solution set is **{–2, –1}**. To check, verify that $y = -2$ and $y = -1$ satisfy the original equation. The check is left for you.

Example 2

Raymond said $x = -2$ and $x = 1$ are the roots of the quadratic equation $x^2 - 3x = -2$ because

$$x^2-3x=-2$$
$$x(x-3)=-2$$
$$x=-2 \text{ or } x-3=-2$$
$$\Big| \quad x=-2+3$$
$$\Big| \quad x=1$$

Explain why Raymond's answer is correct or incorrect.

Solution: Raymond's solution is incorrect. If $x(x-3)=-2$, it is *not* correct to set each factor equal to –2. When the product of two numbers is equal to –2, each number may be equal to a number other than –2. For example, one number could be –0.001 and the other number 2000 since $-0.001 \times 2000 = -2$. Only when the product of two numbers is 0 can we assume that at least one of these numbers is 0. You can also tell that –2 is not a root by checking –2 in the original equation.

Example 3

Find the solution set and check: $28 - x^2 = 3x$.

Solution: Collect all of the nonzero terms on the left side of the equation:

$$-x^2 - 3x + 28 = 0$$

- Make the coefficient of the x^2 term positive by multiplying each member of the equation by –1. This is equivalent to changing the sign of each nonzero term of the equation to its opposite:

$$x^2 + 3x - 28 = 0$$

- Factor and solve:

$$\begin{aligned}(x+7)(x-4)&=0\\ x+7=0 \quad &or\quad x-4=0\\ x=-7 \quad &\Big|\quad x=4\end{aligned}$$

- Write the solution set: **{–7, 4}.**

The check is left for you.

Example 4

Find the solution set: $\dfrac{4}{x+3}=\dfrac{x-5}{5}$.

Solution: Set the cross-products equal:

$$\begin{aligned}(x+3)(x-5)&=20\\ x^2-2x-15&=20\\ x^2-2x-35&=0\\ (x+5)(x-7)&=0\\ x+5=0 \quad &or\quad x-7=0\\ x=-5 \quad &\Big|\quad x=7\end{aligned}$$

The solution set is **{–5, 7}**. The check is left for you.

Example 5

Manny is unable to read a quadratic equation he wrote in his notebook but remembers that its roots are 2 and –5, and its leading coefficient is 1. What is a possible equation?

Solution: Work backward. If 2 and –5 are roots, then $x - 2$ and $x-(-5)$ or, equivalently, $x + 5$ are the factors of the quadratic expression that is equal to 0:

$$\begin{aligned}(x-2)(x+5)&=0\\ x^2-2x+5x-10&=0\\ x^2+3x-10&=0\end{aligned}$$

One possible equation is $\boldsymbol{x^2 + 3x - 10 = 0}$. Equivalent forms of this equation such as $x^2 + 3x = 10$ and $x^2 - 10 = 3x$ are also possible.

Nonfactorable Quadratic Equations

After attempting to solve $x^2 + 3x - 1 = 0$ by factoring, you will soon realize that not all quadratic trinomials are factorable using *rational* numbers. Nonfactorable quadratic equations have either irrational roots or nonreal roots. Algebraic methods for solving these equations will be studied in Section 7.5.

Check Your Understanding of Section 5.6

A. *Multiple Choice.*

1. If $(x-3)(x+2)=0$, which is the greater of the two roots?

(1) –2 (2) 2 (3) 3 (4) –3

2. If $x^2 + 11x + 30 = 0$, which is the smaller of the two roots?

(1) –6 (2) –5 (3) 5 (4) 6

3. If $A=\{x: 5(x-4)=12x+1\}$ and $B=\{x: x^2-x=12\}$, what is the intersection of sets A and B?

(1) –3 (2) { } (3) 3 (4) 4

4. If one of the roots of the equation $x^2 - x + q = 0$ is 3, what is the other root?

(1) –2 (2) 2 (3) –1 (4) –6

5. What is the solution set of $\frac{x+2}{x-2} = \frac{-3}{x}$?

(1) {–2, 3} (2) {–3, –2} (3) {–1, 6} (4) {–6, 1}

B. *Show how you arrived at your answer.*

6–17. *Find the solution set algebraically and check.*

6. $y^2+3y+2=0$

7. $6-x^2=x$

8. $x^2-4x=x+24$

9. $2x^2=6x+20$

10. $13n-n^2=0$

11. $\frac{3+y}{2y}=\frac{y-1}{y}$

12. $3x=\frac{x^2}{2}$

13. $x(x+2)=3$

14. $-2t^2+20t-42=0$

15. $\frac{9n-n^2}{2}=10$

16. $x^2=9(10-x)$

17. $3y^2=9y+84$

18–23. Find the solution set algebraically and check.

18. $\frac{1}{x}=\frac{x+1}{6}$

19. $\frac{y+3}{y}=\frac{y-1}{6}$

20. $\frac{x}{x+3}=\frac{5}{x+7}$

21. $\frac{k+4}{k+2}=\frac{3}{k}$

22. $\frac{x+5}{x+1}=\frac{x-1}{4}$

23. $\frac{x-3}{x-2}=\frac{x+3}{2x}$

24–27. Solve algebraically.

24. The side of a certain square is 3 feet longer than that of another square. If the sum of their areas is 117 square feet, find the length of a side of the *smaller* square.

25. Tamara has two sisters. One of the sisters is 7 years older than Tamara. The other sister is 3 years younger than Tamara. If the product of Tamara's sisters' ages is 24, how many years old is Tamara?

26. The length of a rectangular box is 7. The width of the box is 2 more than the height. The volume of the solid is 105. Find the width and the height of the solid.

27. Barb pulled the plug in her bathtub and it started to drain. The amount of water in the bathtub as it drains can be represented by the equation $L=-t^2-19t+120$, where L represents the number of liters of water in the bathtub and t represents the amount of time, in minutes, since the plug was pulled.

a. How many liters of water were in the bathtub when Barb pulled the plug?

b. Determine the number of minutes it takes for all the water in the bathtub to drain.

28. The sum, S, of a list of consecutive integers beginning with 1 can be determined by evaluating the formula $S=\frac{n}{2}(n+1)$, where n is the number of integers in the list. Using this formula, determine the number of consecutive positive integers beginning with 1 that must be added together so that the sum is 703.

29–31. Find the solution set algebraically and check.

29. $2r^2=5r+3$

30. $3p^2+14p=5$

31. $\frac{x+4}{2}=\frac{5x+4}{3x}$

32. Alexi throws a pebble into an unused well. The distance, d, the pebble falls after t seconds is given by the equation $d = 16t^2 + 56t$, where d is measured in feet. If the water in the well is 240 feet below ground level, how many seconds will it take for the pebble to hit the water?

5.7 SOLVING WORD PROBLEMS WITH QUADRATIC EQUATIONS

KEY IDEAS

The relationships between quantities described in some word problems may lead to quadratic equations. The solutions of these quadratic equations should be checked to make sure they fit the conditions of the word problems.

Area-Related Problems

When solving area-related word problems, carefully label diagrams with the key facts of the problem. Reject any negative solutions that represent dimensions of a figure.

Example 1

A rectangular photograph is 7 inches in length and 3 inches in width. The photograph is enlarged by increasing its length and width by the same amount. If the area of the enlarged photograph is 96 square inches, what are the new dimensions of the photograph?

Solution 1: Solve algebraically. If the length and width are each increased by x inches, then

$$\overbrace{(7+x)}^{\text{new length}} \times \overbrace{(3+x)}^{\text{new width}} = \overbrace{96}^{\text{new area}}$$

This gives $x^2 + 10x + 21 = 96$, so $x^2 - 10x - 75 = 0$, which makes $(x+15)(x-5)=0$. Since the only positive root of this equation is $x = 5$, the length and width are each increased by 5 inches.
The new length is $7 + x = 7 + 5 =$ **12** inches and the new width is $3 + x = 3 + 5 =$ **8** inches.

Solution 2: Work backward.

- The original photograph is 7 in. long and 3 in. wide. Since the length and width are increased by the same amount, the dimensions of the enlarged photograph must also differ by 4 in.
- Work backwards from the new area of 96 in.^{2} by thinking of two positive integers whose product is 96 and whose difference is 4.
- Test different pairs of factors of 96 until you find that $12 \times 8 = 96$ and $12 - 8 = 4$. The length of the enlarged photograph is **12** inches and the width is **8** inches.

Example 2

Deborah built a box open at the top by cutting 3-inch squares from the corners of a rectangular sheet of cardboard, as shown in the accompanying figure, and then folding the sides up. The volume of the box is 150 cubic inches, and the longer side of the box measures 5 inches more than the shorter side. Find the length, in inches, of the shorter side of the *original* sheet of cardboard.

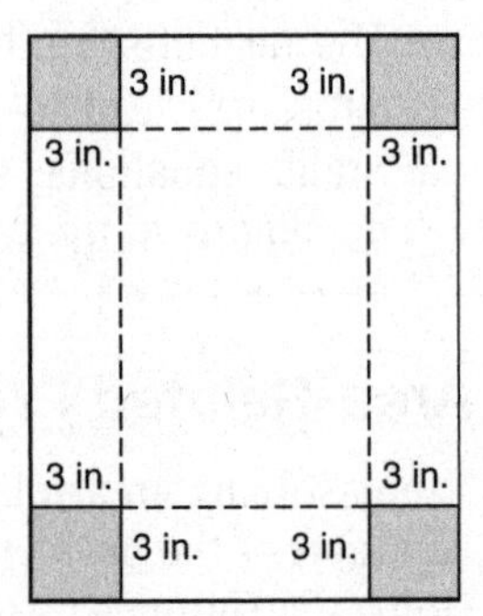

Solution: If x represents the number of inches in the shorter side of the original sheet of cardboard, then the length of the longer side is $x + 5$ inches.

- After 3-inch corners are cut from the original sheet of cardboard, the new dimensions of the sheet are $(x+5)-3-3=x-1$ inches and $x - 3 - 3 = x - 6$ inches.
- When the sides are folded up to form an open box, the height of the box is 3 inches.
- Because the volume of the box formed in this way is 150 cubic inches:

$$\begin{aligned}
\text{length}\times\text{width}\times\text{height} &= \text{volume}\\
(x-1)\times(x-6)\times 3 &= 150\\
(x-1)\times(x-6) &= \frac{150}{3}\\
x^2-7x+6 &= 50\\
x^2-7x-44 &= 0\\
(x-11)(x+4) &= 0\\
x-11=0 \quad &or \quad x+4=0\\
x=11 \quad &| \quad x=-4
\end{aligned}$$

The length of the shorter side of the original sheet is **11** inches.

Example 3

The length of a rectangular garden is twice its width. The garden is surrounded by a rectangular concrete walk having a uniform width of 4 feet. If the area of the garden and the walk is 330 square feet, what are the dimensions of the garden?

Solution: If x = width of the garden, then $2x$ = length of the garden. In the accompanying diagram, the smaller rectangle represents the garden.

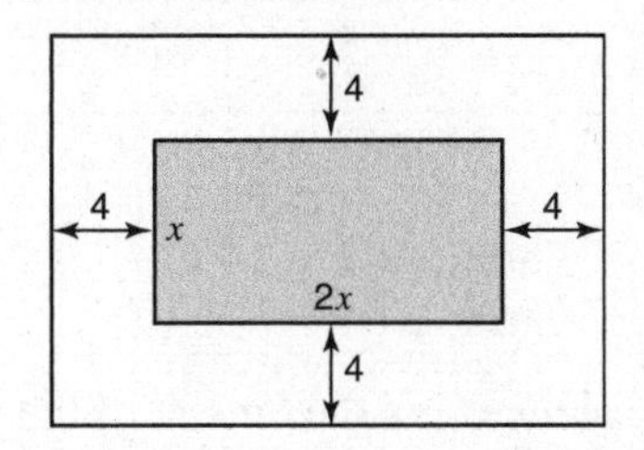

- Since the walk has a uniform width, the width of the larger rectangle is $4 + x + 4 = x + 8$.
- The length of the larger rectangle is $4 + 2x + 4 = 2x + 8$.
- Since the area of the larger rectangle is given as 330 ft^2:

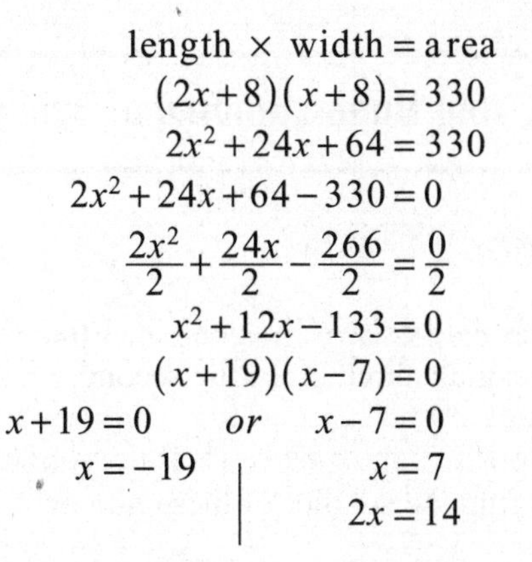

$$\text{length} \times \text{width} = \text{area}$$
$$(2x+8)(x+8) = 330$$
$$2x^2 + 24x + 64 = 330$$
$$2x^2 + 24x + 64 - 330 = 0$$
$$\frac{2x^2}{2} + \frac{24x}{2} - \frac{266}{2} = \frac{0}{2}$$
$$x^2 + 12x - 133 = 0$$
$$(x+19)(x-7) = 0$$
$$x+19 = 0 \quad or \quad x-7 = 0$$
$$x = -19 \quad \Big| \quad x = 7$$
$$2x = 14$$

Reject −19 since the width must be a positive number. The width of the garden is **7** feet and the length is **14** feet.

Number-Type Problems

Some word problems involve the relationships between two or more numbers that can be represented in terms of the same variable.

Example 4

Find the largest of three consecutive odd integers if the product of the first and the third integers is 6 more than three times the second integer.

Solution: If the three consecutive odd integers are represented by x, $x + 2$, and $x + 4$, then

$$\overbrace{x(x+4)}^{\text{product of the first and third integers}} = \overbrace{3(x+2)}^{\text{3 times the second integer}} \overbrace{+6}^{\text{by 6}}$$

$$x^2+4x=3x+6+6$$
$$x^2+4x=3x+12$$
$$x^2+4x-3x-12=0$$
$$x^2+x-12=0$$
$$(x-3)(x+4)=0$$

$x-3=0$ *or* $x+4=0$

$x=3$ | $x=-4$ ← reject since x must be odd

Then $x+2=5$

and $x+4=7$

Hence, 7 is the largest of the three consecutive odd integers.

Check Your Understanding of Section 5.7

Show how you arrived at your answer.

1. Find two consecutive positive integers such that the square of the first decreased by 25 equals three times the second.

2. The sum of two positive integers is 31. If the sum of the squares of these numbers is 625, find the smaller of these numbers.

3. If the second of three positive consecutive integers is added to the *product* of the first and the third, the result is 71. Find the three integers.

4. A positive number is 1 more than twice another number. If the difference of their squares is 40, find the larger of the two numbers.

5. Find three positive consecutive integers such that the product of the first and second is 2 more than 9 times the third.

6. What is the largest of three consecutive odd integers if the product of the first and third integers is 6 more than three times the second integer?

7. Find three positive consecutive odd integers such that the product of the first and second is 25 more than 10 times the third.

8. Find four consecutive positive integers such that the product of the first and fourth is four less than twice the first multiplied by the fourth.

9. A rectangular picture 30 cm wide and 50 cm long is surrounded by a frame having a uniform width. If the combined area of the picture and the frame is 2016 cm^2, what is the width of the frame?

10. The art staff at Central High School is determining the dimensions of paper to be used in the senior yearbook. The area of each sheet is to 432 cm^2. The staff has agreed on margins of 3 cm on each side and 4 cm on top and bottom. If the printed matter is to occupy 192 cm^2 on each page, what must be the overall length and width of the paper?

11. A rectangular picture 24 inches by 32 inches is surrounded by a border of uniform width. If the area of the border is 528 square inches less than the area of the picture, find the width of the border.

12. A rectangular flower garden has an area of 180 square feet. If the width of the garden is 3 feet less than the length, what is the minimum number of feet of fencing needed to enclose the garden completely?

13. The height of a rectangular box is 2 inches and the length of the box exceeds its width by 5 inches. If the volume of the box is 208 cubic inches, find the number of inches in the length and the width of the box.

14. A builder needs to extend a rectangular floor measuring 6 feet by 8 feet so that the number of square feet in the area of the floor increases by 72 square feet. If each of the original dimensions is increased by the same number of feet, what are the dimensions of the new floor?

15. A rectangular plot of land is subdivided into a square flower garden, a rectangular vegetable garden, and a rectangular patio, as shown in the accompanying diagram. The length of the longer side of the vegetable garden exceeds the length of the flower garden by 4 feet, and the length of the longer side of the patio is twice the length of the flower garden. If the total area of the original plot of land is 360 square feet, find the numbers of square feet in the areas of each garden and the patio.

Flower garden	Vegetable garden	Patio

Exercise 15

16. A rectangular piece of cardboard is twice as long as it is wide. From each of its four corners, a square piece 3 inches on a side is cut out. The flaps at each corner are then turned up to form an open box. If the volume of the box is 168 cubic inches, what were the original dimensions of the piece of cardboard?

17. If the length of one side of a square garden is increased by 3 feet, and the length of an adjacent side is increased by 2 feet, the area of the garden increases to 72 square feet. What is the length of a side of the original garden?

18. The length of a rectangle is 7 inches more than its width. If the width is doubled and the length is increased by 2 inches, the area is increased by 42 square inches. Find the dimensions of the original rectangle.

19. The perimeter of a rectangle is 28 inches. When the width of the rectangle is doubled and the length is increased by 1 inch, the area of the rectangle is increased by 55 square inches. Find the possible dimensions of the original rectangle.

CHAPTER 6

RATIONAL EXPRESSIONS AND EQUATIONS

6.1 SIMPLIFYING RATIONAL EXPRESSIONS

KEY IDEAS

A fraction is in **lowest terms** when its numerator and denominator have no factors in common other than 1 or –1. To write $\frac{12}{15}$ in lowest terms, factor the numerator and factor the denominator. Then divide out or "cancel" any factor that appears in both the numerator and the denominator since their quotient is 1:

$$\frac{12}{15} = \frac{4 \times \overset{1}{\cancel{3}}}{5 \times \cancel{3}} = \frac{4}{5} \times 1 = \frac{4}{5}$$

An algebraic fraction is written in lowest terms in much the same way.

Replacement Sets for Rational Expressions

An **algebraic fraction**, also called a **rational expression**, is the quotient of two polynomials as in $\frac{x}{3}$, $\frac{5xy}{x-4}$, and $\frac{2x+3y}{x^2-y^2}$. Because division by 0 is undefined, always assume that the replacement sets for the variables in a rational expression do not include any numbers that make the denominator evaluate to 0. The denominator of $\frac{5xy}{x-4}$ is 0 when $x = 4$ so the replacement set for x does not include 4.

Example 1

For what values of x is $\frac{x-2}{x^2-2x-10}$ undefined?

Solution: To find the values of x for which the rational expression is not defined, set the denominator equal to 0 and solve for x:

$$x^2 - 2x - 10 = 0$$
$$(x+2)(x-5) = 0$$
$$x+2=0 \quad or \quad x-5=0$$
$$x=-2 \quad | \quad x=5$$

The rational expression is undefined when $\boldsymbol{x = -2}$ or $\boldsymbol{x = 5}$.

Writing Rational Expressions in Lowest Terms

To write a rational expression such as $\frac{4x+12}{x^2+3x}$ in lowest terms, factor both the numerator and the denominator completely:

$$\frac{4x+12}{x^2+2x}=\frac{4(x+3)}{x(x+3)}$$

Then divide out any pairs of common factors:

$$=\frac{4\overset{1}{\cancel{(x+3)}}}{x\cancel{(x+3)}}$$

$$=\frac{4}{x}$$

Example 2

Write $\frac{10a^2-15ab}{4a^2-9b^2}$ in lowest terms.

Solution: Rewrite the numerator and the denominator in factored form:

$$\frac{10a^2-15ab}{4a^2-9b^2}=\frac{5a(2a-3b)}{(2a-3b)(2a+3b)}$$

$$=\frac{5a\overset{1}{\cancel{(2a-3b)}}}{\cancel{(2a-3b)}(2a+3b)}$$

$$=\mathbf{\frac{5a}{2a+3b}}$$

Example 3

Write $\frac{18a^4-30a^3}{3a^2}$ in lowest terms.

Solution 1: Factor the numerator. Then divide out any pairs of common factors.

$$\frac{18a^4-30a^3}{3a^2}=\frac{6a^3(3a-5)}{3a^2}$$

$$=\frac{\cancel{6}a^3}{\cancel{3}a^2}\cdot(3a-5)$$

$$=2a^{3-2}\cdot(3a-5)$$

$$=\mathbf{2a(3a-5)}\text{ or }\mathbf{6a^2-10a}$$

Solution 2: Divide each term in the numerator by the denominator:

$$\frac{18a^4-30a^3}{3a^2}=\frac{18a^4}{3a^2}-\frac{30a^3}{3a^2}=\mathbf{6a^2-10a}$$

Example 4

When working on her homework, Carla wrote

$$\frac{\overset{1}{\cancel{3a}}+b^2}{\cancel{3a}}=1+b^2$$

Did Carla simplify the fraction correctly? Explain your answer.

Solution: Carla made a mistake just like it would be a mistake to simplify $\frac{3+4}{3}$ by writing

$$\frac{\overset{1}{\cancel{3}}+4}{\cancel{3}}=1+4=5 \quad \leftarrow \textit{Wrong!}$$

Only *factors* of a *product* common to both the numerator and the denominator can be cancelled. Since in $\frac{3a+b^2}{3a}$, $3a$ is being *added* to b^2, it is not a *factor* of the numerator so it cannot be divided out.

Rewriting a Difference

Sometimes it is necessary to reverse the order of terms in a difference by factoring out -1 as in $b-a=-(a-b)$.

Example 5

Write $\frac{1-y^2}{y^2+4y-5}$ in lowest terms.

Solution: Factor the numerator and the denominator:

$$\frac{1-y^2}{y^2+4y-5}=\frac{\mathbf{(1-y)}(1+y)}{\mathbf{(y-1)}(y+5)}$$

Rewrite $1-y$ as $-(y-1)$: $$=\frac{-(y-1)(1+y)}{(y-1)(y+5)}$$

Divide out common factors: $= -\dfrac{\overset{1}{\cancel{(y-1)}}(1+y)}{\cancel{(y-1)}(y+5)}$

Write the remaining factors: $= -\dfrac{y+1}{y+5}$

Check Your Understanding of Section 6.1

A. *Multiple Choice.*

1. What is the value of $\dfrac{x+1}{x^2-1}$ when $x = 1.002$?

(1) 500 (2) 0.005 (3) 200.2 (4) 1005

2. Expressed in simplest form, $\dfrac{x^2-x-6}{x^2-9}$, $x \neq \pm 3$, is equivalent to

(1) $\dfrac{x+2}{x-3}$ (2) $\dfrac{x+2}{x+3}$ (3) $\dfrac{x-2}{x-3}$ (4) $\dfrac{x-2}{x+3}$

3. Which fraction is expressed in simplest form?

(1) $\dfrac{x-1}{x^2-1}$ (2) $\dfrac{x-1}{x^2-2x+1}$ (3) $\dfrac{x+1}{x^2-1}$ (4) $\dfrac{x+1}{x^2+1}$

4. For what value(s) of x is the fraction $\dfrac{x+4}{x^2-2x-3}$ not defined?

(1) −1,3 (2) 1,−3 (3) −3,−1 (4) − 4

5. Written in simplest form, $\dfrac{3x^3-27xy^2}{12x^2+36xy}$ is

(1) $\dfrac{x+3y}{4}$ (2) $\dfrac{x+4y}{3}$ (3) $\dfrac{x-3y}{4}$ (4) $\dfrac{x-3y}{4(x+3y)}$

B. *Show how you arrived at your answer.*

6–9. *Find the value or values of x for which the rational expression is undefined.*

6. $\dfrac{5x}{3x-12}$ **7.** $\dfrac{8x}{7x+3}$ **8.** $\dfrac{x+1}{8x^2-4x}$ **9.** $\dfrac{2x-1}{x^2+x-6}$

10–27. Simplify by writing each fraction in lowest terms.

10. $\frac{3-b}{2b-6}$

11. $\frac{2x-16}{x^2-64}$

12. $\frac{a^2-16}{3a+12}$

13. $\frac{2ab^2-2a^2b}{4ab}$

14. $\frac{0.48xy-0.16y}{0.8y}$

15. $\frac{ab-b^2}{5ab-5a^2}$

16. $\frac{21r^2s-7r^3s}{14rs}$

17. $\frac{10xy+30x^2}{xy^2-9x^3}$

18. $\frac{3y-12}{y^2+y-20}$

19. $\frac{10-5x}{x^2-x-2}$

20. $\frac{x^2-y^2}{(x-y)^2}$

21. $\frac{p^2-4p-21}{p^2+5p+6}$

22. $\frac{n^2-n-42}{n^2-6n-7}$

23. $\frac{2x^2-50}{2x^2+14x+20}$

24. $\frac{12p^2-27q^2}{9q^2+6pq}$

25. $\frac{s^3-4r^2s}{8r-4s}$

26. $\frac{4x^2-9y^2}{10x^2y+15xy^2}$

27. $\frac{m^2-4n^2}{m^2+4mn+4n^2}$

6.2 MULTIPLYING AND DIVIDING RATIONAL EXPRESSIONS

KEY IDEAS

To multiply arithmetic fractions, divide out any common factors in the numerators and the denominators of the fractions before doing the multiplication. Then multiply:

$$\frac{4}{9} \times \frac{3}{10} = \frac{\overset{2}{\cancel{4}}}{\underset{3}{\cancel{9}}} \times \frac{\overset{1}{\cancel{3}}}{\underset{5}{\cancel{10}}} = \frac{2\times 1}{3\times 5} = \frac{2}{15}$$

To divide one fraction by a second fraction, change to a multiplication example by inverting the second fraction. Multiplying and dividing rational expressions are handled in much the same way.

Multiplying Rational Expressions

Before multiplying rational expressions, factor the numerators and the denominators completely so that any pairs of common factors can be divided out.

Example 1

Write the product $\frac{12y^2}{x^2+7x} \cdot \frac{x^2-49}{2y^5}$ in lowest terms.

Solution: Simplify before doing the multiplication.

- Factor where possible: $$\frac{12y^2}{x^2+7x} \cdot \frac{x^2-49}{2y^5} = \frac{12y^2}{x(x+7)} \cdot \frac{(x+7)(x-7)}{2y^5}$$

- Divide out pairs of common factors: $$= \frac{\overset{6}{\cancel{12y^2}}}{x\cancel{(x+7)}} \cdot \frac{\overset{1}{\cancel{(x+7)}}(x-7)}{\underset{y^3}{\cancel{2y^5}}}$$

- Multiply: $$= \frac{\mathbf{6(x-7)}}{\mathbf{xy^3}}$$

Dividing Rational Expressions

To divide rational expressions, invert the second fraction and then multiply.

Example 2

Write the quotient $\frac{8m^2}{3} \div \frac{6m^3}{3m-12}$ in lowest terms.

Solution: First change to a multiplication example by inverting the divisor:

$$\frac{8m^2}{3} \div \frac{6m^3}{3m-12} = \frac{8m^2}{3} \times \frac{3m-12}{6m^3}$$

- Factor where possible: $$= \frac{8m^2}{3} \times \frac{3(m-4)}{6m^3}$$

- Divide out common factors: $$= \frac{\overset{4}{\cancel{8}}\,\cancel{m^2}}{\cancel{3}} \times \frac{\overset{1}{\cancel{3}}(m-4)}{\underset{3m}{\cancel{6}\,\cancel{m^3}}}$$

- Multiply: $$= \frac{\mathbf{4(m-4)}}{\mathbf{3m}}$$

Example 3

Write the quotient $\frac{x^2-2x-8}{x^2-25} \div \frac{x^2-4}{2x+10}$ in lowest terms.

Solution: First change to a multiplication example by inverting the divisor:

$$\frac{x^2-2x-8}{x^2-25} \div \frac{x^2-4}{2x+10} = \frac{x^2-2x-8}{x^2-25} \cdot \frac{2x+10}{x^2-4}$$

- Factor where possible: $$= \frac{(x-4)(x+2)}{(x-5)(x+5)} \cdot \frac{2(x+5)}{(x+2)(x-2)}$$

- Divide out common factors: $$= \frac{(x-4)\overset{1}{\cancel{(x+2)}}}{(x-5)\cancel{(x+5)}} \cdot \frac{2\overset{1}{\cancel{(x+5)}}}{\cancel{(x+2)}(x-2)}$$

- Multiply: $$= \frac{\mathbf{2(x-4)}}{\mathbf{(x-5)(x-2)}}$$

Check Your Understanding of Section 6.2

A. *Multiple Choice.*

1. When $\frac{y^2-9y}{2y}$ is multiplied by $\frac{6y}{y^2-8y-9}$, the product is

(1) $\frac{3}{y+1}$ (2) $\frac{3y}{y+1}$ (3) $\frac{3}{y-1}$ (4) $3y$

2. When $\frac{x+4}{2}$ is divided by $\frac{x^2-16}{8}$, the quotient is

(1) $\frac{x-4}{4}$ (2) $\frac{1}{x}$ (3) $\frac{1}{x-1}$ (4) $\frac{4}{x-4}$

B. *Show how you arrived at your answer.*

3–12. *For all values of the variables for which the expressions are defined, perform the indicated operation and express the result in simplest form.*

3. $\frac{3b}{4a} \cdot \frac{8a^2-4a}{9b^2}$

4. $\frac{2x+6}{8xy} \div \frac{x+3}{2y^2}$

5. $\frac{32x^2k}{x^2-k^2} \cdot \frac{2x+2k}{8xk^2}$

6. $\frac{c^2-c-6}{3c-9} \cdot \frac{12}{c+2}$

7. $\dfrac{x^2-9}{x^2+6x+9} \cdot \dfrac{x^2+3x}{x^2}$

8. $\dfrac{m^2+3m-4}{5m-5} \cdot \dfrac{10m^2-40m}{m^2-16}$

9. $\dfrac{r^2-7r+10}{5r-r^2} \div \dfrac{r^2-4}{25r^3}$

10. $\dfrac{y^2-49}{y^2-3y-28} \cdot \dfrac{3y+12}{y^2+5y-14}$

11. $\dfrac{x^2-3x}{x^2+2x} \div \dfrac{x^2-5x+6}{x^2-4}$

12. $\dfrac{n^2+4n+4}{n^2+3n+2} \div \dfrac{n^2-3n-10}{n^2-6n-7}$

6.3 COMBINING RATIONAL EXPRESSIONS

KEY IDEAS

To add (or subtract) fractions that have the same denominators, write the sum (or difference) of the numerators over the common denominator, as in

$$\frac{2}{7}+\frac{3}{7}=\frac{2+3}{7}=\frac{5}{7}$$

If the fractions have different denominators, first change each fraction to an equivalent fraction having the LCD (least common denominator) as its denominator.

Combining Fractions with the Same Denominator

Adding and subtracting algebraic fractions are handled in much the same way as fractions in arithmetic.

Example 1

Write the difference $\dfrac{5a+b}{10ab}-\dfrac{3a-b}{10ab}$ in simplest form.

Solution: Write the difference of the numerators over the common denominator. Then combine like terms and simplify.

$$\begin{aligned}\frac{5a+b}{10ab}-\frac{3a-b}{10ab}&=\frac{5a+b-(3a-b)}{10ab}\\&=\frac{5a+b-3a+b}{10ab}\\&=\frac{2a+2b}{10ab}\end{aligned}$$

$$= \frac{\overset{1}{\cancel{2}}(a+b)}{\underset{5}{\cancel{10}}\,ab}$$

$$= \mathbf{\frac{a+b}{5ab}}$$

Example 2

Write the sum $\frac{9x-1}{x^2-4} + \frac{5x+7}{4-x^2}$ as a single fraction in simplest form.

Solution: Rewrite the second denominator as $-(x^2-4)$:

$$\frac{9x-1}{x^2-4} + \frac{5x+7}{4-x^2} = \frac{9x-1}{x^2-4} - \frac{5x+7}{x^2-4}$$

$$= \frac{9x-1-(5x+7)}{x^2-4}$$

$$= \frac{4x-8}{x^2-4}$$

$$= \frac{4\cancel{(x-2)}}{\underset{1}{\cancel{(x-2)}}(x+2)}$$

$$= \mathbf{\frac{4}{x+2}}$$

Combining Fractions with Different Denominators

When adding and subtracting algebraic fractions with unlike denominators, first determine their LCD. Then use the multiplication property of 1 to change each fraction into an equivalent fraction with the LCD as its denominator.

Example 3

Express the sum of $\frac{3}{10x}$ and $\frac{4}{15x}$ as a single fraction in simplest form.

Solution: The LCD of $10x$ and $15x$ is $30x$ since $30x$ is the smallest expression into which both $10x$ and $15x$ divide evenly. Change each fraction into an equivalent fraction that has $30x$ as its denominator by multiplying the first fraction by $\frac{3}{3}$ and multiplying the second fraction by $\frac{2}{2}$. Thus,

$$\frac{3}{10x}+\frac{4}{15x}=\frac{3}{3}\cdot\left(\frac{3}{10x}\right)+\frac{2}{2}\cdot\left(\frac{4}{15x}\right)$$
$$=\frac{9}{30x}+\frac{8}{30x}$$
$$=\frac{9+8}{30x}$$
$$=\mathbf{\frac{17}{30x}}$$

Example 4

Write $\frac{2x+1}{6y}+\frac{3x-5}{9y}+\frac{1}{18y}$ as a single fraction in simplest form.

Solution: The LCD is $18y$ since $18y$ is the smallest expression into which $6y$, $9y$, and $18y$ divide evenly. Change the first two fractions into equivalent fractions that have $18y$ as their denominator by multiplying the first fraction by $\frac{3}{3}$ and multiplying the second fraction by $\frac{2}{2}$. Thus,

$$\frac{2x+1}{6y}+\frac{3x-5}{9y}+\frac{1}{18y}=\frac{3}{3}\cdot\left(\frac{2x+1}{6y}\right)+\frac{2}{2}\cdot\left(\frac{3x-5}{9y}\right)+\frac{1}{18y}$$
$$=\frac{3(2x+1)+2(3x-5)+1}{18y}$$
$$=\frac{12x-6}{18y}$$
$$=\frac{\cancel{6}(2x-1)}{\underset{3}{\cancel{18}}y}$$
$$=\mathbf{\frac{2x-1}{3y}}$$

Check Your Understanding of Section 6.3

A. *Multiple Choice.*

1. Expressed as a single fraction, what is $\frac{x-7}{2}+\frac{x+2}{6}$?

(1) $\frac{2x-5}{8}$ (2) $\frac{4x-19}{6}$ (3) $\frac{8x-5}{12}$ (4) $\frac{x^2-14}{12}$

2. Which expression is equivalent to $\frac{a}{x}+\frac{3b}{2x}$?

(1) $\frac{2a+3b}{3x}$ (2) $\frac{2a+6b}{3x}$ (3) $\frac{2a+3b}{2x}$ (4) $\frac{5ab}{2x}$

3. What is the sum of $\frac{x-2}{3}$ and $\frac{x-3}{2}$?

(1) $\frac{2x-5}{5}$ (2) $\frac{5x-5}{6}$ (3) $\frac{2x-5}{6}$ (4) $\frac{5x-13}{6}$

4. What is $\frac{x+3}{3x}+\frac{x-1}{5x}$, $x \neq 0$, expressed as a single fraction?

(1) $\frac{2x+2}{8x}$ (2) $\frac{8x+12}{15x^2}$ (3) $\frac{2x+2}{15x}$ (4) $\frac{8x+12}{15x}$

5. The probability that an event A will occur is $\frac{x}{4}$ where x can vary from 0 to 4. What is the probability that event A will *not* occur?

(1) $\frac{1-x}{4}$ (2) $\frac{4-x}{4}$ (3) $\frac{4-x}{x}$ (4) $\frac{4}{x}$

6. The sum of $\frac{y-4}{2y}$ and $\frac{3y-5}{5y}$ is

(1) $\frac{11y-30}{7y}$ (2) $\frac{4y-9}{10y}$ (3) $\frac{11y-30}{10y}$ (4) $\frac{4y-9}{7y}$

B. *Show how you arrived at your answer.*

7–21. *Express each sum or difference as a single fraction in simplest form.*

7. $\frac{4b}{5x}-\frac{3b}{10x}$

8. $\frac{3y-5}{5xy}-\frac{1}{10x}$

9. $\frac{b-5}{10b}+\frac{b+10}{15b}$

10. $\frac{3x-1}{7x}+\frac{x+9}{14x}$

11. $\frac{10d+1}{14d}-\frac{d+5}{21d}$

12. $\frac{5w+6}{3w-1}-\frac{8w+5}{3w-1}$

13. $\frac{2c+5}{6c}+\frac{5c-7}{21c}$

14. $\frac{2a-3}{4z}-\frac{3a}{8z}+\frac{a+8}{6z}$

15. $\frac{3y+4}{9}-\frac{2y^2+1}{6y}$

16. $\frac{10ab+1}{5a^2b}-\frac{a+6}{3a}$

17. $\frac{3}{10xy}-\frac{10x-y}{5xy^2}$

18. $\frac{p+1}{p^2-1}+\frac{3p-1}{1-p^2}$

19. $\frac{x-5y}{9x^2-4y^2}+\frac{2x+7y}{9x^2-4y^2}$

20. $\frac{3(m+1)}{14m}-\frac{2(m-3)}{21m}$

21. $\frac{k^2+4}{k^2-25}+\frac{7k-6}{25-k^2}$

6.4 RATIONAL EQUATIONS AND INEQUALITIES

KEY IDEAS

A **rational equation** is an equation that contains rational expressions. To solve a rational equation that has fractional terms, eliminate the fractions by multiplying *both sides* of the equation by the *least common denominator* (LCD) of all of the fractions. An inequality with fractional terms is solved in much the same way.

Eliminating Fractions

When clearing an equation or inequality of any fractional coefficients or terms, multiply each term on *both* sides by the LCD of *all* the denominators.

Example 1

Solve for y: $\frac{3y}{4}+\frac{7}{12}=\frac{y}{6}$.

Solution: The LCD is 12 since 12 is the least positive integer into which 4, 12, and 6 divide evenly. Eliminate the fractions by multiplying each term on both sides of the equation by 12:

$$12\left(\frac{3y}{4}\right)+12\left(\frac{7}{12}\right)=12\left(\frac{y}{6}\right)$$

$$\overset{3}{\cancel{12}}\left(\frac{3y}{\cancel{4}}\right)+\overset{1}{\cancel{12}}\left(\frac{7}{\cancel{12}}\right)=\overset{2}{\cancel{12}}\left(\frac{y}{\cancel{6}}\right)$$

$$9y + 7 = 2y$$

$$9y-2y=-7$$

$$\frac{7y}{7}=\frac{-7}{7}$$

$$\mathbf{y=-1}$$

Example 2

Solve $\frac{4}{3} = \frac{-(3x+13)}{3x} + \frac{5}{6x}$ for x.

Solution: Remove the parentheses by taking the opposite of each term inside the parentheses:

$$\frac{4}{3} = \frac{-3x-13}{3x} + \frac{5}{6x}$$

Eliminate the fractions by multiplying each term by $6x$, the lowest common multiple of all the denominators:

$$\overset{2x}{\cancel{6x}}\left(\frac{4}{\cancel{3}}\right) = \overset{2}{\cancel{6x}}\left(\frac{-3x-13}{\cancel{3x}}\right) + \overset{1}{\cancel{6x}}\left(\frac{5}{\cancel{6x}}\right)$$

$$2x(4) = 2(-3x - 13) + 5$$

$$8x = -6x - 26 \quad + 5$$

$$8x + 6x = -21$$

$$\frac{14x}{14} = \frac{-21}{14}$$

$$x = -\frac{3}{2}$$

Example 3

Solve the equation $\frac{2(n-1)}{3} - \frac{3(n+1)}{4} = \frac{n+3}{2}$ for n.

Solution: Eliminate the fractional terms by multiplying each term of the equation by 12, the lowest common multiple of all the denominators:

$$\overset{4}{\cancel{12}}\left[\frac{2(n-1)}{\cancel{3}}\right] - \overset{3}{\cancel{12}}\left[\frac{3(n+1)}{\cancel{4}}\right] = \overset{6}{\cancel{12}}\left(\frac{n+3}{\cancel{2}}\right)$$

$$4 \cdot 2(n - 1) - 3 \cdot 3(n + 1) = 6(n + 3)$$

$$8n - 8 - 9n - 9 = 6n + 18$$

$$-n - 17 = 6n + 18$$

$$-n - 6n = 18 + 17$$

$$\frac{-7n}{-7} = \frac{35}{-7}$$

$$n = -5$$

Example 4

Find the solution set of $\frac{1}{4}x+1 \leq 7+x$.

Solution: Eliminate the fractional coefficient by multiplying both sides of the inequality by 4:

Divide both sides by −3 and and reverse the inequality:

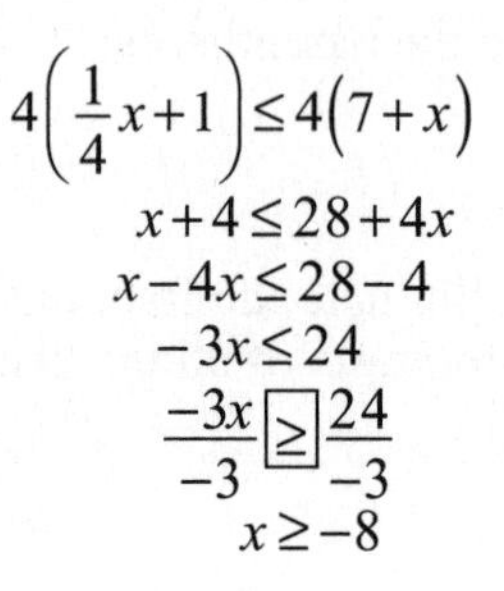

$$4\left(\frac{1}{4}x+1\right) \leq 4(7+x)$$
$$x+4 \leq 28+4x$$
$$x-4x \leq 28-4$$
$$-3x \leq 24$$
$$\frac{-3x}{-3} \boxed{\geq} \frac{24}{-3}$$
$$x \geq -8$$

The solution set is $\{\boldsymbol{x} \mid \boldsymbol{x} \geq -\mathbf{8}\}$.

Check Your Understanding of Section 6.4

1–15. *Solve for the variable and check.*

1. $\frac{x}{3}+\frac{x}{18}=\frac{7}{6}$
2. $\frac{3}{4}+\frac{1}{x}=\frac{2}{3}$
3. $\frac{8}{x}-2=\frac{2}{3}$
4. $\frac{1}{3x}+1=\frac{7}{6}$
5. $\frac{3}{4}w+8=\frac{1}{3}w-7$
6. $\frac{2x}{5}-\frac{x-2}{10}=2$
7. $\frac{y}{5}+7=\frac{y}{2}-2$
8. $\frac{x+4}{2}+\frac{2x}{3}=9$
9. $\frac{3b}{4}=\frac{b}{5}-\frac{11}{10}$
10. $\frac{6}{x}+\frac{x-3}{2x}=2$
11. $\frac{n+6}{3}=2+\frac{n-3}{2}$
12. $\frac{y-2}{2}+\frac{2y-1}{20}=\frac{y}{4}$
13. $\frac{7}{36}+\frac{1}{4x}=\frac{5}{6x}$
14. $\frac{x+1}{4}-\frac{2x}{3}=\frac{x}{12}$
15. $\frac{8}{3x}-\frac{x-1}{12}=\frac{1}{6x}$

16. Solve the formula $C=\frac{5}{9}(F-32)$ for F in terms of C.

17–19. Find the solution set.

17. $\frac{1}{2}x+3<2x-6$ **18.** $2x+4\geq 11-\frac{x}{3}$ **19.** $\frac{1}{3}x-11<7-\frac{1}{3}x$

20. If one-half of a number is 8 less than two-thirds of the number, what is the number?

21. If the reciprocal of a number is multiplied by 3, the result exceeds the reciprocal of the original number by $\frac{1}{3}$. Find the number.

22. Solve $\frac{k}{6}-\frac{3(k+1)}{4}=\frac{k+5}{2}$ for k.

23. Solve $\frac{3}{2}=\frac{-(5m-3)}{3m}+\frac{7}{12m}$ for m.

24. The president of a school board assigns each of the other members of the board to either the budget committee or the curriculum committee. The budget committee has $\frac{2}{3}x-3$ members, and the curriculum committee has $\frac{x}{4}+2$ members. If the two committees have the same number of members and no board member is on both committees, how many people serve on the school board?

25. On Tuesday, CityEx delivered 7 less than two times the number of packages it had delivered on Monday. The number of packages delivered on Wednesday was equal to 10 more than the average number of packages delivered on Monday and Tuesday. If the total number of packages that was delivered over the three days was 661, how many packages were delivered on Monday?

CHAPTER 7
RADICALS AND RIGHT TRIANGLES

7.1 SQUARE AND CUBE ROOTS

KEY IDEAS

Because $(+7)\times(+7)=49$ and $(-7)\times(-7)=49$, both +7 and −7 are *square roots* of 49. The *principal* or positive square root of 49 is +7. The cube root of a number is one of three identical factors of the number. For example, the cube root of 8 is 2 since $2\times2\times2=8$.

Principal Square Root

A **square root** of a nonnegative number is one of two equal factors of the number. Every positive number has two square roots. The positive square root is called the **principal square root**. Although the two square roots of 9 are +3 and −3, the notation $\sqrt{9}$ refers only to the positive or *principal square root* of 9, as shown in Figure 7.1. The symbol $\sqrt{\ }$ is a **radical sign**, and the number underneath it is the **radicand**. Square roots of negative numbers do *not* represent real numbers because the product of two identical real numbers cannot be negative.

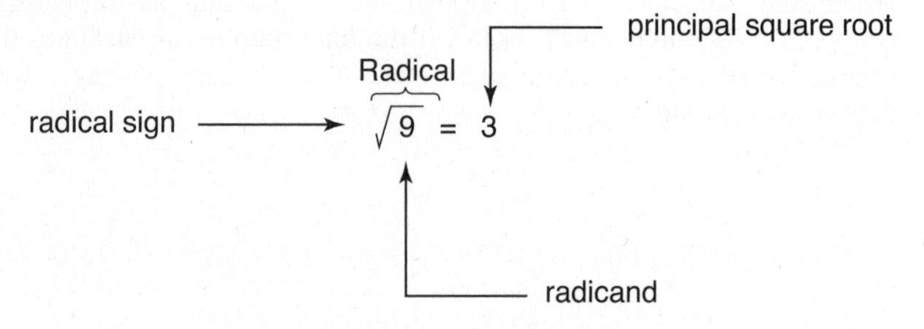

Figure 7.1 Square Root Notation and Terms

Example 1

Evaluate: a. $\sqrt{0.49}+\sqrt{\frac{1}{4}}$ b. $-2\sqrt{\frac{25}{36}}$

Solution:

a. $\sqrt{0.49}+\sqrt{\frac{1}{4}}=0.7+\frac{1}{2}$
$=0.7+0.5$
$=\mathbf{1.2}$

b. $-2\sqrt{\frac{25}{36}}=-\cancel{2}\left(\frac{5}{\cancel{6}_3}\right)=-\mathbf{\frac{5}{3}}$

Comparing *x* and $\sqrt{x}$

The size relationship between x and $\sqrt{x}$ changes according to whether x is between 0 and 1, equal to 0 or 1, or greater than 1.

- If x is between 0 and 1, then $x<\sqrt{x}$. For example,

$$0.64<\sqrt{0.64} \text{ because } \sqrt{0.64}=0.8$$

- If $x=0$ or $x=1$, then $x=\sqrt{x}$. For example,

$$\sqrt{0}=0 \text{ and } \sqrt{1}=1$$

- If $x>1$, then $x>\sqrt{x}$. For example,

$$9>\sqrt{9} \text{ because } \sqrt{9}=3$$

Perfect Squares

A square root may represent a rational or irrational number.

- Numbers such as 0.36, $\frac{16}{25}$, 1, 4, and 9 are *perfect squares* because their square roots are rational numbers. A **perfect square** is a positive real number whose square root is rational.
- The square roots of numbers such as 2, 3, and 79 are decimal numbers that have more decimal digits than can fit in any calculator display. The numbers $\sqrt{2}$, $\sqrt{3}$, and $\sqrt{79}$ are irrational because their decimal representations are nonending and do not have a repeating set of digits. The square root of a positive real number that is not a perfect square is irrational and, as a result, does *not* have an exact decimal equivalent.

Evaluating Cube Roots

The cube root of 64 is 4 since

$$\underbrace{4\times4\times4}_{\text{3 identical factors of 64}} = 64$$

The **cube root** of a number is one of three equal factors of the number. Using radical notation, $\sqrt[3]{64}=4$, where $\sqrt[3]{64}$ is read "the cube root of 64."

Using Technology: Finding Roots

To use a graphing calculator to evaluate the square root of a positive number, use $\frac{1}{2}$ or 0.5 as the exponent. When evaluating the cube root of a number, use $\frac{1}{3}$ as the exponent.

- To find $\sqrt{49}$, press

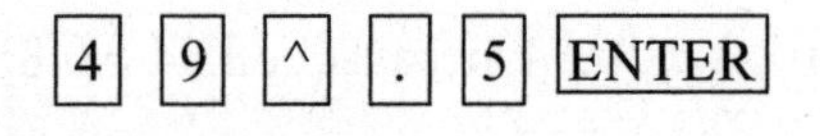

The display should read 7. You can also use the square root key by pressing

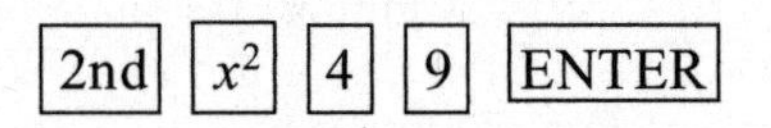

- To find $\sqrt[3]{64}$, press

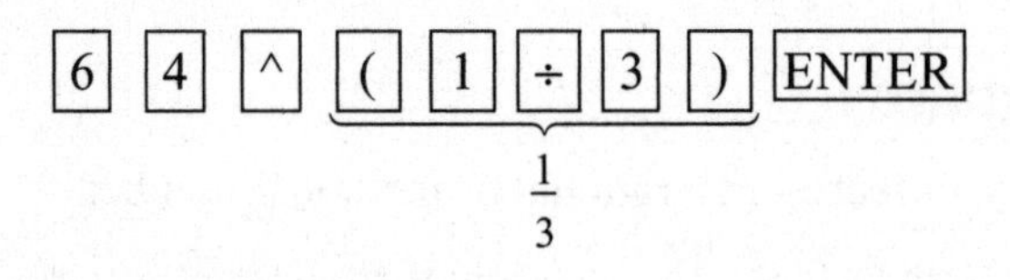

The display should read 4. You can also use the Math key to find the cube root of 64 by pressing

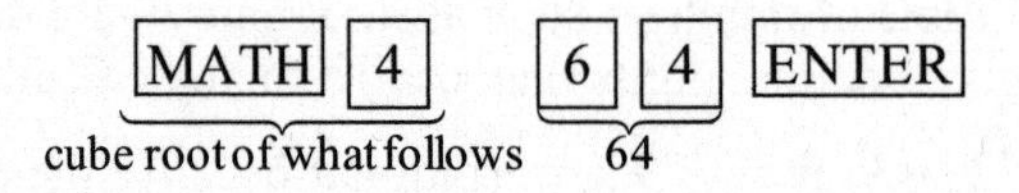

Example 2

Which list of numbers is in order from the smallest value to the largest value?

(1) $\sqrt{10}, \frac{22}{7}, \pi, 3.1$

(2) $3.1, \frac{22}{7}, \pi, \sqrt{10}$

(3) $\pi, \frac{22}{7}, 3.1, \sqrt{10}$

(4) $3.1, \pi, \frac{22}{7}, \sqrt{10}$

Solution: Each answer choice includes the same set of numbers. Locate these numbers on a number line using decimal approximations for the irrational values:

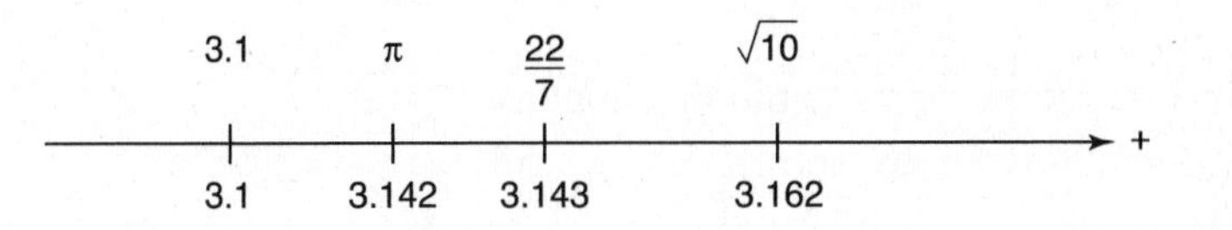

Examine each of the answer choices, in turn, until you find the one that lists the numbers in the same order, from left to right, as on the number line: 3.1, π, $\frac{22}{7}$, and $\sqrt{10}$. The correct choice is **(4)**.

Check Your Understanding of Section 7.1

A. *Multiple Choice.*

1. If $k = 4^3$, then $\sqrt{k} =$
(1) 8 (2) 2 (3) 4 (4) 32

2. Which number is irrational?
(1) $\sqrt{0.64}$ (2) 0.131313 . . . (3) $\sqrt{40}$ (4) $\frac{29}{17}$

3. Which list shows the numbers $|-0.12|$, $\sqrt{\frac{1}{82}}$, $\frac{1}{8}$, and $\frac{1}{9}$ in order from smallest to largest?
(1) $|-0.12|, \frac{1}{8}, \frac{1}{9}, \sqrt{\frac{1}{82}}$
(2) $\frac{1}{8}, \frac{1}{9}, \sqrt{\frac{1}{82}}, |-0.12|$
(3) $\sqrt{\frac{1}{82}}, |-0.12|, \frac{1}{9}, \frac{1}{8}$
(4) $\sqrt{\frac{1}{82}}, \frac{1}{9}, |-0.12|, \frac{1}{8}$

4. Which inequality statement is true when $0 < x < 1$?
(1) $0 < \sqrt{x} < x$ (2) $0 < x < x^2$ (3) $0 < x^2 < x^3$ (4) $0 < x < \sqrt{x}$

5. The amount of time, t, in seconds, it takes an object to fall a distance, d, in meters, is expressed by the formula $t = \sqrt{\frac{d}{4.9}}$. Approximately how long will it take an object to fall 75 meters.
(1) 0.26 sec (2) 2.34 sec (3) 3.9 sec (4) 7.7 sec

6. If $t^2 < t < \sqrt{t}$, then t could be

(1) $-\frac{1}{4}$ (2) 0 (3) $\frac{1}{4}$ (4) 4

7. If $x^3 < x < \frac{1}{x}$, then x could be equal to

(1) 1 (2) 5 (3) $\frac{6}{5}$ (4) $\frac{1}{5}$

B. *Show how you arrived at your answer.*

8–11. *Evaluate.*

8. $\sqrt{0.25}$ **9.** $\sqrt{\frac{1}{3}+\frac{1}{9}}$ **10.** $-2\sqrt{\frac{49}{100}}$ **11.** $\frac{\sqrt{64}}{4}$

12. If $y = \sqrt{b^2 - a^3}$, what is the value of y when $a = -4$ and $b = 6$?

13. What is the value of $\sqrt{y^2 - \left(x^3 - 3\right)}$ when $x = -2$ and $y = 5$?

14. Chris plans to install a fence around the perimeter of his yard. His yard is shaped like a square and has an area of 40,000 square feet. The company that he hires charges $2.50 per foot for the fencing and $50.00 for the installation fee. What will be the cost of the fence, in dollars?

7.2 OPERATIONS WITH RADICALS

KEY IDEAS

Since radicals are real numbers, they can be multiplied and divided using the properties of real numbers. A radical is in simplest form when the radicand does not contain any perfect square factors other than 1 and does not contain any fractions.

Multiplying Radicals

To multiply square root radicals, use the rule

$$a\sqrt{x} \cdot b\sqrt{y} = ab\sqrt{xy}$$

provided x and y are nonnegative numbers. For example,

$$3\sqrt{7}\cdot 2\sqrt{5}=(3\times 2)\sqrt{7\times 5}=6\sqrt{35}$$

Dividing Radicals

To divide square root radicals, use the rule

$$\frac{a\sqrt{x}}{b\sqrt{y}}=\frac{a}{b}\sqrt{\frac{x}{y}}$$

provided x and y are nonnegative numbers, and the denominator of each fraction is not 0. For example,

- $\dfrac{20\sqrt{6}}{4\sqrt{2}}=\left(\dfrac{20}{4}\right)\left(\sqrt{\dfrac{6}{2}}\right)=5\sqrt{3}$

- $\dfrac{28\sqrt{5}}{7\sqrt{30}}=\left(\dfrac{28}{7}\right)\left(\sqrt{\dfrac{5}{30}}\right)=4\left(\sqrt{\dfrac{1}{6}}\right)=\dfrac{4}{\sqrt{6}}$

Multiplying Identical Radicals

The product of two identical square root radicals is their radicand, as in

$$\sqrt{9}\cdot\sqrt{9}=\sqrt{81}=9$$

In general, $\sqrt{N}\times\sqrt{N}=N$, provided $N\geq 0$.

Example 1

Write $\left(3y\sqrt{7}\right)^2$ in simplest form.

Solution:
$$\begin{aligned}\left(3y\sqrt{7}\right)^2&=\left(3y\sqrt{7}\right)\times\left(3y\sqrt{7}\right)\\&=(3y\cdot 3y)\times\left(\sqrt{7}\cdot\sqrt{7}\right)\\&=\ 9y^2\ \times\ 7\\&=\mathbf{63y^2}\end{aligned}$$

Example 2

Solve for n: $\sqrt{2n}=4\sqrt{5}$.

Solution: Eliminate the square root radicals by raising both sides of the equation to the second power:

$$\left(\sqrt{2n}\right)^2 = \left(4\sqrt{5}\right)^2$$
$$2n = 16 \cdot 5$$
$$\frac{2n}{2} = \frac{80}{2}$$
$$\mathbf{n = 40}$$

Simplifying Square Root Radicals

To simplify $\sqrt{N}$, factor N so that one of the two factors is its greatest perfect square factor.

Example 3

Simplify $\sqrt{108}$.

Solution: The greatest perfect square factor of 108 is 36. Thus,

$$\begin{aligned}\sqrt{108} &= \sqrt{36 \cdot 3}\\ &= \sqrt{36} \cdot \sqrt{3}\\ &= \mathbf{6\sqrt{3}}\end{aligned}$$

Beginning with a different perfect square factor other than 36 leads to some additional work. If you began by factoring 108 as 4×27, then

$$\sqrt{108} = \sqrt{4 \cdot 27} = \sqrt{4} \cdot \sqrt{27} = 2\sqrt{27}$$

Now $\sqrt{27}$ must be simplified:

$$\begin{aligned}\sqrt{108} &= 2\sqrt{27}\\ &= 2\sqrt{9} \cdot \sqrt{3}\\ &= 2 \cdot 3 \cdot \sqrt{3}\\ &= \mathbf{6\sqrt{3}}\end{aligned}$$

Example 4

Express the product of $3\sqrt{14}$ and $5\sqrt{2}$ in simplest form.

Solution: $3\sqrt{14}\times 5\sqrt{2}=(3\times 5)\times(\sqrt{14}\times\sqrt{2})$

$$=15\sqrt{28}$$
$$=15\sqrt{4}\cdot\sqrt{7}$$
$$=15\cdot 2\sqrt{7}$$
$$=\mathbf{30\sqrt{7}}$$

Solving $ax^2 + c = k$

To solve a quadratic equation that does not have an x term, isolate the x^2 term. Then take the square root of both sides of the equation:

$$2x^2-47=3$$
$$2x^2=50$$
$$\frac{2x^2}{2}=\frac{50}{2}$$
$$x^2=25$$
$$x=\pm\sqrt{25}=+5 \text{ or } -5$$

Check Your Understanding of Section 7.2

A. Multiple Choice.

1. If the length of a rectangle is $5\sqrt{2}$ and the width is $2\sqrt{3}$, what is the area of the rectangle?

(1) $10\sqrt{6}$ (2) $7\sqrt{6}$ (3) $7\sqrt{5}$ (4) $10\sqrt{5}$

2. If $\sqrt{7-2x}$ represents a rational number, then x cannot be equal to

(1) 1 (2) $\frac{3}{2}$ (3) 3 (4) -1

3. If $4x^2-48=x^2$, then $x=$

(1) -4 (2) -2 (3) ± 4 (4) $+4, -2$

4. If $\frac{1}{2}\sqrt{96}=k\sqrt{6}$, then k is equal to

(1) 8 (2) 2 (3) 3 (4) 4

5. What is the value of $\sqrt{\frac{x+4}{2}}$ when $x = \frac{1}{2}$?

(1) $\frac{3}{2}$ (2) 2 (3) $\frac{3}{4}$ (4) $\frac{5}{2}$

6. If the dimensions of a rectangular box are $\sqrt{2}$, $\sqrt{3}$, and $4\sqrt{6}$, what is the volume of the box in simplest form?

(1) 24 (2) $4\sqrt{11}$ (3) $4\sqrt{30}$ (4) $8\sqrt{3}$

7. If $(x-3)^2 = 5$, then x is equal to

(1) $3 \pm \sqrt{5}$ (2) $-3 \pm \sqrt{5}$ (3) $-\sqrt{5} \pm 3$ (4) $\frac{\pm\sqrt{5}}{3}$

B. *Show how you arrived at your answer.*

8-11. *Write each expression in simplest form.*

8. $\sqrt{0.04} + \sqrt{0.36}$ **9.** $5\sqrt{48}$ 10. $\frac{1}{9}\sqrt{63}$ **11.** $\frac{1}{2}\sqrt{192}$

12–20. *Perform the indicated operations and write the answer in simplest radical form.*

12. $\sqrt{6} \cdot \sqrt{15}$ **15.** $\left(3\sqrt{5}\right)^2$ **18.** $\frac{12\sqrt{54}}{4\sqrt{3}}$

13. $\frac{3\sqrt{252}}{12\sqrt{7}}$ **16.** $\sqrt{2} \cdot \sqrt{6} \cdot \sqrt{21}$ **19.** $\frac{40\sqrt{10x}}{5\sqrt{2x}}$

14. $\left(7\sqrt{6}\right)\left(3\sqrt{18}\right)$ **17.** $\left(\frac{3}{4}\sqrt{75}\right)\left(\frac{1}{5}\sqrt{24}\right)$ **20.** $\sqrt{6}\left(\sqrt{6}\,y + \sqrt{8}\right)$

21. Solve for y: $y^2 + 29 = 4y^2 - 7$. Express the roots in simplest radical form.

22–24. *Solve for the variable.*

22. $\sqrt{3y} = 4\sqrt{21}$ **23.** $3\sqrt{2n} - 1 = 8$ **24.** $\sqrt{2m-1} = 5\sqrt{7}$

7.3 COMBINING RADICALS

KEY IDEAS

Square root radicals with the same radicand can be combined using the distributive property in much the same way that like monomials are combined:

Monomials	Radical Expressions
$2x+3x=(2+3)x=5x$	$2\sqrt{7}+3\sqrt{7}=(2+3)\sqrt{7}=5\sqrt{7}$
$9y-7y=(9-7)y=2y$	$9\sqrt{3}-7\sqrt{3}=(9-7)\sqrt{3}=2\sqrt{3}$

Before square root radicals with different radicands can be combined, they must first be changed into like radicals.

Combining Like Radicals

Square root radicals such as $7\sqrt{3}$ and $5\sqrt{3}$ are **like radicals** since they have the same radicand. The sum or difference of like radicals can be expressed as a single radical by combining their rational coefficients, as in

$$7\sqrt{3}+5\sqrt{3}=(7+5)\sqrt{3}=12\sqrt{3}$$

Example 1

Expand $\left(\sqrt{3}+2\right)^2$ and write the result in simplest form.

Solution: Rewrite $\left(\sqrt{3}+2\right)^2$ as $\left(\sqrt{3}+2\right)\left(\sqrt{3}+2\right)$ and then use FOIL:

$$\left(\sqrt{3}+2\right)\left(\sqrt{3}+2\right)=\overset{\text{F}}{\left(\sqrt{3}\cdot\sqrt{3}\right)}+\underbrace{\overset{\text{O}}{\left(2\sqrt{3}\right)}+\overset{\text{I}}{\left(2\sqrt{3}\right)}}+\overset{\text{L}}{\left(2\cdot 2\right)}$$

$$= 3 + 4\sqrt{3} + 4$$

$$= 4\sqrt{3}+7$$

Combining Unlike Radicals

Unlike radicals cannot be combined into a single radical unless they can be changed into like radicals. To add $5\sqrt{3}$ and $\sqrt{48}$, first simplify $\sqrt{48}$ so its radicand is 3. Then combine the like radicals:

$$\begin{aligned} 5\sqrt{3}+\sqrt{48} &= 5\sqrt{3}+\overbrace{\sqrt{16}\cdot\sqrt{3}}^{\sqrt{48}} \\ &= 5\sqrt{3}+4\sqrt{3} \\ &= 9\sqrt{3} \end{aligned}$$

It is not always possible to combine radicals into a single radical, as is the case with $\sqrt{8}+\sqrt{5}$.

Example 2

Express $2\sqrt{28}-3\sqrt{63}$ as a single radical in simplest form.

Solution: Simplify each radical so that their radicands are the same.

$$\begin{aligned} 2\sqrt{28}-3\sqrt{63} &= 2\sqrt{4\cdot 7}-3\sqrt{9\cdot 7} \\ &= 2\times 2\sqrt{7}-3\times 3\sqrt{7} \\ &= 4\sqrt{7}-9\sqrt{7} \\ &= \mathbf{-5\sqrt{7}} \end{aligned}$$

Example 3

If $3\sqrt{20}-\sqrt{5}+\frac{1}{2}\sqrt{80}=x\sqrt{5}$, what is the value of x?

Solution: Simplify the first and third radicals:

$$\begin{aligned} 3\sqrt{20}-\sqrt{5}+\frac{1}{2}\sqrt{80} &= 3\sqrt{4}\cdot\sqrt{5}-\sqrt{5}+\frac{1}{2}\sqrt{16}\cdot\sqrt{5} \\ &= (3\cdot 2)\sqrt{5}-\sqrt{5}+\left(\frac{1}{2}\cdot 4\right)\sqrt{5} \\ &= 6\sqrt{5}-\sqrt{5}+2\sqrt{5} \\ &= 7\sqrt{5} \end{aligned}$$

Since it is given that $3\sqrt{20}-\sqrt{5}+\frac{1}{2}\sqrt{80}=x\sqrt{5}$, $x\sqrt{5}=7\sqrt{5}$ so $\mathbf{x=7}$.

Check Your Understanding of Section 7.3

A. Multiple Choice.

1. The expression $2\sqrt{2} + \sqrt{50}$ is equivalent to

   (1) $2\sqrt{52}$ (2) $3\sqrt{52}$ (3) $7\sqrt{2}$ (4) $27\sqrt{2}$

2. The expression $5\sqrt{2} - \sqrt{18}$ is equivalent to

   (1) $2\sqrt{2}$ (2) $-2\sqrt{2}$ (3) $8\sqrt{2}$ (4) $-8\sqrt{2}$

3. The expression $4\sqrt{12} + 2\sqrt{27}$ is equivalent to

   (1) $6\sqrt{39}$ (2) $34\sqrt{3}$ (3) $14\sqrt{3}$ (4) $14\sqrt{6}$

4. The expression $\sqrt{90}\cdot\sqrt{40} - \sqrt{8}\cdot\sqrt{18}$ simplifies to

   (1) 22.9 (2) 48 (3) 864 (4) 3,456

5. If the length of a rectangle is $\sqrt{27}$ and the width is $\sqrt{12}$, what is the perimeter in simplest radical form?

   (1) $5\sqrt{3}$ (2) $10\sqrt{3}$ (3) $5\sqrt{6}$ (4) $26\sqrt{3}$

B. Show how you arrived at your answer.

6–17. Perform the indicated operations and write the answer in simplest radical form.

6. $\sqrt{11} - 7\sqrt{11} + \sqrt{44}$

7. $2\sqrt{8} + 8\sqrt{2}$

8. $\sqrt{75} + \left(\sqrt{3}\right)^3$

9. $\left(\sqrt{5} + 1\right)^2$

10. $\left(\sqrt{3} - 4\right)^2$

11. $\frac{1}{2}\sqrt{20} + \frac{1}{3}\sqrt{45}$

12. $\dfrac{\sqrt{50} - \sqrt{8}}{\sqrt{2}}$

13. $\sqrt{2}\left(3\sqrt{2} - \sqrt{8}\right)$

14. $\left(5 - 3\sqrt{2}\right)^2$

15. $5\sqrt{12}-\sqrt{48}+\frac{2}{3}\sqrt{108}$ **16.** $\frac{\sqrt{27}+7\sqrt{3}}{\sqrt{12}}$

17. $\frac{3\sqrt{32}+4\sqrt{50}}{\sqrt{96}}$

18. If $\sqrt{18}-\sqrt{200}+\sqrt{72}=x\sqrt{2}$, what is the value of x?

19. Write $\frac{9\sqrt{40}+6\sqrt{10}}{\sqrt{120}}$ as a single radical in simplest form without a radical denominator.

20. If $3\sqrt{54}-5\sqrt{24}+\frac{1}{2}\sqrt{96}=k\sqrt{6}$, what is the value of k?

7.4 THE PYTHAGOREAN THEOREM

KEY IDEAS

In a right triangle, the side opposite the right or 90° angle is the **hypotenuse**, and the other two sides are the **legs**, as shown in the figure at the right. The Pythagorean Theorem relates the lengths of the three sides of a right triangle:

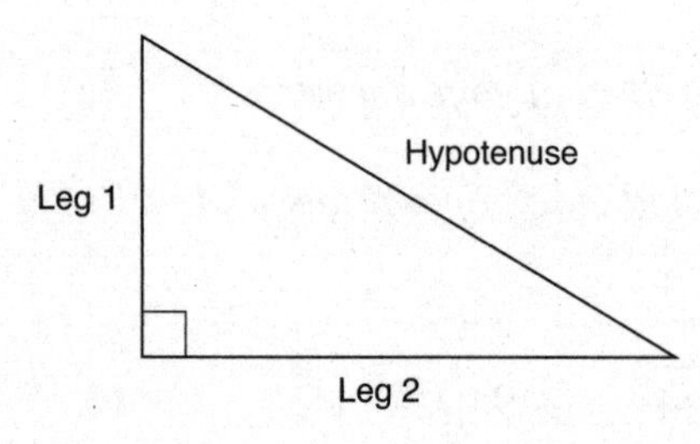

$$(\text{Leg1})^2+(\text{Leg2})^2=(\text{Hypotenuse})^2$$

Applying the Pythagorean Theorem

The Pythagorean relationship can be used to find the length of *any* side of a right triangle when the lengths of the other two sides are known. Referring to the right triangle in Figure 7.2,

- If $a = 3$ and $b = 5$, then

$$\begin{aligned} a^2 + b^2 &= c^2 \\ 3^2 + 5^2 &= c^2 \\ 9 + 25 &= c^2 \\ c^2 &= 34 \\ c &= \sqrt{34} \end{aligned}$$

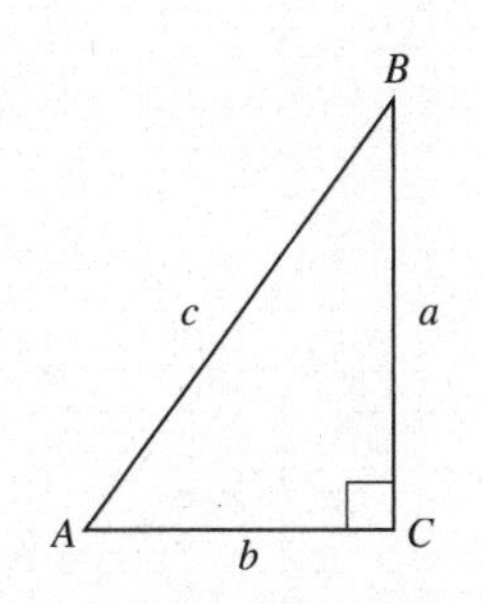

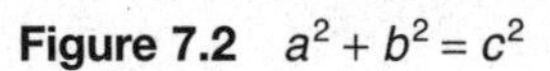

Figure 7.2 $a^2 + b^2 = c^2$

- If $a = \sqrt{13}$ and $c = 7$, then

$$\begin{aligned} a^2 + b^2 &= c^2 \\ \left(\sqrt{13}\right)^2 + b^2 &= 7^2 \\ 13 + b^2 &= 49 \\ b^2 &= 49 - 13 \\ b &= \sqrt{36} = 6 \end{aligned}$$

Example 1

Ray wants to build a square garden in which the distance between opposite corners is *at least* 18 feet. What is the shortest possible side length of the square garden correct to the *nearest foot*?

Solution: If x represents the length of a side of the square garden, then

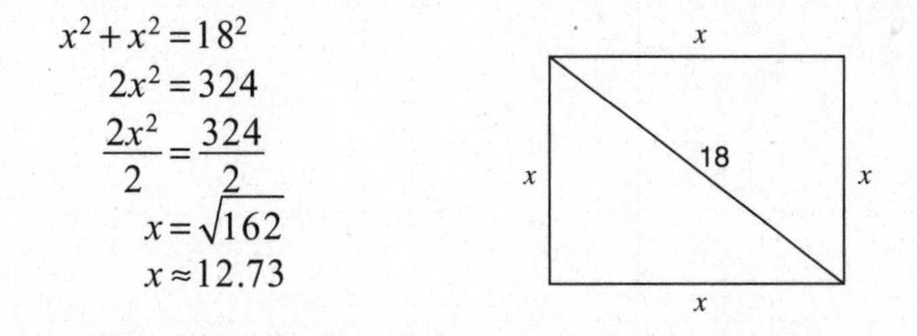

$$\begin{aligned} x^2 + x^2 &= 18^2 \\ 2x^2 &= 324 \\ \frac{2x^2}{2} &= \frac{324}{2} \\ x &= \sqrt{162} \\ x &\approx 12.73 \end{aligned}$$

Because the side length must be greater than 12.73 feet, the shortest possible side length, to the *nearest foot*, is **13 feet**.

Example 2

Katie hikes 5 miles north, 7 miles east, and then 3 miles north again. *To the nearest tenth of a mile*, how far, in a straight line, is Katie from her starting point?

Solution: The four key points on Katie's trip are labeled A through D in the accompanying diagram.

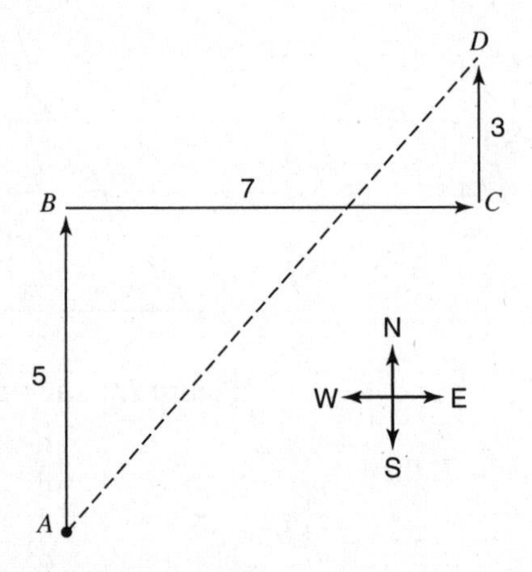

- To determine how far, in a straight line, Katie is from her starting point at A, find the length of $\overline{AD}$.
- Form a right triangle in which $\overline{AD}$ is the hypotenuse by completing rectangle $BCDE$ as shown in the accompanying diagram.
- Because opposite sides of a rectangle have the same length, $ED = BC = 7$, and $BE = CD = 3$. Thus, $AE = 5 + 3 = 8$.

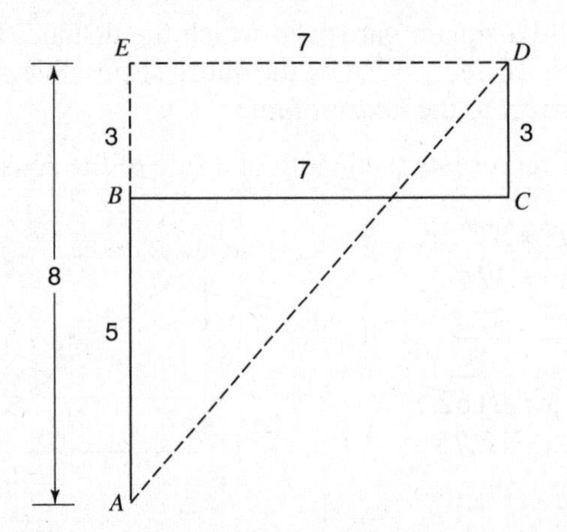

- As $\triangle AED$ is a right triangle, use the Pythagorean theorem to find AD:

$$\begin{aligned}(AD)^2 &= (AF)^2 + (FD)^2\\ &= 8^2 \quad + 7^2\\ &= 64 \quad + 49\\ &= 113\\ AD &= \sqrt{113} \approx 10.63\end{aligned}$$

Correct to the *nearest tenth of a mile*, Katie is **10.6** miles from her starting point.

Pythagorean Triples

A **Pythagorean triple** is a set of three positive integers $\{x, y, z\}$ that satisfy the relationship $x^2 + y^2 = z^2$. Here are some commonly encountered Pythagorean triples that should be memorized:

{3,4,5}, **{5,12,13}**, and **{8,15,17}**

Multiplying each member of a Pythagorean triple by the same whole number produces another Pythagorean triple. For example, multiplying each member of $\{3, 4, 5\}$ by 3 forms a 9-12-15 Pythagorean triple:

$$\{\underline{3}\times 3,\ \underline{4}\times 3,\ \underline{5}\times 3\} = \{9, 12, 15\}$$

Recognizing Pythagorean triples can make problems easier to solve.

Example 3

A ladder 13 feet in length leans against the side of a building, as shown in the accompanying diagram. The foot of the ladder is 5 feet from the building. How far up the building does the ladder reach?

Solution: Assume the ladder reaches x feet up the building.

- The ladder serves as the hypotenuse of a right triangle whose side lengths form a 5-12-13 Pythagorean triple where $x = 12$.

- If you did not recognize that the lengths of the sides form a Pythagorean triple, use the Pythagorean theorem to find x:

$$\begin{aligned} x^2 + 5^2 &= 13^2 \\ x^2 + 25 &= 169 \\ x^2 &= 169 - 25 \\ x &= \sqrt{144} \\ &= \mathbf{12} \end{aligned}$$

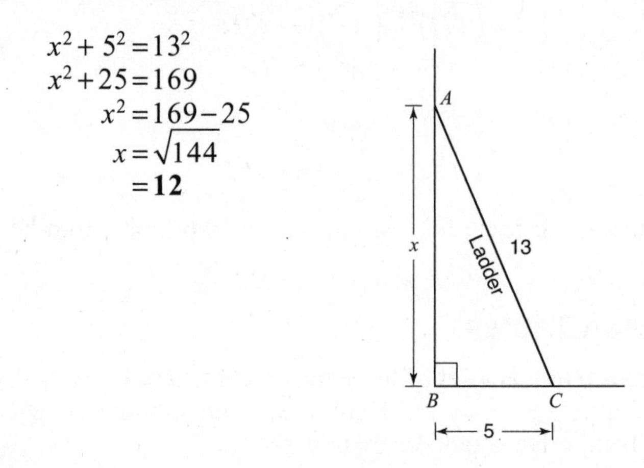

Example 4

A diagonal of a rectangle *ABCD* measures 20 inches and the width measures 12 inches. What is the number of square inches in the area of the rectangle?

Solution 1: Draw rectangle *ABCD*. The lengths of the sides of right triangle *DAB* are a multiple of the basic 3-4-5 Pythagorean triple since $20 = \underline{5} \times 4$ and $12 = \underline{3} \times 4$ so $AB = \underline{4} \times 4 = 16$.

The area of rectangle *ABCD* is, therefore, $16 \times 12 = \mathbf{192}$ square inches.

Solution 2: If you did not recognize that a Pythagorean triple was involved, use the Pythagorean theorem to find the length of leg $\overline{AB}$:

$$\begin{aligned} 12^2 + (AB)^2 &= 20^2 \\ (AB)^2 &= 400 - 144 \\ AB &= \sqrt{256} \\ &= 16 \end{aligned}$$

The area of the rectangle can now be calculated as before.

A Rhombus Application

A rectangle is a special type of *parallelogram*. A **parallelogram** is a quadrilateral whose opposite sides are parallel (and equal in length). A **rhombus** is a parallelogram in which all four sides have the same length. Like all parallelograms, the diagonals of a rhombus bisect each other. Additionally, a rhombus has the special property that its diagonals intersect at right angles.

Example 5

If the lengths of the diagonals of a rhombus are 18 and 24, what is the length of a side of the rhombus?

Solution: Draw rhombus $ABCD$, as shown in the accompanying figure. Because diagonals $\overline{AC}$ and $\overline{BD}$ bisect each other and intersect at right angles, $\triangle AED$ is a right triangle whose legs measure 9 and 12. Since $9 = \underline{3} \times 3$ and $12 = \underline{4} \times 3$, the lengths of the sides of this right triangle form a 9-12-15 Pythagorean triple in which the length of hypotenuse $\overline{AD}$ is $\underline{5} \times 3 = \mathbf{15}$.

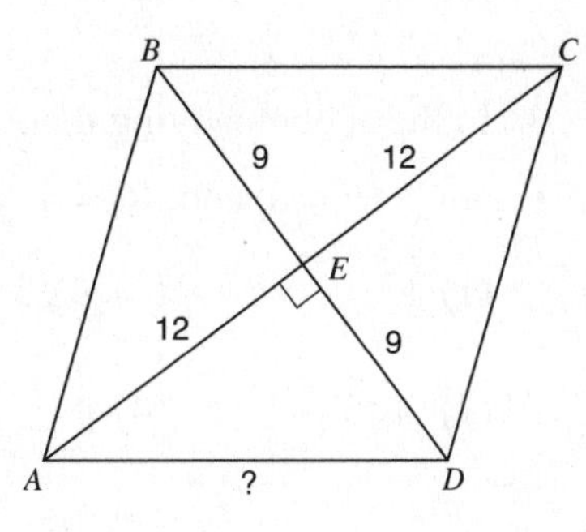

Proving a Triangle Is a Right Triangle

If the lengths of the three sides of a triangle satisfy the Pythagorean relationship, then the triangle is a right triangle. To determine if the triangle whose sides measure $\sqrt{7}$, 3, and 4 is a right triangle, test whether the square of the largest of the three numbers is equal to the sum of the squares of the other two numbers:

$$\begin{aligned} 4^2 &\boxed{?} \left(\sqrt{7}\right)^2 + 3^2 \\ 16 &\boxed{?} \ 7 + 9 \\ 16 &= 16 \checkmark \end{aligned}$$

Hence, this triangle is a right triangle.

Check Your Understanding of Section 7.4

A. Multiple Choice.

1. The lengths of the two legs of a right triangle are 2 and $2\sqrt{3}$. What is the length of the hypotenuse?

(1) $4\sqrt{3}$ (2) 10 (3) $\sqrt{14}$ (4) 4

2. The length of the hypotenuse of a right triangle is $\sqrt{15}$, and the length of one leg is 3. What is the length of the other leg?

(1) 6 (2) 9 (3) $\sqrt{6}$ (4) $3\sqrt{2}$

3. In the accompanying diagram of $\triangle ABC$, a right angle is at C, $AB = 8$, and $AC = 4$. What is the length of $\overline{BC}$?

(1) 12 (2) $4\sqrt{3}$

(3) $4\sqrt{5}$ (4) 4

B
8
C 4 A

Exercise 3

4. What is the length of a diagonal of a square whose perimeter is 32?

(1) $\sqrt{2}$ (2) $2\sqrt{2}$ (3) $4\sqrt{2}$ (4) $8\sqrt{2}$

5. Which set of numbers do *not* represent the lengths of the sides of a right triangle?

(1) $\{12, 16, 20\}$ (2) $\{2, 4, 2\sqrt{5}\}$ (3) $\{\sqrt{12}, 6, 7\}$ (4) $\{1, 1, \sqrt{2}\}$

6. What is the length of a diagonal of a rectangle in which the lengths of two adjacent sides are $\sqrt{13}$ and 6?

(1) 7 (2) 19 (3) $5\sqrt{3}$ (4) $\sqrt{23}$

7. If the lengths of the diagonals of a rhombus are 6 and 8, the perimeter of the rhombus is

(1) 14 (2) 20 (3) 28 (4) 40

8. In rectangle *MATH*, $AT = 8$ and $TH = 12$. What is the length of diagonal $\overline{HA}$ to the *nearest tenth*?

(1) 14.4 (2) 11.7 (3) 8.9 (4) 7.2

9. What is the perimeter of a square whose diagonal measures $6\sqrt{2}$?

(1) 36 (2) 24 (3) $4\sqrt{6}$ (4) $24\sqrt{2}$

10. A carpenter is building a rectangular deck with dimensions of 16 feet by 30 feet. To ensure that the adjacent sides form 90° angles, what should each diagonal measure?

(1) 26 ft (2) 30 ft (3) 34 ft (4) 46 ft

11. At 9:00 AM a car starts at point A and travels north for 1 hour at an average rate of 60 miles per hour. Without stopping, the car then travels east for 2 hours at an average rate of 45 miles per hour. At 12:00 PM, what is the best approximation of the distance, in miles, of the car from point A?

(1) 100 (2) 105 (3) 108 (4) 115

B. *Show how you arrived at your answer.*

12. The accompanying diagram shows a kite that has been secured to a stake in the ground with a 20-foot string. The kite is located 12 feet from the ground, directly over point X. What is the distance, in feet, between the stake and point X?

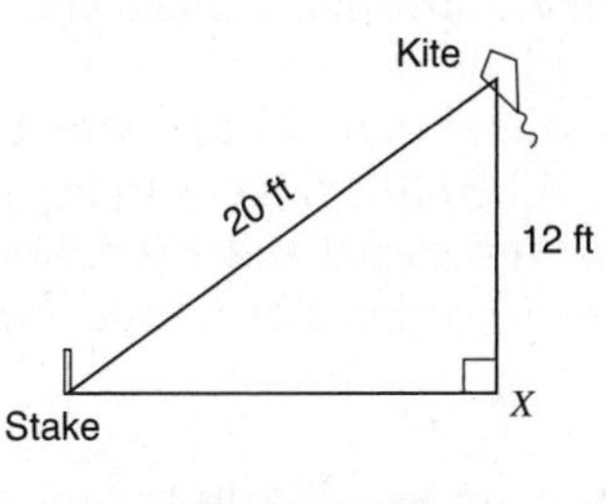

Exercise 12

13. The NuFone Communications Company must run a telephone line between two poles at opposite ends of a lake, as shown in the accompanying diagram. The length and width of the lake are 75 feet and 30 feet, respectively. What is the distance between the two poles to the *nearest tenth* of a foot.

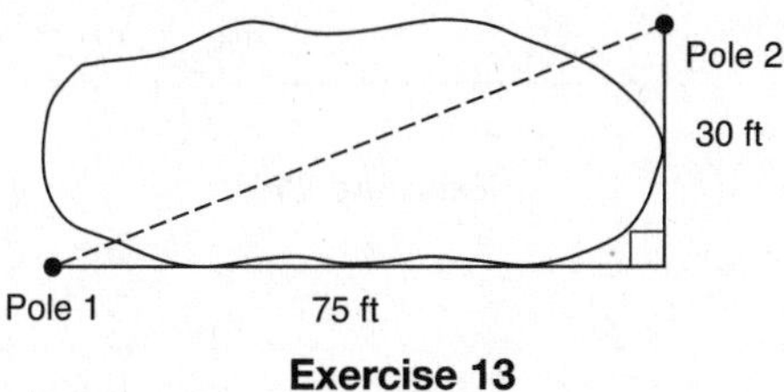

Exercise 13

14. A baseball diamond is in the shape of a square with a side length of 90 feet. What is the distance from home plate to second base, correct to the *nearest tenth of a foot*?

15. A diagonal of a rectangle $ABCD$ measures 15 inches, and the width measures 9 inches. What is the number of square inches in the area of the rectangle?

16. The perimeter of a rhombus is 100 centimeters, and the length of the longer diagonal is 48 centimeters. What is the number of centimeters in the length of the shorter diagonal?

17. What is the exact length of a diagonal of a square whose area is 81 square units?

18. The length and width of a rectangle are in the ratio of 3 : 4. If the length of the diagonal of the rectangle is 60, what are the length and width of the rectangle?

19. Two hikers started at the same location. One traveled 2 miles east and then 1 mile north. The other traveled 1 mile west and then 3 miles south. At the end of their hikes, how many miles apart were the two hikers?

20. To get from his high school to his home, Jamal travels 5.0 miles east and then 4.0 miles north. When Sheila goes to her home from the same high school, she travels 8.0 miles east and 2.0 miles south. What is the shortest distance, to the *nearest tenth* of a mile, between Jamal's home and Sheila's home?

21. In the accompanying diagram of right triangles ABC and DBC, $AB = 5$, $AD = 4$, and $CD = 1$. Find the length of $\overline{BC}$, to the *nearest tenth*.

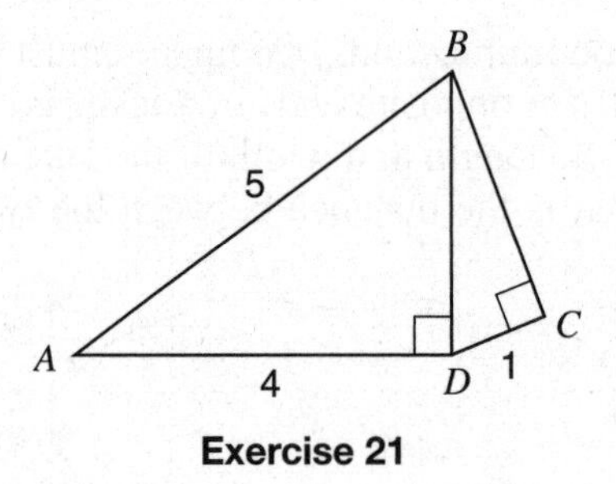

Exercise 21

22. Mary and Martin start at the same time from point A shown on the accompanying grid. Mary walks at a constant average rate of 3.5 miles per hour from point A to point R to point S to point C. Martin walks at a constant average rate of 3 miles per hour directly from point A to point C on $\overline{AC}$. Which person reaches point C first and how many minutes later does the second person reach point C?

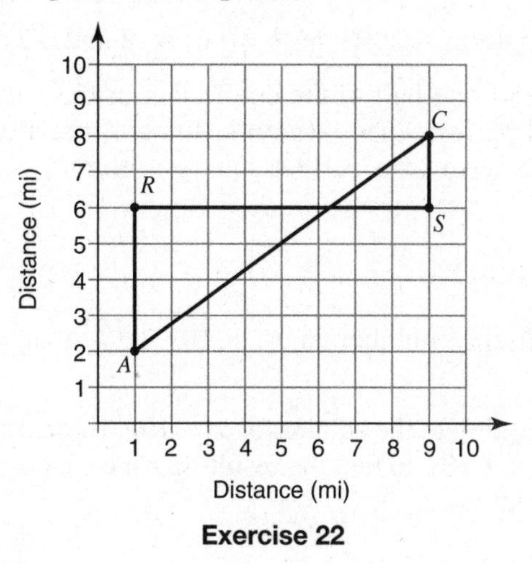

Exercise 22

7.5 GENERAL METHODS FOR SOLVING QUADRATIC EQUATIONS*

KEY IDEAS

If a quadratic equation can be rewritten so that one side is the square of a binomial and the other side is a constant, as in $(x-3)^2 = 16$, then the equation can be solved by taking the square root of both sides:

$$(x-3)^2 = 16$$
$$\sqrt{(x-3)^2} = \pm\sqrt{16}$$
$$x - 3 = +4 \quad \text{or} \quad x - 3 = -4$$
$$x = 3 + 4 \qquad\qquad x = 3 - 4$$
$$x = 7 \qquad\qquad x = -1$$

By using this method, a general formula can be obtained for solving any quadratic equation that has the form $ax^2 + bx + c = 0$.

*This is an optional topic and is not required for the Integrated Algebra Regents examination.

Completing the Square

A **perfect square trinomial** is a trinomial that can be written as the square of a binomial. For example, $x^2 - 6x + 9$ is a perfect square trinomial:

$$x^2 - 6x + 9 = (x - 3)(x - 3) = (x - 3)^2$$

In $x^2 - 6x + 9$, what is the relationship between the constant term, 9, and the coefficient of the x-term, -6? Because $\frac{1}{2}(-6) = -3$ and $(-3)^2 = 9$, the constant term is the square of one-half of the coefficient of the x-term. This relationship is true for all perfect square trinomials; therefore, it can be used to discover the constant term of a perfect square trinomial when the other two terms are known.

Example 1

Determine the missing number in $x^2 + 10x + \boxed{?}$ that will complete the square.

Solution: To complete the square for $x^2 + 10x$, figure out the number that must be added to $x^2 + 10x$ so that the resulting trinomial can be written as the square of a binomial.

- Take one-half of the coefficient of the x-term: $\frac{1}{2}(10) = 5$
- Square that number: $5^2 = 25$
- Add the result to the first two terms: $x^2 + 10x + 25$
- Check your answer by factoring:

$$x^2 + 10x + 25 = (x + 5)(x + 5) = (x + 5)^2$$

Solving Quadratic Equations By Completing the Square

Any quadratic equation, including those that are not factorable, can be solved by putting them in the form $(x + p)^2 = k$ and then taking the square root of both sides of the equation. To solve $x^2 - 8x + 9 = 0$ by completing the square:

- Rewrite the equation as $x^2 - 8x = -9$ so that only terms involving x are on one side of the equation.
- Complete the square on the left side of the equation. Since $\frac{1}{2}(-8) = -4$ and $(-4)^2 = 16$, add 16 to both sides of the equation and simplify:

$$x^2 - 8x + 16 = -9 + 16$$
$$x^2 - 8x + 16 = 7$$
$$(x - 4)^2 = 7$$

- Take the square roots of both sides of the equation:

$$\sqrt{(x-4)^2} = \pm\sqrt{7}$$

$$x - 4 = +\sqrt{7} \qquad or \qquad x - 4 = -\sqrt{7}$$

$$x = 4 + \sqrt{7} \qquad\qquad x = 4 - \sqrt{7}$$

MATH FACTS

Any quadratic equation, including those that have irrational roots, can be solved by completing the square. To use this method, follow these steps:

- Rewrite the quadratic equation in the form $x^2 + bx = k$.
- Add the number that completes the square to both sides of the equation.
- Take the square root of both sides of the equation and solve for x.

Example 2

Solve $4x^2 + 8x - 5 = 0$ by completing the square.

Solution: Begin by transposing the constant term to the right side of the equation, which gives $4x^2 + 8x = 5$. Then complete the square:

- Make the coefficient of the x^2-term 1 by dividing each member of the equation by 4:

$$\frac{4x^2}{4} + \frac{8x}{4} = \frac{5}{4}$$
$$x^2 + 2x = \frac{5}{4}$$

- Since $\frac{1}{2}(2) = 1$ and $(1)^2 = 1$, complete the square by adding 1 to both sides of the equation:

$$x^2 + 2x + 1 = 1 + \frac{5}{4}$$

$$(x+1)^2 = \frac{4}{4} + \frac{5}{4}$$

$$(x+1)^2 = \frac{9}{4}$$

- Solve for x by taking the square root of both sides of the equation:

$$\sqrt{(x+1)^2} = \pm\sqrt{\frac{9}{4}}$$

$$x+1=\frac{3}{2} \qquad or \qquad x+1=-\frac{3}{2}$$

$$x=-1+\frac{3}{2} \qquad\qquad x=-1-\frac{3}{2}$$

$$x=\frac{1}{2} \qquad\qquad x=-\frac{5}{2}$$

Example 3

Solve $y^2 + 6y + 2 = 0$.

Solution: The quadratic trinomial $y^2 + 6y + 2$ is not factorable. Solve the equation by completing the square.

- Rewrite the equation as $y^2 + 6y = -2$.
- To complete the square, add $\left(\frac{6}{2}\right)^2 = 3^2 = 9$ to both sides of the equation:

$$y^2 + 6y + 9 = -2 + 9$$
$$(y + 3)^2 = 7$$

- Take the square root of both sides of the equation:

$$y + 3 = \pm\sqrt{7}$$

- Solve for y:

$$y = -3 + \sqrt{7} \qquad or \qquad y = -3 - \sqrt{7}$$

In Example 3, although we could use a graphing calculator to estimate the roots of this equation, solving the equation by completing the square allows us to obtain an exact representation of the irrational roots in radical form.

The Quadratic Formula

If $ax + b = c$, solving for x gives $x = \frac{c-b}{a}$. Any linear equation can be solved by writing it in the form $ax + b = c$ and then substituting the values for a, b, and c in the formula $x = \frac{c-b}{a}$. For example, if $2x + 3 = 9$, then $a = 2$, $b = 3$, and $c = 9$ so $x = \frac{9-3}{2} = \frac{6}{2} = 3$. Similarly, if $ax^2 + bx + c = 0$, by completing the square x can be solved for in terms of the coefficients a, b, and c. The result is called the **quadratic formula**:

$$x = \frac{-b \pm \sqrt{b^2 - 4ac}}{2a}$$

Any quadratic equation can be solved by first writing it in the form $ax^2 + bx + c = 0$, and then substituting the numerical values of a, b, and c into the quadratic formula. The roots of a quadratic equation may be rational, irrational, or nonreal (when the radicand is negative).

Example 4

Find the roots of $2x^2 - 3x = 1$ in radical form.

Solution: Rewrite the quadratic equation so all of the nonzero terms are on the same side of the equation, as in $2x^2 - 3x - 1 = 0$. Use the quadratic formula where $a = 2$, $b = -3$, and $c = -1$:

$$\begin{aligned} x &= \frac{-b \pm \sqrt{b^2 - 4ac}}{2a} \\ &= \frac{-(3) \pm \sqrt{(-3)^2 - 4(2)(-1)}}{2(2)} \\ &= \frac{3 \pm \sqrt{9+8}}{4} \\ &= \frac{3 \pm \sqrt{17}}{4} \end{aligned}$$

The two roots are $\frac{3+\sqrt{17}}{4}$ and $\frac{3-\sqrt{17}}{4}$.

Check Your Understanding of Section 7.5

A. *Multiple Choice.*

1. Brian correctly used the method of completing the square to solve the equation $x^2 + 7x - 11 = 0$. Brian's first step was to rewrite the equation as $x^2 + 7x = 11$. He then added a number to both sides of the equation. Which number did he add?

(1) $\frac{7}{2}$ (2) $\frac{49}{4}$ (3) $\frac{49}{2}$ (4) 49

2. If $x^2 + 2 = 6x$ is solved by completing the square, an intermediate step would be

(1) $(x + 3)^2 = 7$
(2) $(x - 3)^2 = 7$
(3) $(x - 3)^2 = 11$
(4) $(x - 6)^2 = 34$

B. *Show all work.*

3–5. *Solve each equation by completing the square, and express the roots in simplest form.*

3. $x^2 + 15 = 8x$

4. $3m^2 = 21 + 6m$

5. $6 - 2x^2 = 20x$

6–8. *Solve each equation using the quadratic formula, and express the roots in simplest radical form.*

6. $2x^2 = 5x + 1$

7. $-x^2 + 5x + 9 = 0$

8. $3x^2 - 2x - 4 = 0$

CHAPTER 8
SIMILAR TRIANGLES AND TRIGONOMETRY

8.1 TRIGONOMETRIC RATIOS

KEY IDEAS

A *trigonometric ratio* relates the measures of *two* sides and one of the acute angles of a *right* triangle.

Opposite Sides and Adjacent Legs

Associated with each acute angle of a right triangle is an *opposite side* and an *adjacent side*, both of which are legs of the right triangle.

- With respect to $\angle A$ in Figure 8.1(a), $\overline{BC}$ is the opposite side and $\overline{AC}$ is the adjacent side since it is one of the sides of $\angle A$.
- With respect to $\angle B$ in Figure 8.1(b), $\overline{AC}$ is the opposite side and $\overline{BC}$ is the adjacent side because it is one of the sides of $\angle B$.

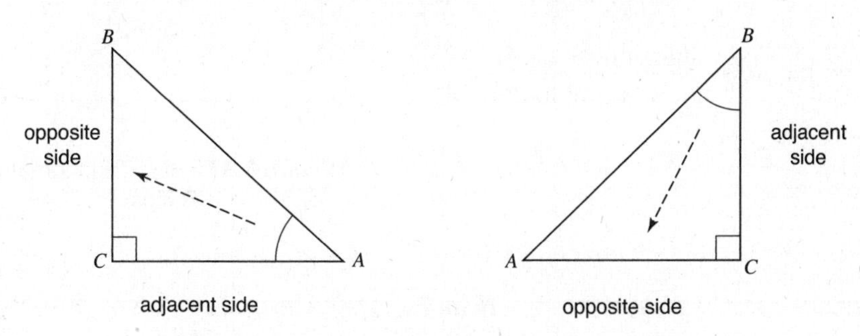

Figure 8.1 Identifying Opposite and Adjacent Sides

The Three Basic Trigonometric Ratios

In Figure 8.2, $\triangle ABC$, $\triangle ADE$, and $\triangle AFG$ each contain a right angle and have acute $\angle A$ in common. The ratio of the length of the side opposite $\angle A$ to the length of the hypotenuse is the same for each right triangle:

$$\frac{\text{side opposite } \angle A}{\text{hypotenuse}} = \frac{BC}{AC} = \frac{DE}{AE} = \frac{FG}{AG}$$

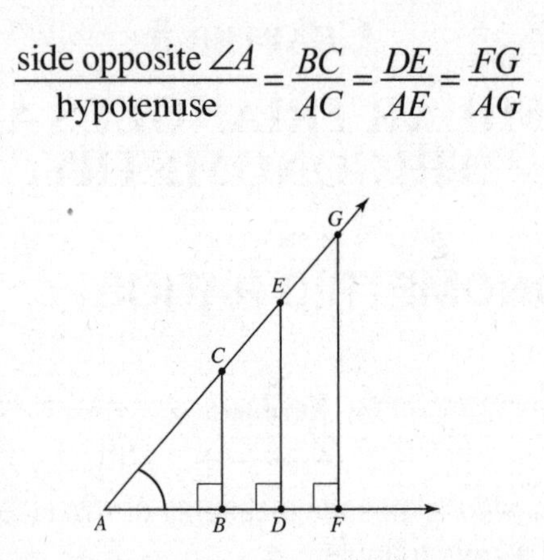

Figure 8.2 Comparing Ratios in Right Triangles

This constant ratio is called the **sine** of $\angle A$. Other ratios named **cosine** and **tangent** can also be defined in a right triangle. The abbreviations for sine, cosine, and tangent ratios are sin, cos, and tan, respectively. In right triangle *ABC* shown in Figure 8.3,

- $\sin A = \dfrac{\text{side opposite } \angle A}{\text{hypotenuse}} = \dfrac{3}{5}$
- $\cos A = \dfrac{\text{side adjacent to } \angle A}{\text{hypotenuse}} = \dfrac{4}{5}$
- $\tan A = \dfrac{\text{side opposite } \angle A}{\text{side adjacent to } \angle A} = \dfrac{3}{4}$

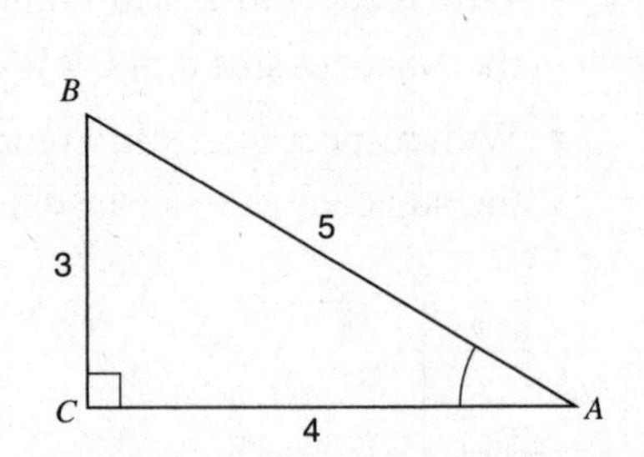

Figure 8.3 Finding Trigonometric Ratios

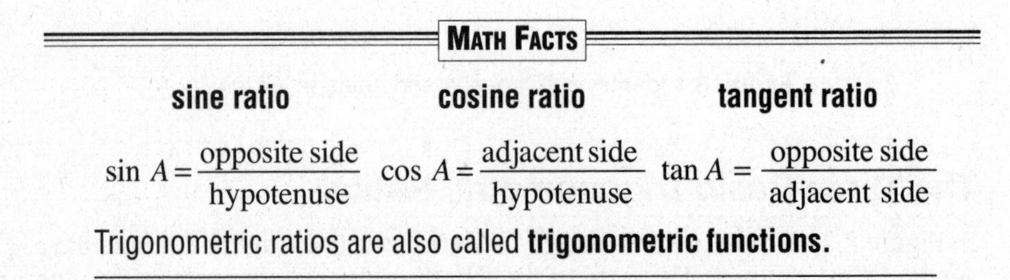

MATH FACTS

sine ratio **cosine ratio** **tangent ratio**

$$\sin A = \frac{\text{opposite side}}{\text{hypotenuse}} \quad \cos A = \frac{\text{adjacent side}}{\text{hypotenuse}} \quad \tan A = \frac{\text{opposite side}}{\text{adjacent side}}$$

Trigonometric ratios are also called **trigonometric functions.**

Example 1

In right triangle ABC, $\angle C$ is the right angle, $AC = 12$, and $AB = 13$. What is the value of $\sin A$, $\cos A$, and $\tan A$?

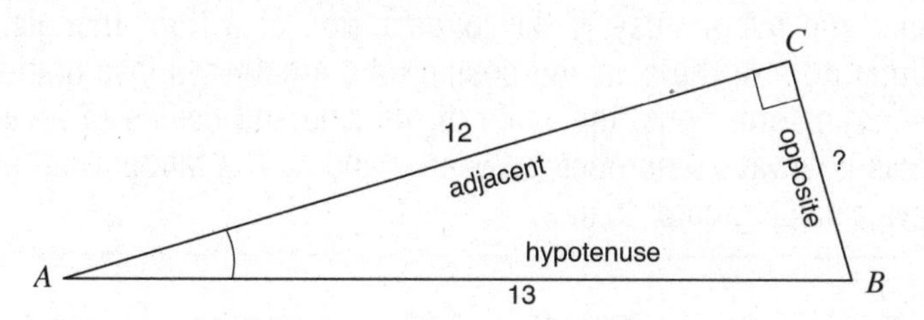

Solution: Since the lengths of the sides of right triangle ABC form a 5-12-13 Pythagorean triple, $BC = 5$. Thus,

$$\sin A = \frac{\text{opposite side}}{\text{hypotenuse}} = \frac{BC}{AB} = \mathbf{\frac{5}{13}}$$

$$\cos A = \frac{\text{adjacent side}}{\text{hypotenuse}} = \frac{AC}{AB} = \mathbf{\frac{12}{13}}$$

$$\tan A = \frac{\text{opposite side}}{\text{adjacent side}} = \frac{BC}{AC} = \mathbf{\frac{5}{12}}$$

Using Technology: Finding Trigonometric Function Values

To use your graphing calculator to find the value of a trigonometric function of an angle:

- Press either the [SIN], [COS], or [TAN] keys, depending on the trigonometric function that is required.
- Enter the number of degrees of the given angle.
- Press [ENTER]. You should verify that

$$\sin 54° \approx 0.8090169944, \cos 60° = 0.5, \text{ and } \tan 73° = 3.270852618$$

If your calculator does not produce the expected results, it may be because the angular mode of your calculator is not set to degrees. To set a graphing calculator to degree mode, press [MODE], highlight **DEGREE**, and press [ENTER]. Then return to the home screen by pressing [2nd] [MODE].

MATH FACTS

- The values of $\sin x$, and $\tan x$ do not change in different right triangles in which the measure of acute angle x is the same.
- Since the hypotenuse is the longest side of a right triangle, the numerators of the sine and cosine ratio are always less than their denominators. Thus, the value of the sine and cosine of an acute angle is always a number between 0 and 1. The tangent ratio may have a value greater than 1.

Using Technology: Finding Inverse Trigonometric Function Values

If $\tan x = \frac{5}{12}$, then "x is the angle whose tangent is $\frac{5}{12}$." This statement can be expressed using a special mathematical notation as, $x = \tan^{-1}\left(\frac{5}{12}\right)$. Your graphing calculator has $\tan^{-1}$ printed above the TAN key, $\sin^{-1}$ printed above the SIN key, and $\cos^{-1}$ printed above the COS key. The -1 is *not* an exponent. It indicates that these are *inverse* functions used to find the measure of an angle when the value of a trigonometric function of that angle is known. To find the degree measure of angle x, press

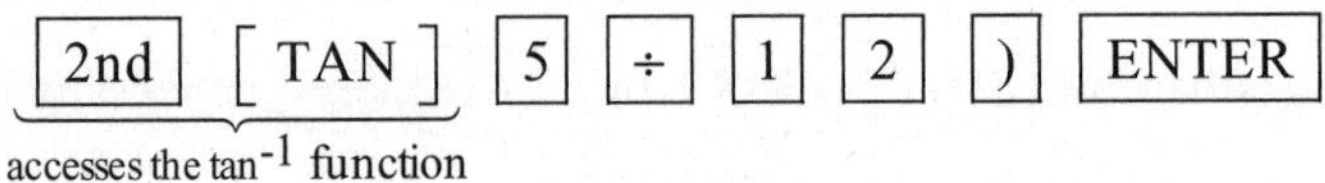

which gives $x \approx 22.61986495°$. Pressing the rightmost parentheses key,), is optional.

Example 2

Find the value of x: a. $\sin x = \frac{4}{9}$ b. $\cos x = 0.1529$

Solution: a. If $\sin x = \frac{4}{9}$, then "x is the angle whose sine is $\frac{4}{9}$" so $x = \sin^{-1}\left(\frac{4}{9}\right)$. To find the degree measure of angle x, press:

2nd [SIN] 4 ÷ 9) ENTER

which gives $x \approx$ **26.38779996°**.

b. If $\cos x = 0.1529$, then "x is the angle whose cosine is 0.1529" so $x = \cos^{-1}(0.1529)$.

To find the degree measure of angle x, press

which gives $x \approx$ **81.20497661°**.

Solving a Right Triangle

When the measures of any two sides, or any side and an acute angle, of a right triangle are given, the measures of any of the remaining sides and angles of the right triangle can be determined using an appropriate trigonometric ratio or the Pythagorean Theorem.

Example 3

In the right triangle JKL, $\angle K$ is a right angle, $\angle J = 56°$, and $JL = 14$. What is the length of $\overline{JK}$ to the *nearest tenth*?

Solution:

- Draw a right triangle. Label the right angle K and the opposite side, $\overline{JL}$.
- Mark the diagram with the given dimensions. Represent the length of $\overline{JK}$ by x.

  To help identify the required trigonometric ratio, label the two sides involved in the problem relative to their position to $\angle J$, as was done in the accompanying figure.

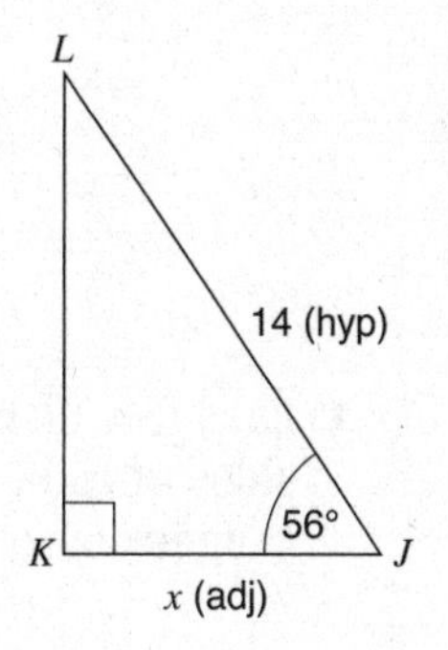

- Select the trigonometric ratio that is defined in terms of the given and the required parts of the right triangle. As the problem involves the side *adjacent* to $\angle J$ and the *hypotenuse*, use the **cosine** ratio:

$$\cos J = \frac{\text{adjacent side }(JK)}{\text{hypotenuse }(JL)}$$

$$\cos 56° = \frac{x}{14}$$

- Solve for x *before* evaluating cos 56°:

$$x = 14 \cos 56°$$

- Perform the indicated multiplication using the full power/display of your calculator:

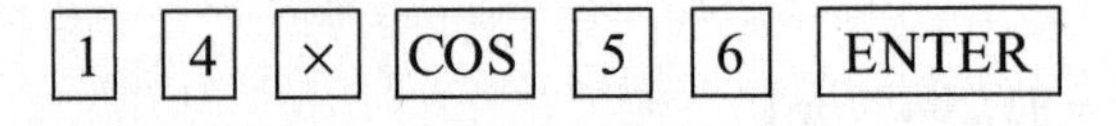

The calculator display window shows 7.828700649. As the final step, round off the answer in the display window.

The length of $\overline{JK}$ is **7.8** to the *nearest tenth*.

Example 4

The lengths of a pair of adjacent sides of rectangle *RECT* are 8 and 13. Find to the *nearest tenth of a degree* the angle that a diagonal makes with the longer side of the rectangle.

Solution: Draw rectangle *RECT* and diagonal $\overline{ET}$, which is the hypotenuse of right triangle *ERT*.

- Mark the diagram with the given dimensions where x represents the measure of $\angle RTE$.

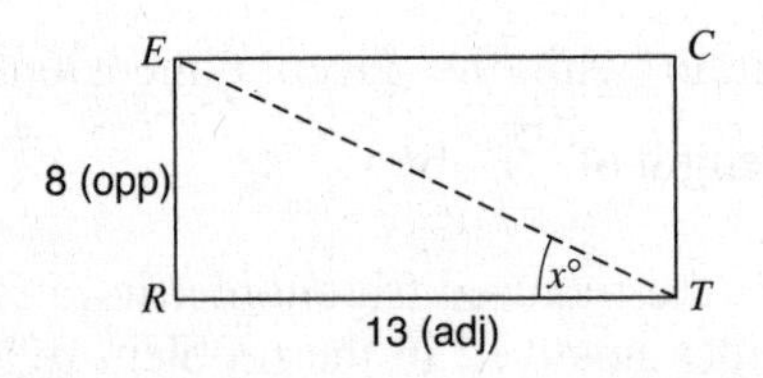

- Decide which trigonometric ratio to use. As the lengths of the given sides of right triangle *ERT* are *opposite* and *adjacent* to $\angle RTE$, use the **tangent** ratio:

$$\tan x = \frac{\text{opposite side}(RE)}{\text{adjacent side }(RT)}$$

$$= \frac{8}{13}$$

- Solve for x:

$$x = \tan^{-1}\left(\frac{8}{13}\right)$$

- Use the full power/display of your calculator to find x:

2nd [TAN] 8 ÷ 1 3 ENTER
(8 ÷ 1 3: wait until this step to divide)

The calculator display window shows 31.60750225°. As the last step, round off the answer in the display window.

To the *nearest tenth of a degree*, the diagonal makes an angle of **31.6°** with the longer side of the rectangle.

Check Your Understanding of Section 8.1

A. *Multiple Choice.*

1. In the accompanying diagram of right triangle ABC, $\angle C$ is a right angle, $\angle A = 63°$, and $AB = 10$. If BC is represented by x, which equation can be used to find x?

(1) $\sin 63° = \frac{x}{10}$ (2) $\tan 63° = \frac{x}{10}$ (3) $x = 10 \cos 63°$ (4) $x = \frac{\tan 27°}{10}$

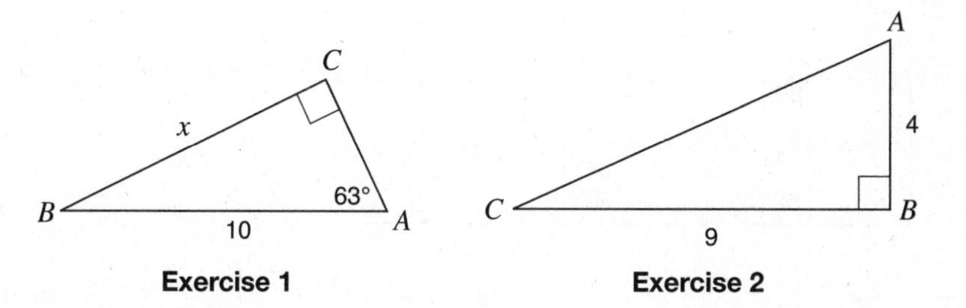

Exercise 1 Exercise 2

2. In the accompanying diagram of right triangle ABC, $AB = 4$ and $BC = 9$. What is the measure of $\angle A$ correct to the *nearest degree*?
(1) 24 (2) 35 (3) 55 (4) 66

3. In right triangle REM, $\angle E$ is a right angle, $RE = 6$, and $EM = 8$. The value of $\cos R$ is
(1) 0.25 (2) 0.6 (3) 0.8 (4) 0.75

4. In the accompanying diagram of right triangle RUN, $m\angle U = 90$, $m\angle N = 37$, and $RN = 21$. What is the length of $\overline{RU}$, expressed to the *nearest tenth*?

(1) 12.6 (2) 15.8 (3) 16.8 (4) 34.9

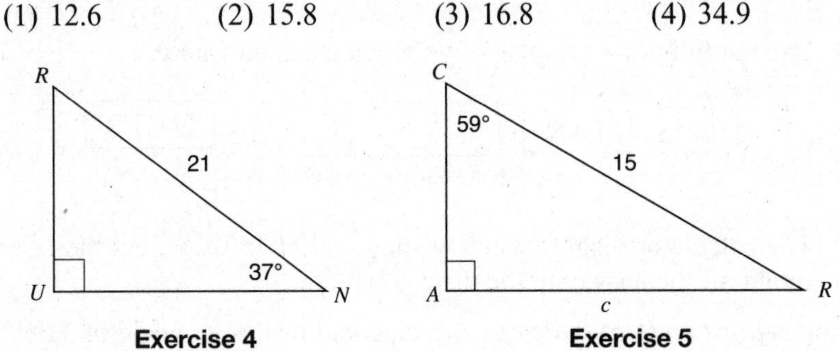

Exercise 4 **Exercise 5**

5. In the accompanying diagram of right triangle CAR, $\angle A$ is a right angle, $m\angle C = 59$, and $CR = 15$. If AR is represented by c, which equation can be used to find c?

(1) $\cos 59° = \frac{c}{15}$ (2) $\tan 59° = \frac{c}{15}$ (3) $\cos 31° = \frac{c}{15}$ (4) $\sin 31° = \frac{c}{15}$

***B.** Show how you arrived at your answer.*

6. In right triangle JKL, $\angle K$ is a right angle, $JK = 7$, and $KL = 24$. Find the value of $\sin J$, $\cos J$, and $\tan J$.

7–9. *Find x to the nearest hundredth. Figures are not drawn to scale.*

7. **8.** **9.**

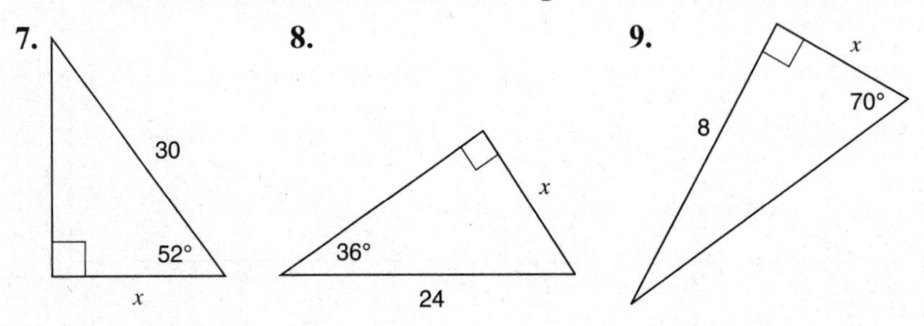

10–12. *Find x to the nearest tenth of a degree. Figures are not drawn to scale.*

10. **11.** **12.**

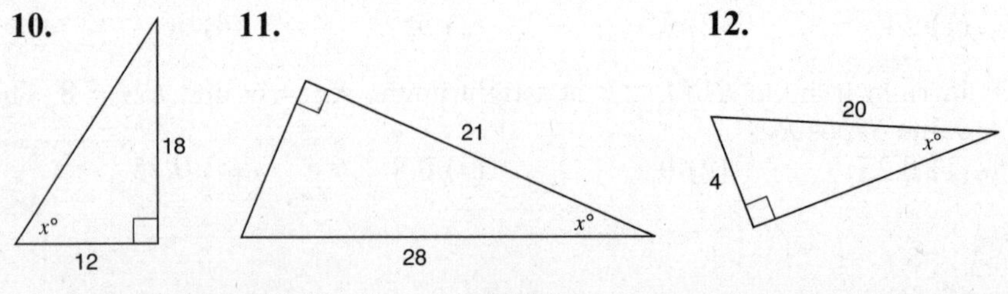

13. The lengths of a pair of adjacent sides of rectangle *JKLM* are 7 and 18. Find to the *nearest tenth of a degree* the angle that a diagonal makes with the shorter side of the rectangle.

14. In right triangle *ABC*, $\angle C$ is a right angle, $AB = 17$, and $AC = 15$. Find to the *nearest degree* the measure of the angle opposite the shorter leg of the right triangle.

15. In rectangle *ABCD*, diagonal $\overline{AC}$ makes an angle of 39° with the longer side. If the length of the diagonal is 20.0 inches, find

a. The length of the longer side to the *nearest tenth* of an inch.
b. The perimeter of the rectangle to the *nearest inch*.

16. As shown in the accompanying diagram, a person can travel from New York City to Buffalo by going north 170 miles to Albany and then west 280 miles to Buffalo.

a. If an engineer wants to design a highway to connect New York City directly to Buffalo, at what angle, *x*, would she need to build the highway? Find the angle to the *nearest tenth* of a degree.
b. To the *nearest mile*, how many miles would be saved by traveling directly from New York City to Buffalo rather than traveling first to Albany and then from Albany to Buffalo?

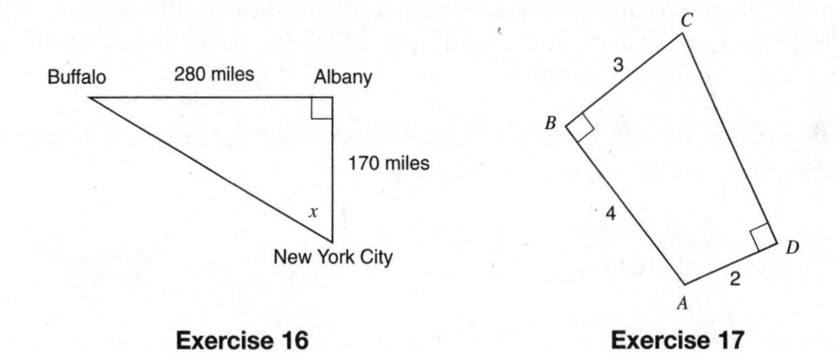

Exercise 16 **Exercise 17**

17. In quadrilateral *ABCD*, $\overline{AB} \perp \overline{BC}$, $\overline{AD} \perp \overline{CD}$, $AB = 4$, $BC = 3$, and $AD = 2$.

a. Find $\angle C$ correct to the *nearest degree*.
b. Find the perimeter of quadrilateral *ABCD* correct to the *nearest tenth*.

8.2 SOLVING PROBLEMS USING TRIGONOMETRY

KEY IDEAS

Trigonometric functions are often used to calculate the measures of sides and angles that represent real-world quantities. Angles of *elevation* and *depression* are special names given to angles that involve the sighting of distant objects relative to a horizontal line of sight.

Indirect Measurement

Trigonometric functions are particularly useful when it is necessary to calculate the measure of a side or an angle of a right triangle that may be difficult, if not impossible, to measure directly.

Example 1

A plane takes off from a runway at point A and climbs while maintaining a constant angle with the ground, as shown in the accompanying diagram. When the plane has traveled 1000 meters, its altitude is 290 meters. Find, correct to the *nearest degree*, the angle x at which the plane has climbed with respect to the horizontal ground.

Solution: Since the lengths of the side opposite the required angle and the hypotenuse are given, use the sine ratio:

$$\sin x = \frac{\text{opposite side}}{\text{hypotenuse}}$$

$$= \frac{290}{1000}$$

$$x = \sin^{-1} 0.2900$$

$$x \approx 16.85795602^\circ$$

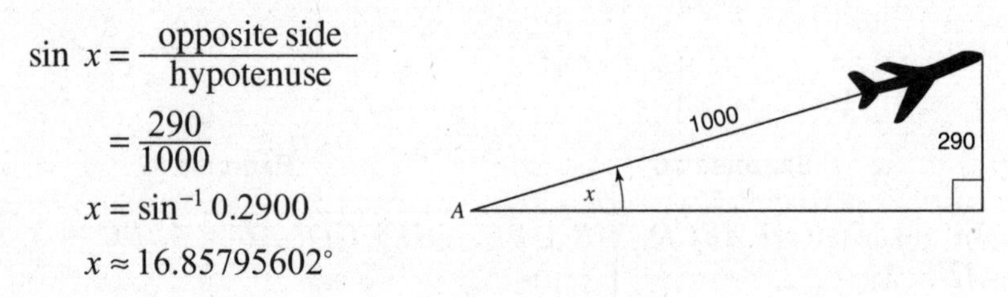

To the *nearest degree*, the plane has climbed at an angle of **17°**.

Example 2

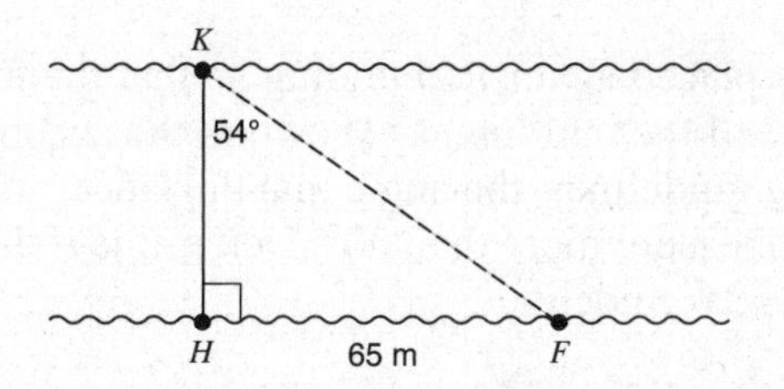

To determine the distance across a river, a surveyor marked two points on one riverbank, H and F, 65 meters apart. She also marked one point, K, on the opposite bank such that $\overline{KH} \perp \overline{HF}$, as shown in the accompanying figure. If $\angle K = 54°$, what is the width of the river, to the *nearest tenth of a meter*?

Solution: Represent the width of the river, KH, by x.

- Because the problem involves the sides opposite and adjacent to the given angle, use the tangent ratio:

$$\tan\angle K = \frac{\text{opposite side } (HF)}{\text{adjacent side } (KH)}$$

$$\tan 54° = \frac{65}{x}$$

- Solve for x before evaluating tan 54°:

$$x = 65 \div \tan 54°$$

- Use the full power/display of your calculator to find x:

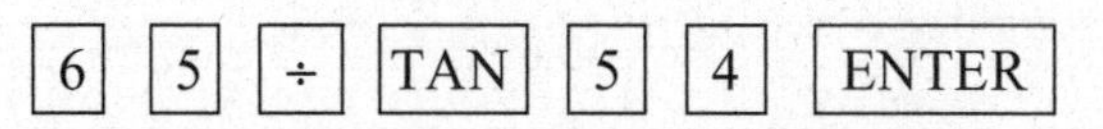

The calculator display window shows 47.22526432.

To the *nearest tenth* of a meter, the width of the river is **47.2 meters**.

Example 3

A 16-foot ladder is placed against a building so that the foot of the ladder is 9 feet from the base of the building, as shown in the accompanying diagram. According to safety guidelines, the angle that the ladder makes with the level ground should not measure more than 60°. Determine if the placement of the ladder meets the safety guidelines.

Solution: Let x represent the measure of the angle formed by the ladder and the ground. Because lengths of the side adjacent to the required angle and the hypotenuse are given, use the cosine ratio:

$$\begin{aligned}\cos x &= \frac{\text{adjacent side}}{\text{hypotenuse}}\\ &= \frac{9}{16}\\ x &= \cos^{-1}\left(\frac{9}{16}\right)\\ x &\approx 55.8^\circ\end{aligned}$$

Because 55.8° < 60°, the **placement of the ladder meets the safety guidelines.**

Angles of Elevation and Depression

The **angle of elevation** represents the angle through which an observer must *lift* his or her horizontal line of sight in order to see an object above ground level, as a birdwatcher might need to do. The **angle of depression** is the angle through which an observer must *lower* his or her horizontal line of sight in order to spot a landmark at ground level, as an airplane pilot might need to do. The angle of elevation, e, falls inside a right triangle while the angle of depression, d, falls outside a right triangle, as shown in Figure 8.4.

Because the horizontal lines of sight are parallel, the measures of the angles of elevation and depression are numerically equal.

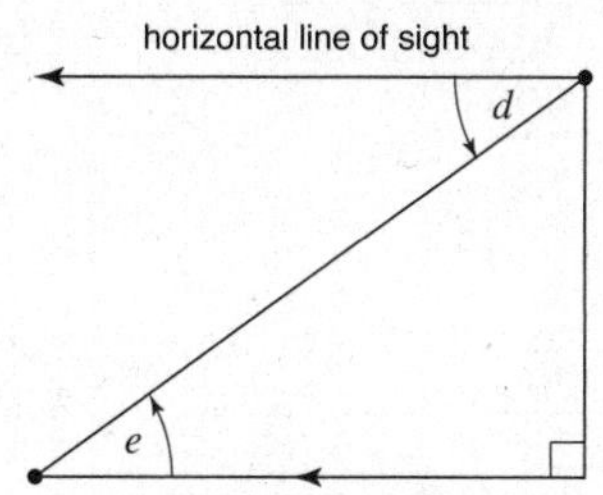

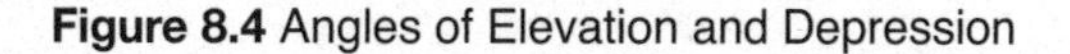

Figure 8.4 Angles of Elevation and Depression

Example 4

A man standing 30 feet from a flagpole observes that the angle of elevation of its top is 48°. Find the height of the flagpole, correct to the *nearest tenth* of a foot.

Solution: Represent the height of the flagpole by x.

- Draw a right triangle and label it with the given information.
- Use the tangent ratio:

$$\tan 48° = \frac{\text{opposite side}}{\text{adjacent side}} = \frac{x}{30}$$

Solve for x:

$$x = 30 \times \tan 48°$$
$$x \approx 33.31837544$$

To the *nearest tenth* of a foot, the height of the flagpole is **33.3 feet**.

Example 5

An airplane pilot observes the angle of depression of a point on a landing field to be 28°. If the plane's altitude at this moment is 900 meters, find the distance from the pilot to the observed point on the landing field, correct to the *nearest meter*.

Solution: Represent the distance from the pilot to the observed point on the landing field by x.

- Draw a right triangle based on the given information.

- Use the sine ratio:

$$\sin 28° = \frac{\text{opposite side}}{\text{hypotenuse}} = \frac{900}{x}$$

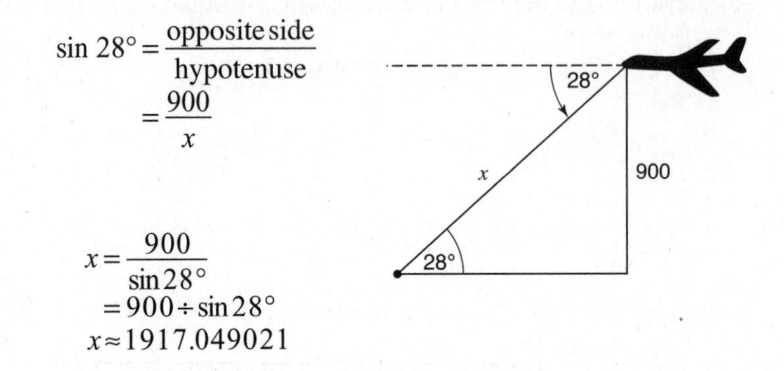

Solve for x:

$$x = \frac{900}{\sin 28°} = 900 \div \sin 28°$$
$$x \approx 1917.049021$$

The pilot's distance to the point on the landing field is, to the *nearest meter*, **1,917 meters.**

Check Your Understanding of Section 8.2

Show how you arrived at your answer.

1. As shown in the accompanying diagram, a kite is flying at the end of a 200 meter straight string. If the string makes an angle of 68° with the level ground, how high, to the *nearest meter*, is the kite?

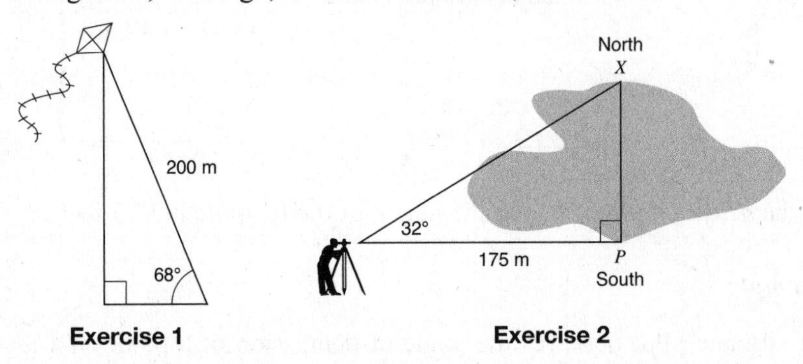

Exercise 1 **Exercise 2**

2. A surveyor needs to determine the distance across the pond shown in the accompanying diagram. She determines that the distance from her points to point P on the south shore of the pond is 175 meters, and the angle from her position to point X on the north shore is 32°. Determine the distance, PX, across the pond, rounded to the *nearest meter*.

3. At Slippery Ski Resort, the beginner's slope is inclined at an angle of 12.3°, while the advanced slope is inclined at an angle of 26.4°. If Rudy skis 1000 meters down the advanced slope while Valerie skis the same distance on the beginner's slope, how much greater was the horizontal distance, to the *nearest tenth* of a meter, that Valerie covered?

4. At noon, a tree having a height of 10 feet casts a shadow 15 feet in length. Find, to the *nearest degree*, the angle of elevation of the sun at this time.

5. As shown in the accompanying diagram, a ship at sea is sighted from the top of a 60-foot lighthouse. If the angle of depression of the ship from the top of the lighthouse measures 15°, find, to the *nearest foot*, how far the ship is from the foot of the lighthouse.

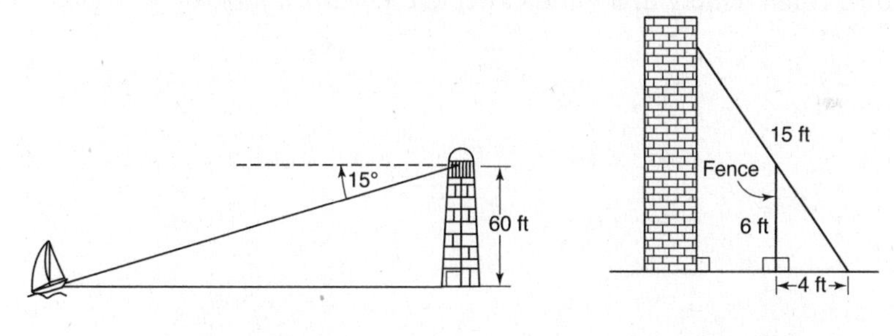

Exercise 5 **Exercise 6**

6. In the accompanying diagram, the base of a 15-foot ladder rests on the ground 4 feet from a 6-foot fence.

a. If the ladder touches the top of the fence and the side of a building, what angle, to the *nearest degree*, does the ladder make with the ground?
b. Using the angle found in part *a*, determine how far the top of the ladder reaches up the side of the building, to the *nearest tenth of a foot*.

7. Find, to the *nearest tenth of a foot*, the number of feet in the height of a building that casts a shadow of 80 feet when the angle of elevation of the sun is 42°.

8. A man observes the angle of depression from the top of a cliff overlooking the ocean to a ship to be 37°. If at this moment the ship is 1,000 meters from the foot of the cliff, find, to the *nearest meter*, the number of meters in the height of the cliff.

9. A ship on the ocean surface detects a sunken ship on the ocean floor at an angle of depression of 48°. The distance between the ship on the surface and the sunken ship on the ocean floor is 200 meters. If the ocean floor is level in this area, how far above the ocean floor, to the *nearest tenth of a meter*, is the ship on the surface?

10. A 10-foot ladder is to be placed against the side of a building. The base of the ladder must be placed at an angle of 72° with the level ground for a secure footing. Find, to the *nearest inch*, how far the base of the ladder should be from the side of the building.

11. Joe is holding his kite string 3 feet above the ground, as shown in the accompanying diagram. The distance between his hand and a point directly under the kite is 95 feet. If the angle of elevation to the kite is 50°, find the height, h, of his kite, to the *nearest foot*.

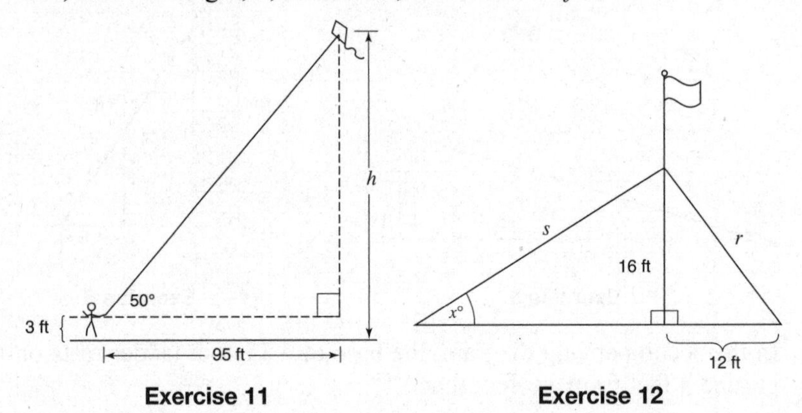

Exercise 11 **Exercise 12**

12. The accompanying diagram shows a flagpole that stands on level ground. Two cables, r and s, are attached to the pole at a point 16 feet above the ground. The combined length of the two cables is 50 feet. If cable r is attached to the ground 12 feet from the base of the pole, what is the measure of the angle, x, to the *nearest degree*, that cable s makes with the ground?

CHAPTER 9
AREA AND VOLUME

9.1 AREAS OF PARALLELOGRAMS AND TRIANGLES

KEY IDEAS

The areas of parallelograms and triangles can be calculated using the appropriate area formulas.

Area of a Parallelogram = Base × Height

A parallelogram can be transformed into a rectangle with the same base and height, as shown in Figure 9.1. Thus, the area, A, of a parallelogram is given by the same formula as the area of a rectangle, $A = bh$.

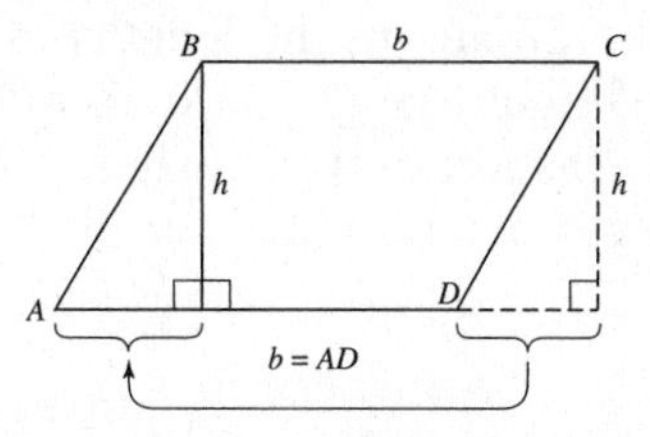

Figure 9.1 Area Parallelogram $ABCD = bh$

Example 1

In the accompanying diagram of parallelogram $ABCD$, $\overline{BH} \perp \overline{AD}$, $AB = 10$, $BC = 23$, and $DH = 17$. Find the number of square units in the area of parallelogram $ABCD$.

Solution:

- Since opposite sides of a parallelogram are equal in length, $AD = 23$ so $AH = 23 - 17 = 6$.
- The lengths of the sides of right triangle AHB form a 6-8-10 Pythagorean triple with altitude $BH = 8$.
- Use the area of a parallelogram formula:

$$\begin{aligned}\text{area of parallelogram } ABCD &= bh\\ &= AD \times BH\\ &= 23 \times 8\\ &= \mathbf{184} \text{ square units}\end{aligned}$$

Area of a Triangle = $\frac{1}{2}$ × Base × Height

Since a diagonal of a parallelogram divides the parallelogram into two triangles of exactly the same size and shape, the area of each triangle is one-half of the area of the parallelogram, as shown in Figure 9.2.

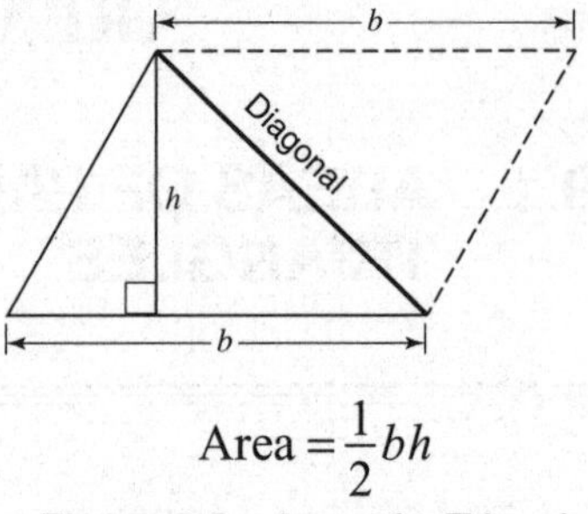

$$\text{Area} = \frac{1}{2}bh$$

Figure 9.2 Area of a Triangle

Example 2

Find the number of square units in the area of $\triangle ABC$ shown in the accompanying figure.

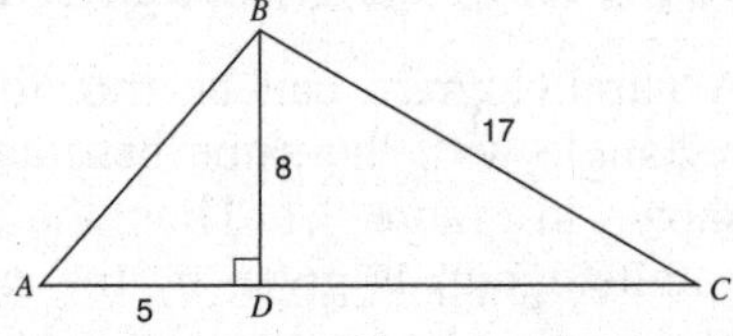

Solution: The lengths of the sides of right triangle *BDC* form a 8-*15*-17 Pythagorean triple, where $CD = 15$ so $AC = 5 + 15 = 20$.

$$\begin{aligned}\text{area of } \triangle ABC &= \frac{1}{2} \times BD \times AC \\ &= \frac{1}{2} \times 8 \times 20 \\ &= \mathbf{80} \text{ square units}\end{aligned}$$

Example 3

Find the number of square units in the area of $\triangle ABC$ shown in the accompanying figure.

Solution: The lengths of the sides of right triangle *BDC* form a *6*-8-10 Pythagorean triple with $BD = 6$.

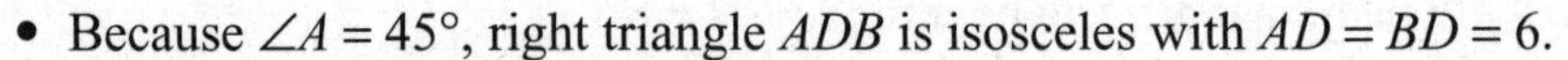

- Because $\angle A = 45°$, right triangle *ADB* is isosceles with $AD = BD = 6$.
- Thus, $AC = AD + CD = 6 + 8 = 14$.
- In $\triangle ABC$, height $BD = 6$ and base $AC = 14$ so

$$\begin{aligned}\text{area of } \triangle ABC &= \frac{1}{2} \times BD \times AC \\ &= \frac{1}{2} \times 6 \times 14 \\ &= \mathbf{42} \text{ square units}\end{aligned}$$

Area of a Right Triangle = $\frac{1}{2}$ × Leg1 × Leg2

In a right triangle, two legs serve as the base and height so that the area of a right triangle is equal to one-half the products of the lengths of its legs.

Example 4

What is the number of square inches in the area of a right triangle in which the hypotenuse measures 25 inches and the shorter leg measures 7 inches?

Solution: Use the Pythagorean theorem to find the length of the longer leg. Referring to the accompanying diagram:

$$\begin{aligned} a^2+b^2&=c^2\\ 7^2+b^2&=25^2\\ 49+b^2&=625\\ b^2&=625-49\\ b&=\sqrt{576}=24 \end{aligned}$$

Thus,

$$\begin{aligned} \text{area}&=\frac{1}{2}ab\\ &=\frac{1}{2}\times 7\times 24\\ &=\mathbf{84\ in.^2} \end{aligned}$$

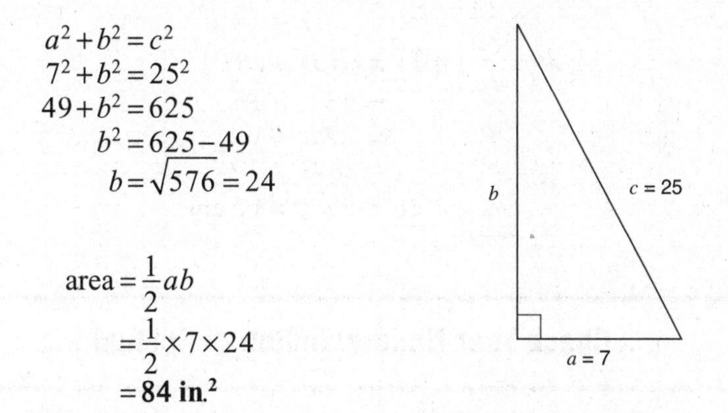

Example 5

The area of right $\triangle ABC$ is 60 cm^2. If the length of leg $\overline{AB}$ is 15 cm, find the length of hypotenuse $\overline{AC}$.

Solution: Use the area of a right triangle formula to find the length of leg $\overline{BC}$.

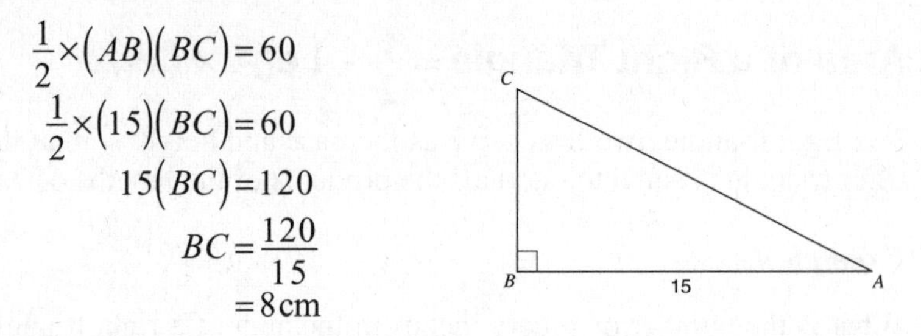

$$\frac{1}{2}\times(AB)(BC)=60$$
$$\frac{1}{2}\times(15)(BC)=60$$
$$15(BC)=120$$
$$BC=\frac{120}{15}$$
$$=8\,\text{cm}$$

The legs of the right triangle measure 8 cm and 15 cm so the lengths of the sides of this right triangle form a 8-15-17 Pythagorean triple, where hypotenuse AC = 17 cm. If you did not recognize this fact, use the Pythagorean theorem to find the length of $\overline{AC}$:

$$(AC)^2=(AB)^2+(BC)^2$$
$$=15^2\ +\ 8^2$$
$$=225\ +64$$
$$=289$$
$$AC=\sqrt{289}=\mathbf{17\ cm}$$

Check Your Understanding of Section 9.1

A. Multiple Choice.

1. What is the number of square inches in the area of a square whose diagonal measures 8 inches?
(1) 16 in.^{2} (2) 24 in.^{2} (3) 32 in.^{2} (4) 64 in.^{2}

2. In right triangle ABC, $\angle C$ is a right angle and AC = 10 inches. If the area of $\triangle ABC$ is 150 square inches, what is the number of inches in the length of $\overline{AB}$.
(1) 15 (2) $10\sqrt{10}$ (3) $10\sqrt{15}$ (4) $20\sqrt{2}$

3. In the accompanying figure of quadrilateral $ABCD$, $\overline{DA}\perp\overline{AB}$, $\overline{DB}\perp\overline{BC}$, $AD = 4$, $AB = 3$, and $CD = 13$. What is the number of square units in the area of quadrilateral $ABCD$?
(1) 28 (3) 36
(2) 32 (4) 42

B. *Show how you arrived at your answer.*

4–5. *Find the number of square units in the area of $\triangle ABC$.*

4. **5.**

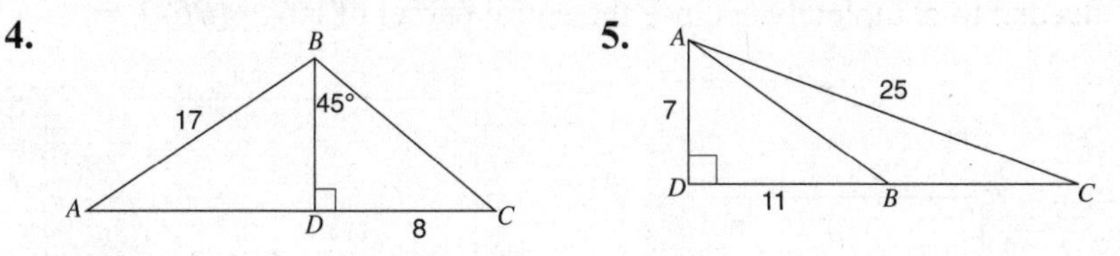

6. Find the area of a right triangle whose hypotenuse measures 14 and whose longest leg measures $7\sqrt{3}$.

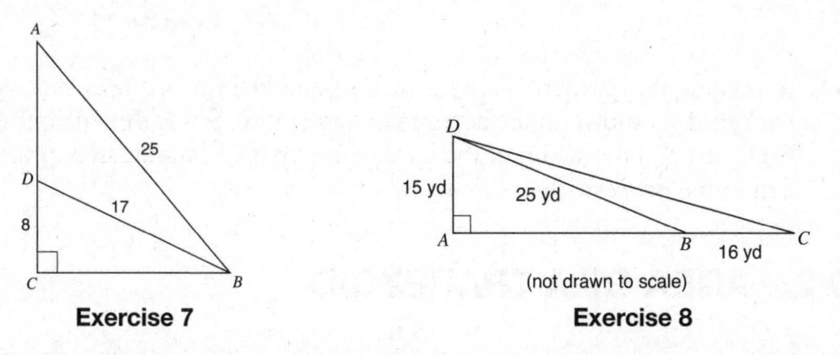

Exercise 7 **Exercise 8**

7. In the accompanying figure of right triangle ABC, $\angle C$ is a right angle, $AB = 25$, $DB = 17$, and $CD = 8$. What is the number of square units in the area of $\triangle ADB$?

8. In the accompanying diagram of right triangle ACD, $\angle A$ is a right angle, $AD = 15$ yards, $BD = 25$ yards, and $BC = 16$ yards. Find the number of square yards in the area of $\triangle ACD$.

9. Mr. Carlin owns a plot of land with boundaries that form right triangle ACD with right angle C, $AC = 9$ yards, and $AD = 15$ yards. He purchases an adjacent plot of land that extends $\overline{CD}$ to B with $AB = 41$ yards. If the additional plot of land costs \$217 per square yard, how many dollars did Mr. Carlin pay for this land?

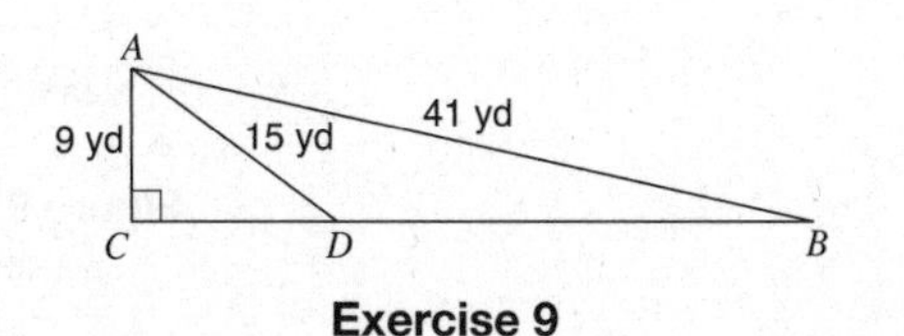

Exercise 9

10. A parcel of land is represented by quadrilateral *ABCD* in the accompanying diagram where *DEBC* is a rectangle. If the area of right triangle *AEB* is 600 square feet, find the minimum number of feet of fence needed to completely enclose the entire parcel of land, *ABCD*.

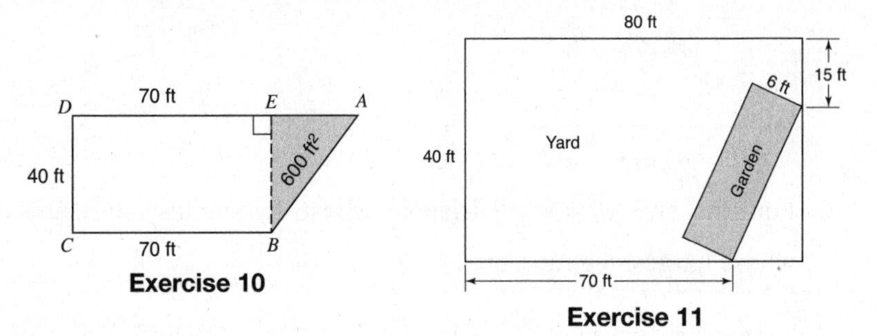

11. A rectangular garden is going to be planted in a person's rectangular backyard, as shown in the accompanying figure. Some dimensions of the backyard and the width of the garden are given. Find the area of the garden to the *nearest square foot.*

9.2 AREA OF A TRAPEZOID

KEY IDEAS

A **trapezoid** is a quadrilateral in which exactly one pair of sides, called the **bases**, are parallel. The nonparallel sides are the **legs**. The area of a trapezoid depends on its height and the lengths of its two bases.

The height of a trapezoid is any segment drawn from one base perpendicular to the other base. To find the area of a trapezoid, multiply the height by one-half the sum of the lengths of the two bases, as shown in Figure 9.3.

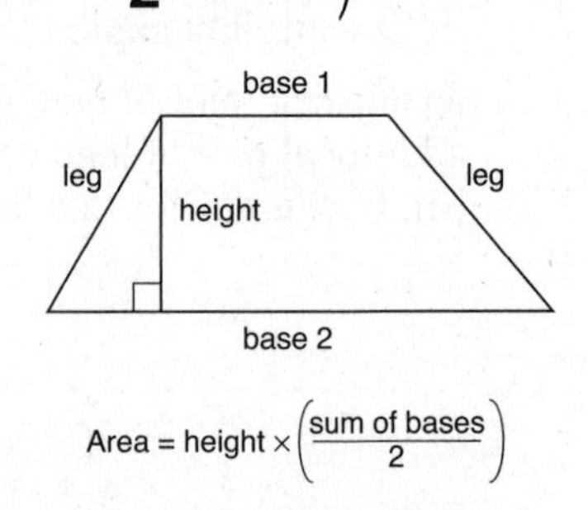

$$\text{Area} = \text{height} \times \left(\frac{\text{sum of bases}}{2}\right)$$

Figure 9.3 Formula for Area of a Trapezoid

Example 1

The length of the longer base of a trapezoid exceeds the length of the shorter base by 8 cm. If the height of the trapezoid is 14 cm and the area is 126 cm^2, find the lengths of the bases.

Solution: Represent the lengths of the two bases by x and $x + 8$ as shown in the accompanying figure. Then apply the formula for the area of a trapezoid:

$$\text{area} = \text{height} \times \left(\frac{\text{base 1} + \text{base 2}}{2}\right)$$

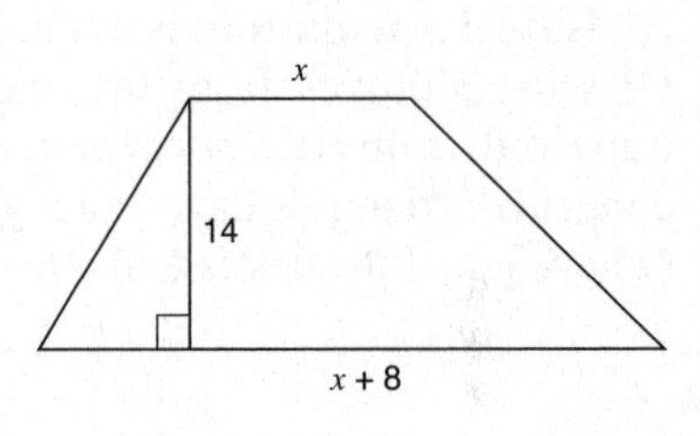

$$\begin{aligned}
126 &= 14\frac{(x+8+x)}{2}\\
126 &= 7(2x+8)\\
126 &= 14x+56\\
-14x &= 56-126\\
\frac{-14x}{-14} &= \frac{-70}{-14}\\
x &= 5 \quad \text{so } x+8=13
\end{aligned}$$

The length of the bases are **5 cm** and **13 cm**.

Example 2

In the accompanying figure of trapezoid $ABCD$, $\overline{AB} \| \overline{CD}$, $\overline{BA} \perp \overline{AD}$, $\overline{CD} \perp \overline{AD}$, $AB =$ 10, $BC = 17$, and $AD = 15$. Find the number of square units in the area of trapezoid $ABCD$.

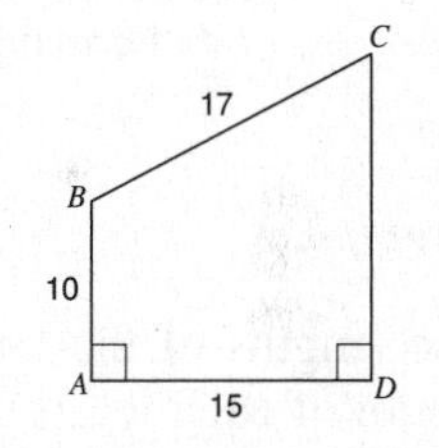

Solution: First, find the length of base $\overline{CD}$ by drawing altitude $\overline{BH}$ to $\overline{CD}$, thereby forming rectangle $ABHD$, as shown in the accompanying figure.

- Since opposite sides of a rectangle are equal in length, $BH = AD = 15$ and $DH = AB = 10$.
- The lengths of the sides of right triangle BHC form an 8-15-17 Pythagorean triple, where $CH = 8$. Thus,

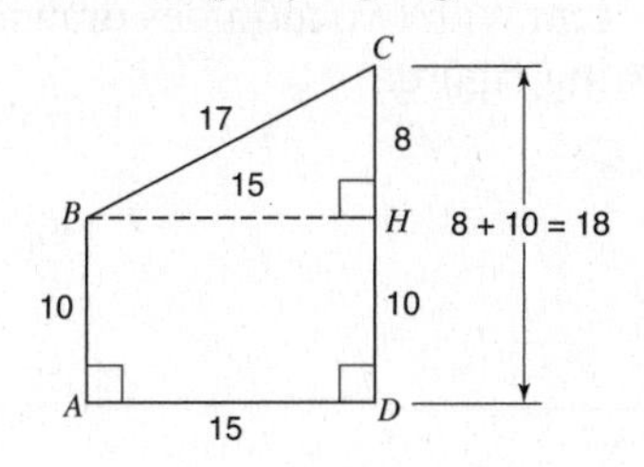

$$\begin{aligned}
CD &= CH + DH\\
&= 8+10\\
&= 18
\end{aligned}$$

- Use the area formula for a trapezoid:

$$\begin{aligned}\text{area of trapezoid } ABCD &= BH \times \left(\frac{AB+CD}{2}\right)\\ &= 15\times\left(\frac{10+18}{2}\right)\\ &= 15(14)\\ &= \mathbf{210} \text{ square units}\end{aligned}$$

Isosceles Trapezoids

An **isosceles trapezoid** is a trapezoid in which the legs have the same length. Drawing altitudes from the endpoints of the shorter base of an isosceles trapezoid, forms two *congruent* right triangles, as shown Figure 9.4. Because congruent triangles have exactly the same measurements, base angles A and D have equal measures and $AE = DF$.

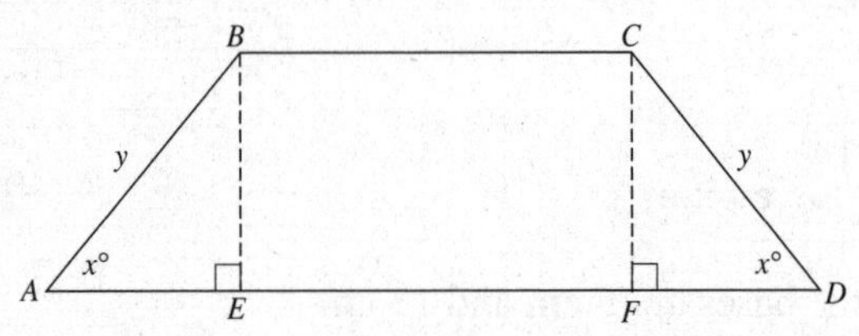

Figure 9.4 Isosceles Trapezoid *ABCD* with Altitudes $\overline{BE}$ and $\overline{CF}$ Forming Right Triangles with the Same Dimensions

Example 3

The lengths of the bases of an isosceles trapezoid are 21 and 45, and the length of each leg is 20. Find the number of square units in the area of the trapezoid.

Solution: Draw an isosceles trapezoid that fits the conditions of the problem with two altitudes drawn to the longer base, as shown in the accompanying figure.

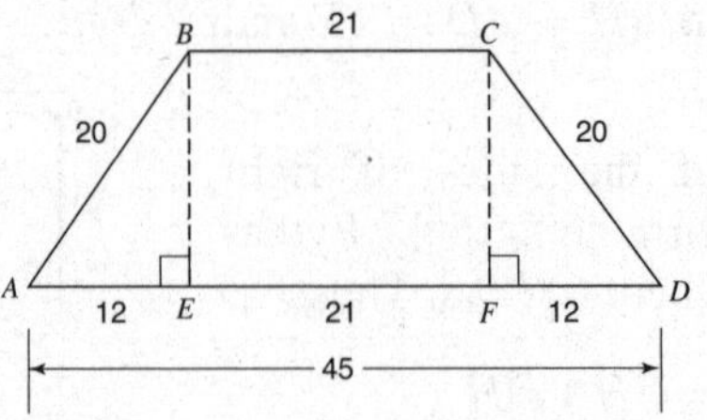

- $EBCF$ is a rectangle so $EF = BC = 21$. Hence, $AE + DF = 45 - 21 = 24$.
- Because the trapezoid is isosceles, $AE = DF = \frac{1}{2} \times 24 = 12$.
- In right triangle AEB, $AB = 20 = 4 \times \underline{5}$ and $AE = 12 = 4 \times \underline{3}$ so $BE = 4 \times \underline{4} = 16$. If you did not recognize that the lengths of the sides of right triangle AEB form a 12-16-20 Pythagorean triple with $BE = 16$, use the Pythagorean theorem to find BE.
- Use the area formula for a trapezoid:

$$\begin{aligned}\text{area of trapezoid } ABCD &= BE \times \left(\frac{BC + AD}{2}\right)\\ &= 16 \times \left(\frac{21 + 45}{2}\right)\\ &= 15(33)\\ &= \mathbf{495} \text{ square units}\end{aligned}$$

Check Your Understanding of Section 9.2

A. Multiple Choice.

1. In the accompanying diagram of trapezoid $ABCD$, $AB = 26$, $BC = 10$, $CD = 18$, $\overline{CB} \perp \overline{AB}$, and altitude $\overline{DE}$ is drawn. What is the number of square units in the area of trapezoid $ABCD$?
(1) 440 (2) 228 (3) 220 (4) 180

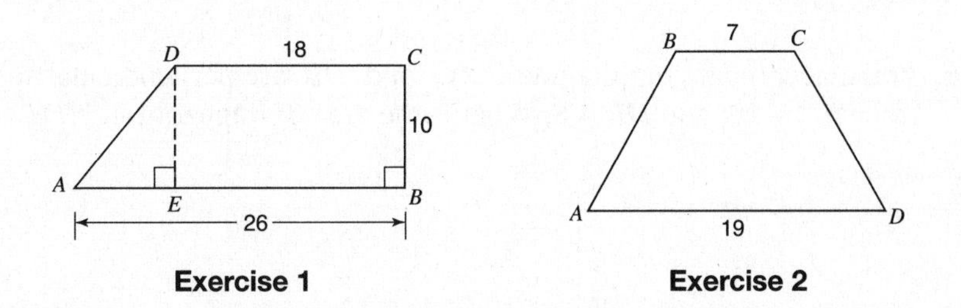

Exercise 1 Exercise 2

2. In the accompanying diagram of isosceles trapezoid $ABCD$, $BC = 7$ centimeters and $AD = 19$ centimeters. If the perimeter of $ABCD$ is 46 centimeters, what is the number of square centimeters in the area of the trapezoid?
(1) 525 cm^2 (2) 104 cm^2 (3) 112 cm^2 (4) 124 cm^2

3. In the accompanying diagram of quadrilateral $ABCD$, $\overline{CD} \perp \overline{BC}$, $\overline{AB} \perp \overline{BC}$, $m\angle A = 45$, $BC = 6$, and $CD = 2$. What is the number of square units in the area of quadrilateral $ABCD$?

(1) 24 (3) 32
(2) 30 (4) 36

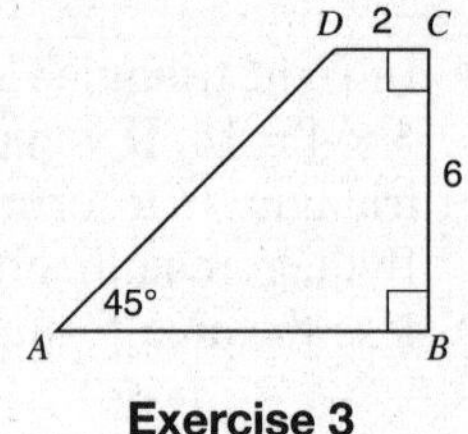

Exercise 3

B. *Show how you arrived at your answer.*

4. In the accompanying diagram of rectangle $ABCD$, E is a point on $\overline{DC}$, $EC = 5$, $BE = 13$, and $AB = 20$. What is the number of square units in the area of trapezoid $ABED$?

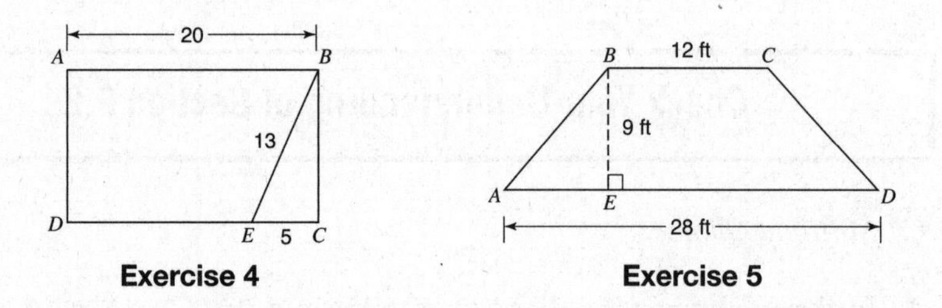

Exercise 4 **Exercise 5**

5. The cross section of an attic is in the shape of an isosceles trapezoid, as shown in the accompanying figure. If the height of the attic is 9 feet, $BC = 12$ feet, and $AD = 28$ feet, find the length of $\overline{AB}$ to the *nearest foot.*

6. In the accompanying diagram, $\overline{AC}$ and $\overline{DE}$ are perpendicular to $\overline{BC}$, $AC = BC = 10$, and $DE = 8$. What is the area of trapezoid $ACED$?

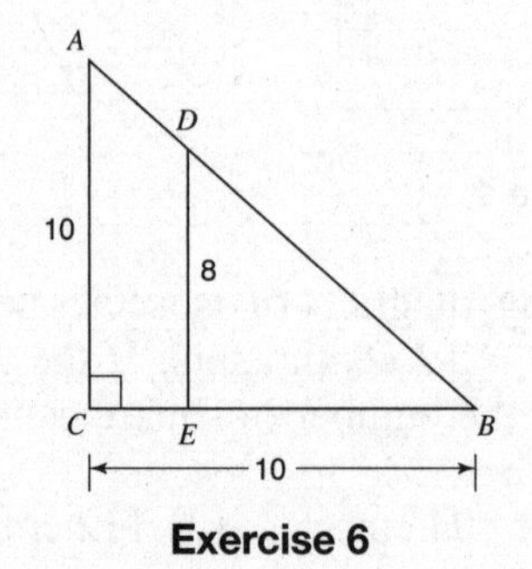

Exercise 6

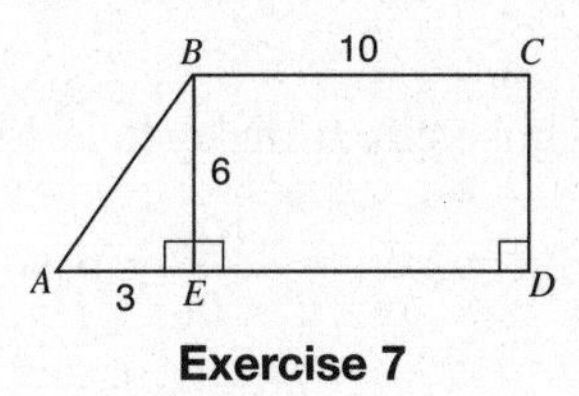

Exercise 7

7. In the accompanying diagram of trapezoid $ABCD$, $\overline{BE}$ is an altitude, $AE = 3$, $BE = 6$, and $BC = 10$.

a. Find AB to the *nearest tenth*.
b. The area of $\triangle AEB$ is what percent of the area of trapezoid $ABCD$? Express your answer correct to the *nearest whole percent*.

9.3 CIRCUMFERENCE AND AREA OF A CIRCLE

KEY IDEAS

The **circumference** of a circle is the distance around it. If d is the diameter of the circle and r is its radius, then the circumference, C, and area, A, of a circle are given by the formulas:

$$C = \pi \times d = 2\pi r \quad \text{and} \quad A = \pi r^2$$

Applying Circle Formulas

When performing a calculation involving the constant π, use the calculator's stored value of π and round off the answer in the last step of the calculation.

Example 1

A wheel has a radius of 5 feet. What is the minimum number of *complete* revolutions that the wheel must make to roll at least 1000 feet?

Solution: In each revolution the wheel turns a distance equal to its circumference which is $2\pi \times 5 = 10\pi$. To roll at least 1000 feet, the wheel must make

$$\frac{1000}{10\pi} = 1000 \div (10 \times \pi) \approx 31.83098862 \text{ revolutions}$$

Hence, the minimum number of *complete* revolutions the wheel must make is **32**.

Example 2

Find the area of a circle whose circumference is 12π.

Solution: Since $C = 12\pi = 2\pi r$, $r = \frac{12\pi}{2\pi} = 6$ so $A = \pi r^2 = \pi(6^2) = \mathbf{36\pi}$

Example 3

A point P on a bicycle wheel is touching the ground. The next time point P hits the ground, the wheel has rolled 3.5 feet in a straight line along the ground. What is the radius of the wheel correct to the *nearest tenth of an inch*.

Solution: Since the distance the wheel rolls is equal to the circumference of the bicycle wheel, 3.5 ft $= 2\pi r$, so

$$r = \frac{3.5\,\text{ft}}{2\pi}$$

Change from feet to inches: $= \frac{3.5 \times 12\,\text{in}}{2\pi}$

$$= 42 \div (2 \times \pi)$$

$$r \approx \mathbf{6.7\ in.}$$

Example 4

Find in terms of π the diameter and the area of a circle whose circumference is $\frac{6}{\pi}$.

Solution: Let C represent the circumference of a circle whose diameter is d.

- $C = \pi d = \frac{6}{\pi}$ so $d = \frac{6}{\pi \times \pi} = \frac{6}{\pi^2}$.
- Because the diameter of the circle is $\frac{6}{\pi^2}$, its radius, r, is $\frac{1}{2} \times \frac{6}{\pi^2} = \frac{3}{\pi^2}$.
- Find the area, A, of the circle whose radius is $\frac{3}{\pi^2}$:

$$A = \pi\left(\frac{3}{\pi^2}\right)^2 = \pi\left(\frac{9}{\pi^4}\right) = \frac{9}{\pi^3}$$

The diameter of the circle is $\boldsymbol{\frac{6}{\pi^2}}$, and its area is $\boldsymbol{\frac{9}{\pi^3}}$.

Semicircles and Quarter Circles

To find the circumference (or area) of a fractional part of a circle, first find the circumference (or area) of the entire circle. Then multiply the result by that fraction.

Example 5

A window is made up of a single piece of glass in the shape of a semicircle and a rectangle, as shown in the diagram to the right. Theresa is decorating for a party and wants to put a string of lights all the way around the outside edge of the window.

To the *nearest foot*, what is the length of the string of lights that Theresa will need to decorate the window?

Solution: The length of the string of lights needed is equal to the circumference of the semicircle plus the lengths of the three sides of the rectangle indicated by solid lines.

- The circumference of the semicircle is one-half the circumference of the circle. The broken line represents the diameter of the circle and has a length of 10 ft. The circumference of the semicircle is, therefore,

$$\frac{1}{2} \times (\pi \times 10) = 5\pi \text{ ft}$$

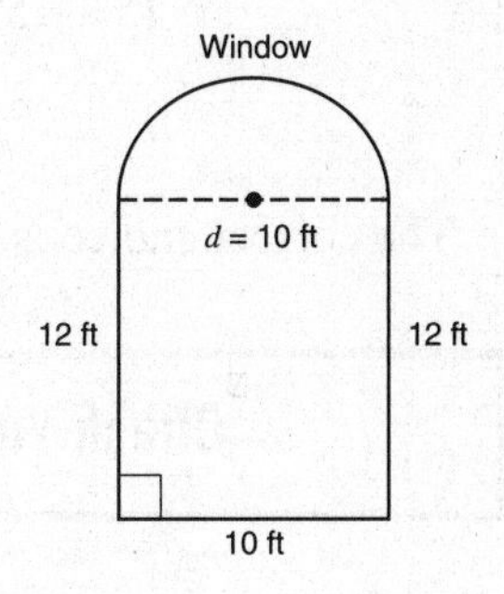

- The length of the string of lights needed is the length of 3 sides of the rectangle plus the circumference of the semicircle:

$$\begin{aligned} 5\pi + 12 + 10 + 12 &= 34 + 5\pi \\ &\approx 34 + 15.70796327 \\ &\approx 49.70796327 \text{ ft} \end{aligned}$$

The length of string Theresa needs, to the *nearest foot*, is **50 feet**.

Example 6

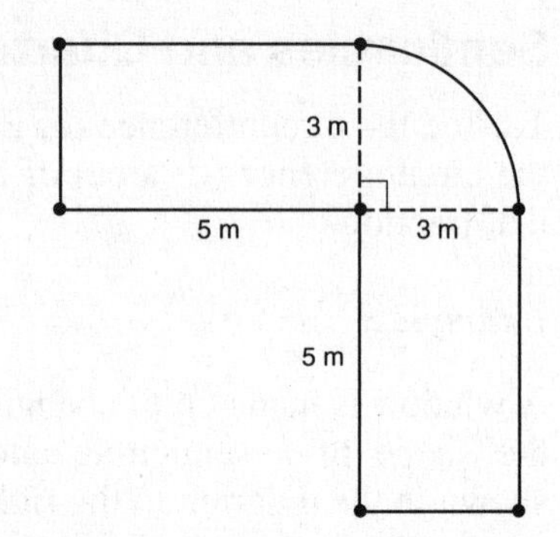

The paved entryway to a house has the shape of two rectangles and a quarter-circle, as shown in the accompanying figure. What is the area of the walkway, to the nearest square meter ?

(1) 33 (3) 44
(2) 37 (4) 58

Solution: The area of the walkway is equal to the sum of the areas of the two rectangles and the area of the quarter circle.

- The area of each rectangle is $5 \times 3 = 15$ m^2.
- The area of the quarter circle is one-fourth the area of the entire circle:
$$\frac{1}{4} \times (\pi \times 3^2) = \frac{9}{4}\pi \text{ m}^2$$
- The number of square meters in the area of the walkway is:
$$\begin{aligned} 15 + 15 + \frac{9}{4}\pi &= 30 + 2.25\pi \\ &\approx 37.06858347 \\ &\approx 37 \text{ m}^2 \end{aligned}$$

The correct choice is **(2)**.

Check Your Understanding of Section 9.3

A. *Multiple Choice.*

1. What is the circumference of a circle whose area is 16π square units?
(1) 4π (2) 8π (3) 16π (4) 64π

2. What is the area of a circle whose circumference is 10π centimeters?
(1) 5π cm^2 (2) 20π cm^2 (3) 25π cm^2 (4) 100π cm^2

3. What is the radius of a circle whose circumference is 7?
(1) $\frac{3.5}{\pi}$ (2) $\frac{\sqrt{3.5}}{\pi}$ (3) $\frac{7}{\pi}$ (4) $\frac{\sqrt{7}}{\pi}$

4. A circular garden has a diameter of 12 feet. How many bags of topsoil must Linda buy to cover the garden if one bag covers an area of 3 square feet?
(1) 13 (2) 38 (3) 40 (4) 151

5. If the number of square units in the area of a circle increases from π to 4π, by what percent does the radius length increase?
(1) 100 (2) 200 (3) 300 (4) 400

6. A figure is made up of a rectangle and a semicircle as shown in the diagram below.

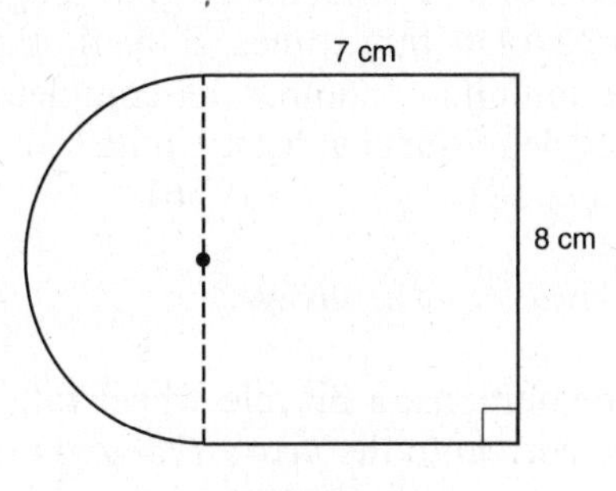

Exercise 6

What is the area of the figure, to the *nearest tenth of a square centimeter*?
(1) 81.1 (2) 106.3 (3) 156.5 (4) 257.1

7. A playground consists of a rectangle and two semicircles, as shown in the diagram below.

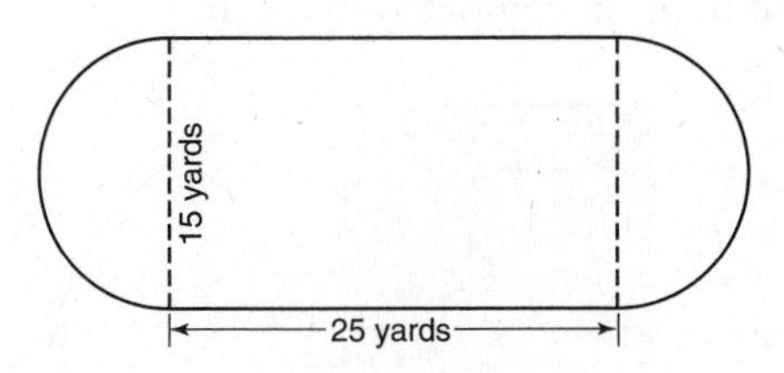

Exercise 7

Which expression represents the amount of fencing, in yards, that would be needed to enclose the playground completely?
(1) $15\pi + 50$ (2) $15\pi + 80$ (3) $30\pi + 50$ (4) $30\pi + 80$

8. In the accompanying figure of square $ABCD$, circle O touches each of the sides of $ABCD$ at one point. If the area of square $ABCD$ is 64 square inches, what is the number of square inches in the area of circle O?

(1) 4π (3) 16π
(2) 8π (4) 18π

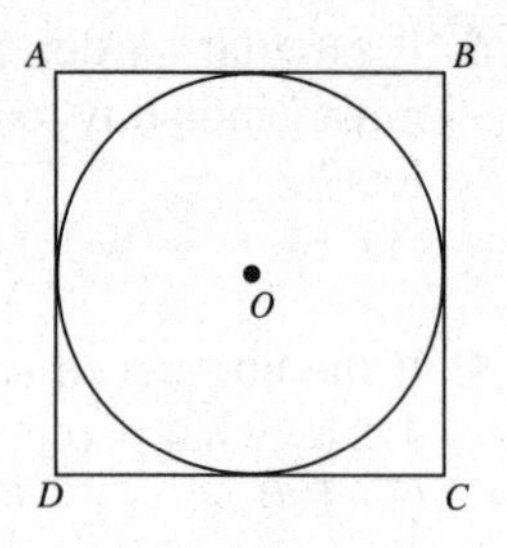

Exercise 8

9. Each time the pedals of a certain bicycle go through one complete circular rotation, the tires rotate three times. If the tires are 24 inches in diameter, what is the minimum number of complete rotations of pedals needed for the bicycle to travel at least 1 mile?

(1) 841 (2) 281 (3) 561 (4) 700.1

B. *Show how you arrived at your answer.*

10. In six complete revolutions, a bicycle wheel rolls 27 feet. What is the number of inches, correct to the *nearest tenth*, in the radius of the bicycle wheel?

11. The ratio of the areas of two circles is 9 to 16. If the length of the radius of the larger circle is 20 centimeters, what is the length of the radius of the smaller circle?

12. A designer created a logo composed of two equal squares and a quarter-circle as shown in the accompanying figure.

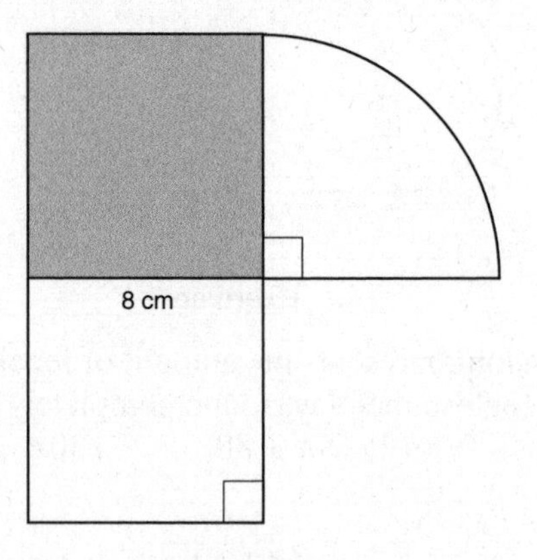

Exercise 12

What is the number of square centimeters in the area of the logo, to the *nearest tenth*?

13. To measure the number of miles in a hiking trail, a worker uses a device with a 2-foot diameter wheel that counts the number of revolutions the wheel makes. If the device reads 0 revolution at the beginning of the trail and 2300 revolutions at the end of the trail, how many miles long, correct to the *nearest tenth of a mile*, is the trail?

9.4 AREAS OF OVERLAPPING FIGURES

KEY IDEAS

When a geometric figure is embedded in another figure, the area of the region between the two figures is the difference between the areas of the original two figures. Circles with the same center and different radii are called **concentric circles**.

Finding Areas Indirectly

To find the area of the region between two embedded figures, subtract their areas.

Example 1

Find in terms of π the number of square units in the area of the shaded region between the two concentric circles with radii of 5 and 8, as shown in the accompanying figure.

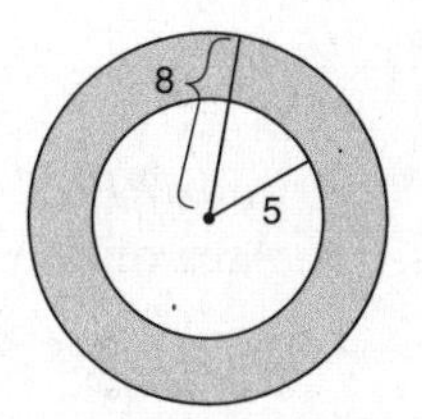

Solution: The shaded region represents the area that the two concentric circles do *not* have in common. To find this area, *subtract* the area of the smaller circle from the area of the larger circle:

$$\begin{array}{rl} \text{area of larger circle} & = \pi \times 8^2 = 64\pi \\ -\ \text{area of smaller circle} & = \pi \times 5^2 = 25\pi \\ \hline \text{area of shaded region} & = \mathbf{39\pi} \end{array}$$

Example 2

In the accompanying diagram, *RECT* is a rectangle, $AH = 4.2$, $RT = 5$, and $TC = 12$.

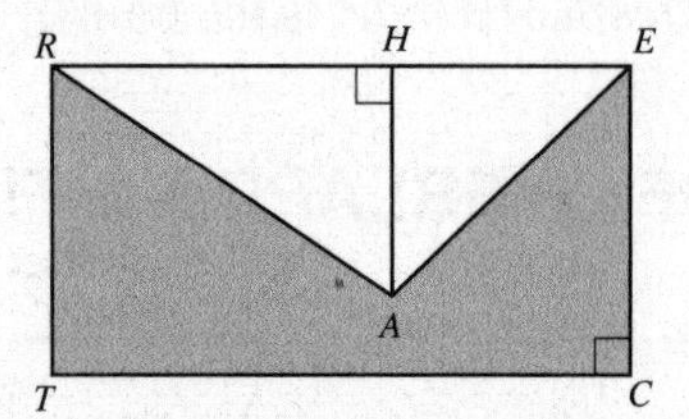

What is the number of square units in the area of polygon *RAECT*?

(1) 9.6 (2) 25.2 (3) 34.8 (4) 60

Solution: To find the area of polygon *RAECT*, subtract the area of triangle *EAR* from the area of rectangle *RECT*.

- Since $AH = 4.2$ and $ER = TC = 12$,

$$\text{Area of } \triangle EAR = \frac{1}{2} \times ER \times AH$$
$$= \frac{1}{2} \times 12 \times 4.2$$
$$= 25.2$$

- Area $RECT = TC \times RT = 12 \times 5 = 60$.
- Area of polygon $RAECT = 60 - 25.2 = 34.8$.

The correct choice is **(3)**.

Example 3

Mr. Petri has a rectangular plot of land with length equal to 20 feet and width equal to 10 feet. He wants to design a flower garden in the shape of a circle with two semicircles at each end of the center circle, as shown in the accompanying diagram. He will fill in the shaded region with wood chips. If one bag of wood chips covers 5 square feet, how many bags must he buy?

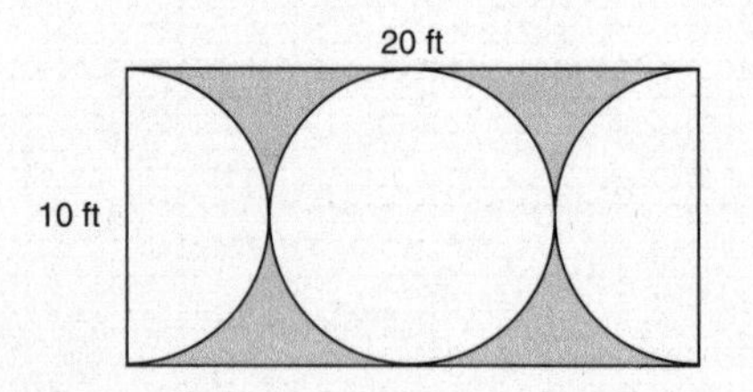

Solution: Calculate the area of the shaded region by subtracting the sum of the areas of the circle and two semicircles from the area of the rectangular plot.

- The area of the rectangular plot of land is $20 \times 10 = 200$ square feet.
- The width of the rectangle is the diameter of the embedded circles. Hence, the radius of each of the two semicircles and the circle is $\frac{10}{2} = 5$ feet. The area of the center circle is $\pi \times (\text{radius})^2 = \pi \times (5)^2 = 25\pi$ square feet. The two semicircles are equivalent to one circle. Hence, the total area enclosed by the two semicircles and the one circle in the center is $25\pi + 25\pi = 50\pi$ square feet.
- Find the area of the shaded region:

$$\begin{aligned}\text{area of shaded region} &= \text{area of rectangle} - \text{areas of circles}\\ &= 200 - 50\pi\\ &= 42.92036732\end{aligned}$$

Since $42.92036732 \div 5 = 8.5841 \approx 9$, Mr. Petri must by **9 bags** of wood chips.

Probability and Area

Finding the probability of an event may involve comparing the areas of two regions. If a point is selected at random from within rectangular region B in Figure 9.5, then the probability that the point lies in circular region A is the ratio

$$\frac{\text{area of region } A}{\text{area of region } B}$$

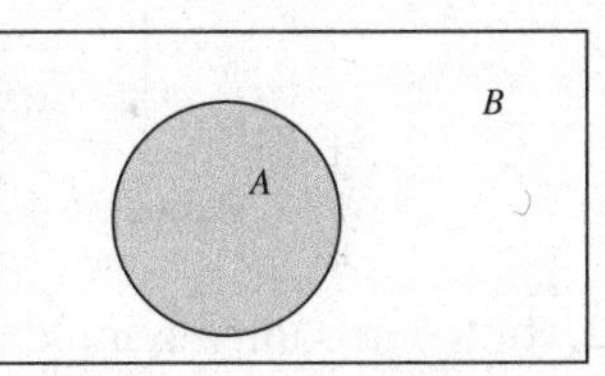

Figure 9.5 Comparing Areas

Example 4

The accompanying diagram shows a square dartboard. Each side of the dartboard measures 30 inches. The square shaded region at the center has a side that measures 10 inches. If darts thrown at the board are equally likely to land anywhere on the board, what is the theoretical probability that a dart does *not* land in the shaded region?

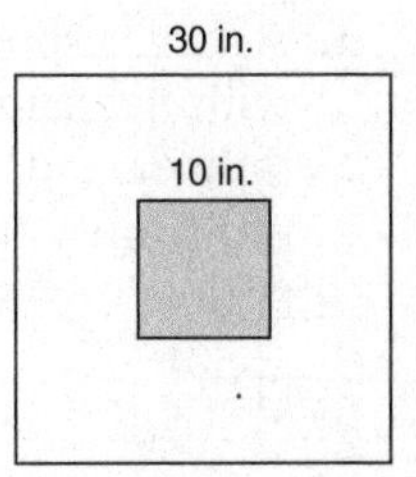

Solution: The probability, P, that the dart does *not* land inside the shaded region is the ratio of the area of the region between the two squares to the area of the large square.

$$
\begin{aligned}
P &= \frac{\text{area between the two squares}}{\text{area of large square}} \\
&= \frac{30^2 - 10^2}{30^2} \\
&= \frac{900 - 100}{900} \\
&= \mathbf{\frac{8}{9}}
\end{aligned}
$$

Check Your Understanding of Section 9.4

A. Multiple Choice.

1. In the accompanying diagram, a circle with radius 4 is inscribed in a square. What is the area of the shaded region?

(1) $64 - 16\pi$ (2) $16\pi - 16$ (3) $64\pi - 8\pi$ (4) $8\pi - 16$

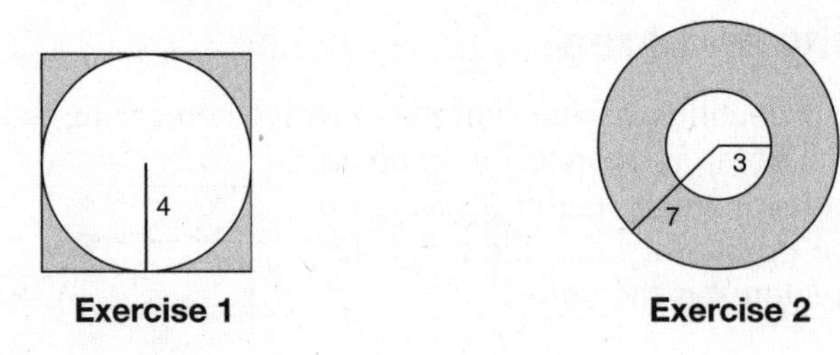

Exercise 1 Exercise 2

2. In the accompanying diagram, the radii of the two concentric circles are 3 and 7, respectively. Approximately what percent of the area of the larger circle is shaded?

(1) 82% (2) 57% (3) 43% (4) 18%

3. A target shown in the accompanying diagram consists of three circles with the same center. The radii of the circles are 3, 7, and 9. A dart is thrown and lands on the target. What is the probability that the dart will land on the shaded region?

(1) $\frac{4}{9}$ (3) $\frac{49}{81}$

(2) $\frac{40}{81}$ (4) $\frac{40}{49}$

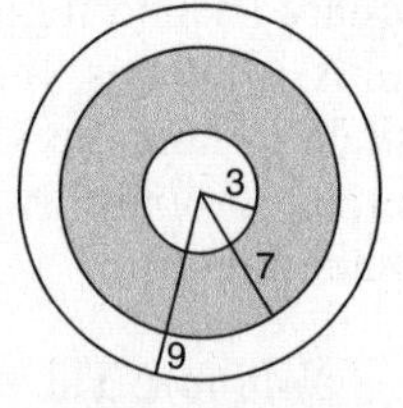

Exercise 3

B. *Show how you arrived at your answer.*

4-5. *Find the area of the shaded region.*

4. Quadrilateral *ABCD* is a square.

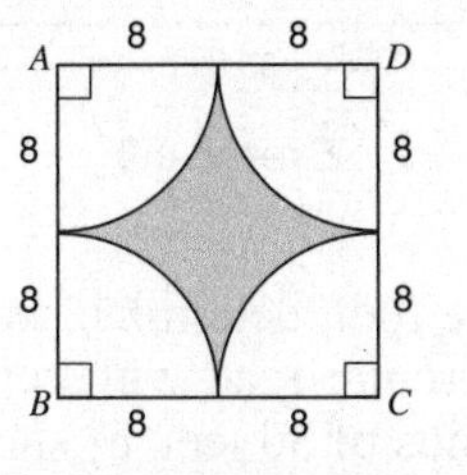

The four quarter-circles have centers at the vertices of the square.

Exercise 4

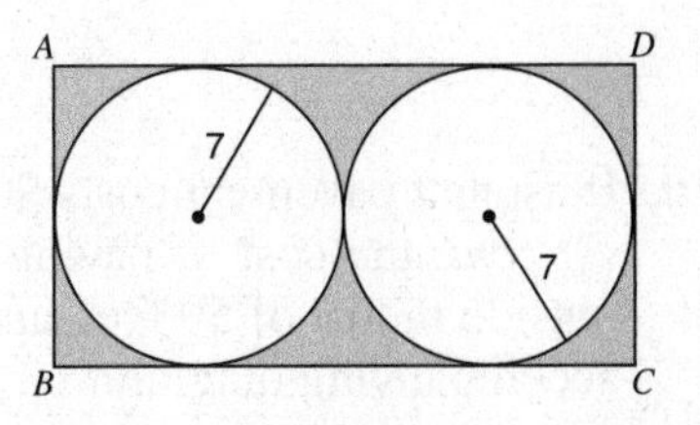

Exercise 5

5. Two circles are tangent to the sides of rectangle *ABCD* and to each other.

6. The target shown in the accompanying diagram consists of three concentric circles with radii of 2 in., 5 in., and 8 in.

a. What is the area of the shaded region to the *nearest tenth of a square inch*?

b. To the *nearest percent*, what percent of the target is not shaded?

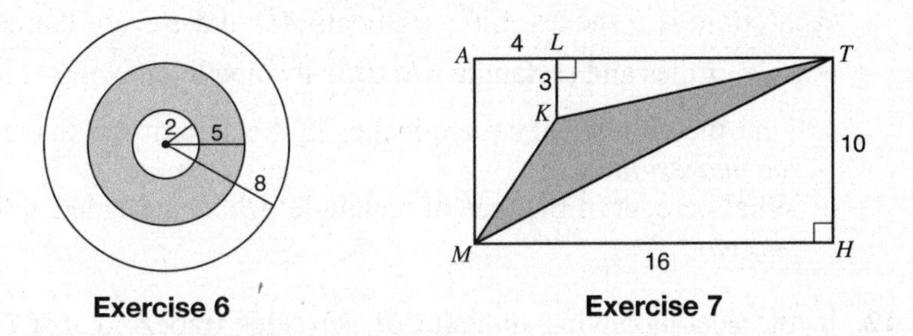

Exercise 6 **Exercise 7**

7. In the accompanying diagram, *MATH* is a rectangle, *KLT* is a right triangle, $AL = 4$, $LK = 3$, $TH = 10$, and $MH = 16$. What is the number of square units in the area of $\triangle MKT$?

8. In the accompanying diagram, *ABCD* is a square with a semicircle having *AD* as its diameter. If $AB = 10$ cm, what is the number of square centimeters in the area of the shaded region to the nearest tenth?

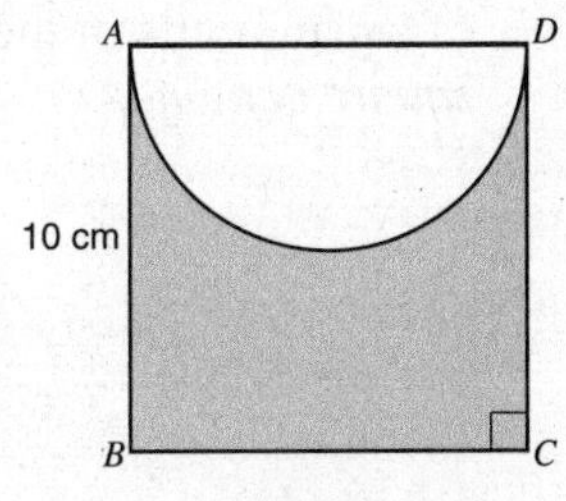

Exercise 8

9. In the accompanying figure, $ABCE$ is a rectangle, $BC = 29$ feet, $CD = 10$ feet, and $AD = 35$ feet. Find the number of square feet in the area of trapezoid $ABCD$.

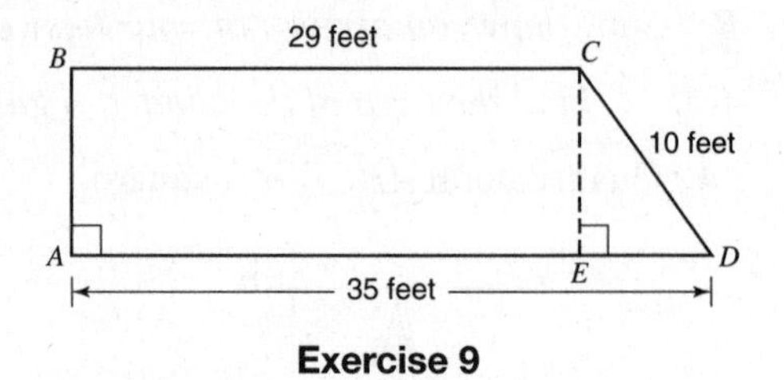

Exercise 9

10. If asphalt pavement costs \$0.78 per square foot, determine, to the *nearest cent*, the cost of paving the shaded circular road with center O, an outside radius of 50 feet, and an inner radius of 36 feet, as shown in the accompanying diagram.

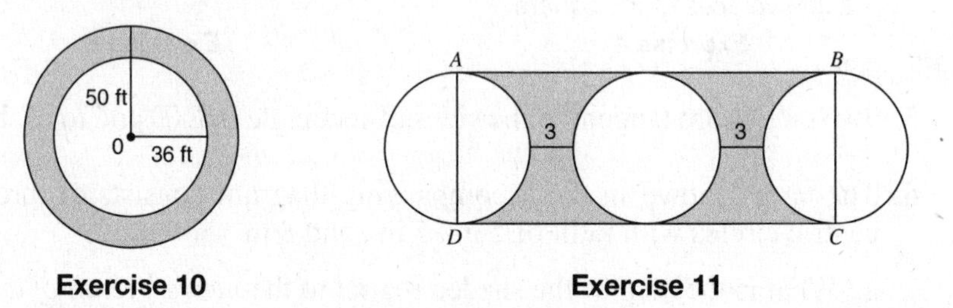

Exercise 10 **Exercise 11**

11. The accompanying diagram shows three equal circles aligned in a row such that the distance between adjacent circles is 3 inches. The radius of each circle is 5 inches. Line segments AD and BC are diameters of the outside circles and rectangle $ABCD$ is formed by drawing $\overline{AB}$ and $\overline{CD}$.

a. Find the number of square inches in the area of the shaded region to the *nearest integer.*

b. What percent of the area of rectangle $ABCD$ is shaded, to the *nearest percent*?

12. In the accompanying diagram of isosceles trapezoid $ABCD$, each leg measures 5 centimeters, the length of $\overline{DC}$ is 6 centimeters longer than the length of $\overline{AB}$, and the perimeter is 36 centimeters. Circle O is inscribed in the trapezoid. Radius $OE = 2$ centimeters. Find the number of square units in the area of the shaded region to the *nearest tenth of a square centimeter*.

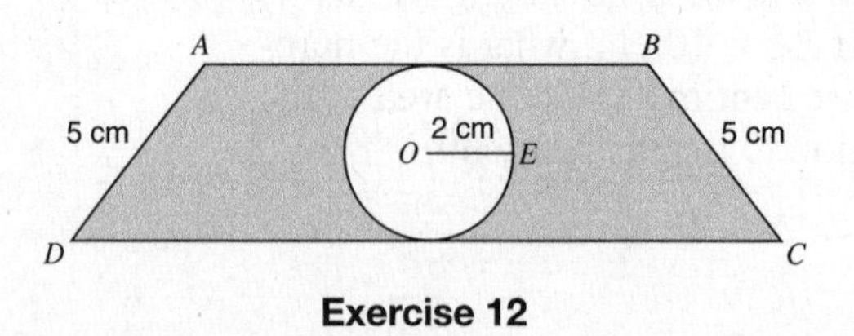

Exercise 12

9.5 SURFACE AREA AND VOLUME

KEY IDEAS

The surface areas of some solids can be computed by adding the areas of the plane (two-dimensional) surfaces that comprise the figure. The volume of a cylinder is the product of the area of its circular base and height.

Surface Area of a Rectangular Box

You already know that the volume of a rectangular box (prism) is the product of its three dimensions. The surface area of a rectangular box is the sum of the areas of its six rectangular sides. In Figure 9.6, the area of the bottom side is $\ell \times w$; the area of the front side is $\ell \times h$; and the area of the right side is $w \times h$.

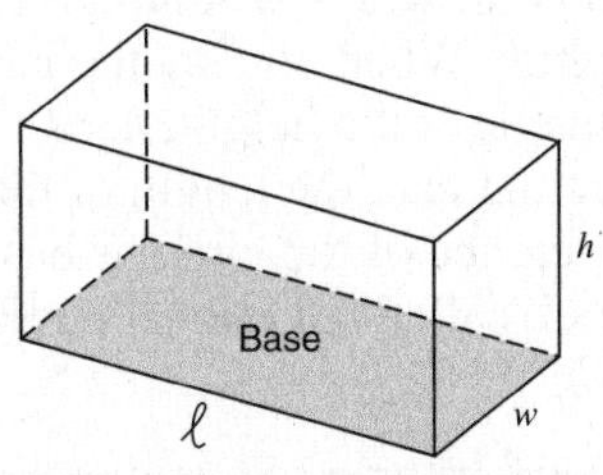

Figure 9.6 Rectangular Box

Because the areas of opposite sides are the same, the areas of the bottom, front, and right sides can be doubled and then added together to obtain the surface area of the rectangular solid:

$$\text{surface area of a rectangular box} = 2(\ell \times w) + 2(\ell \times h) + 2(w \times h)$$

If the length, width, and height of a closed rectangular box are 10 inches, 7 inches, and 4 inches, respectively, then its surface area is

$$\begin{aligned} 2(10\times7)+2(10\times4)+2(7\times4) &= 140\,\text{in.}^2+80\,\text{in.}^2+56\,\text{in.}^2 \\ &= 276\text{ in.}^2 \end{aligned}$$

Surface Area of a Cube

Because the area of each of the six sides of a cube whose edge length is represented by e is e^2, the surface area of a cube is given by the formula

$$\text{surface area of a cube} = 6e^2$$

Volume of a Cylinder

A cylinder has two equal circular bases, as shown in Figure 9.7. If r represents the radius of the base of a cylinder and h its height, then the volume of the cylinder is $\pi r^2 h$.

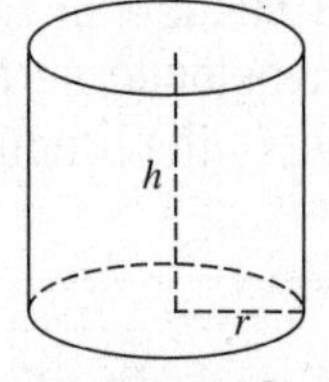

$V_{\text{cylinder}} = \pi r^2 h$

Figure 9.7 Volume of a Cylinder

Lateral Surface Area of a Cylinder

Imagine peeling off a label that wraps around the full height of a cylindrical can with radius r and height h. When you lay the label flat it will be a rectangle. The width of the rectangle is the height, h, of the can. Because the label wraps once around the circular can, the length of the other side of the rectangle is equal to the circumference of the circular base which is $2\pi r$. Thus, the area of the rectangular label, called the **lateral surface area of the cylinder**, is the product of $2\pi r$ and h:

$$\text{lateral surface area of a cylinder} = 2\pi rh$$

Example 1

The circumference of the base of a cylinder is 10π, and the height of the cylinder is 6. Find the volume and lateral surface area of the cylinder in terms of π?

Solution: Let C represent the circumference of the base, r its radius, and h the height of the cylinder.

- Since $C = 10\pi = 2\pi r$, $r = 5$.
- Use the volume formula with $r = 5$ and $h = 6$:

$$\begin{aligned}\text{volume} &= \pi r^2 h \\ &= \pi \times 5^2 \times 6 \\ &= \mathbf{150\pi}\end{aligned}$$

- Apply the formula for lateral surface area:

$$\begin{aligned}\text{lateral surface area} &= 2\pi rh \\ &= 2\pi \times 5 \times 6 \\ &= \mathbf{60\pi}\end{aligned}$$

Surface Area of a Cylinder

The surface area of a cylinder with radius r and height h is the sum of the areas of its two circular bases, each of which has an area of πr^2, and its lateral surface area:

$$\text{surface area of a cylinder} = 2\pi r^2 + 2\pi rh$$

Example 2

The metal used to manufacture a cylindrical can costs \$0.43 per square foot. What is the cost, to the *nearest cent*, of the metal used to manufacture a cylindrical can whose radius is 6 inches and height is 2 feet ?

Solution: The radius of the cylindrical can is 6 inches, which is equivalent to 0.5 feet.

- Find the number of square feet in the surface area of the cylinder:

$$\begin{aligned}\text{surface area} &= 2\pi r^2 + 2\pi rh \\ &= 2\pi(0.5)^2 + 2\pi(0.5)(2) \\ &= 2\pi(0.25) + 2\pi \\ &= 2.5\pi\end{aligned}$$

- Use your calculator to multiply the number of square feet in the surface area of the can by \$0.43:

$$\text{cost} = 2.5 \times \pi \times 0.43 = \mathbf{\$3.38}$$

The formulas for surface area and volume that you need to know are summarized in the accompanying table.

SURFACE AREA AND VOLUME FORMULAS

Figure	Dimensions	Surface Area	Volume
Rectangular solid	ℓ by w by h	$2(\ell \times w) + 2(\ell \times h) + 2(w \times h)$	$\ell \times w \times h$
Cube	edge length e	$6e^2$	e^3
Cylinder	radius r and height h	$2\pi r^2 + \underbrace{2\pi rh}_{\text{lateral surface area}}$	$\pi r^2 h$

Relative Error in Measurement

When using instruments such as a ruler or tape measure, the linear measurements that are taken may have some errors. The **relative error** in a measurement is a number that expresses the magnitude of the error compared with the true value of that measurement:

$$\text{relative error} = \frac{|\text{actual value} - \text{measured value}|}{\text{actual value}}$$

Example 3

Raymond incorrectly measures an edge of a wooden cube as 5.0 cm when the true edge length is 5.1 cm. If Raymond uses his measurement to calculate the volume of the cube, what is the relative error in his volume calculation to the *nearest hundredth*?

(1) 0.05 (2) 0.06 (3) 0.07 (4) 0.08

Solution: To determine the relative error in calculating the volume of a figure when there is an error in the measurement of its edge length, calculate the volume using both actual and measured edge lengths. Then divide the absolute value of their difference by the actual volume.

- Calculate the actual volume using the true edge length:

$$\text{actual volume} = (5.1)^3 = 132.651 \text{ cm}^3$$

- Calculate the measured volume using Raymond's measurement:

$$\text{measured volume} = (5.0)^3 = 125.0 \text{ cm}^3$$

- Calculate the relative error:

$$\text{relative error} = \frac{|\text{actual volume} - \text{measured volume}|}{\text{actual volume}}$$

$$= \frac{|132.651 - 125.0|}{132.651}$$

$$= \frac{7.651}{132.651}$$

$$\approx 0.0576776655$$

The relative error in his volume calculation to the *nearest hundredth* is 0.06.

The correct choice is **(2)**.

Example 4

Carlos measured a rectangular piece of paper to be 21.7 cm by 28.5 cm. The piece of paper is actually 21.6 cm by 28.4 cm.

a. Determine the relative error in calculating the area to the *nearest thousandth*.

b. Is the amount of error significant? Explain.

Solution:

a. Calculate the area using both actual and measured dimensions. Then divide the absolute value of their difference by the actual area.

- Find the actual area using the true measurements:

$$\begin{aligned} \text{actual area} &= 21.6 \text{ cm} \times 28.4 \text{ cm} \\ &= 613.44 \text{ cm}^2 \end{aligned}$$

- Calculate the area using Carlos's measurements:

$$\begin{aligned} \text{Measured area} &= 21.7 \text{ cm} \times 28.5 \text{ cm} \\ &= 618.45 \text{ cm}^2 \end{aligned}$$

- Calculate the relative error:

$$\begin{aligned} \text{relative error} &= \frac{|\text{actual area} - \text{measured area}|}{\text{actual area}} \\ &= \frac{|613.44 - 618.45|}{613.44} \\ &= \frac{5.01}{613.44} \\ &= 0.0081670579 \\ &\approx 0.008 \end{aligned}$$

To the *nearest thousandth*, the relative error is **0.008**.

b. No, a relative error of 0.008 is *not* significant since it represents an error of less than 1% in calculating the area of a notebook-sized piece of paper. This error should be acceptable for all practical purposes in this situation. On the other hand, an error of this magnitude might not be acceptable in critical engineering applications such as the design of computer circuits.

Greatest Possible Error in Measurement

The precision of a measurement typically depends on the measuring instrument used. Suppose a ruler is marked in increments of $\frac{1}{8}$ inch.

- The precision of any measurement made with this ruler must be estimated to the nearest $\frac{1}{8}$ inch.
- The **greatest possible error** in a measurement is one-half the smallest unit of measurement, which in this case is $\frac{1}{2} \times \frac{1}{8} = \frac{1}{16}$ inch or 0.0625 inch.
- To find the relative error in measurement when the true dimensions are not known, calculate the area (or volume) using the measurements taken and recalculate the area (or volume) with each measurement increased by the greatest possible error. Write their difference over the measured area (or volume) as illustrated in Example 5.

Example 5

The dimensions of a rectangular box are measured with a ruler that is precise to 0.1 cm. A rectangular box is measured as 20.0 cm by 12.0 cm by 5.0 cm. Find the relative error in calculating the volume of the box correct to the *nearest tenth of a percent.*

Solution: Since the precision of the ruler is 0.1 cm, the greatest possible error in each linear measurement is $\frac{1}{2} \times 0.1 = 0.05$ cm. Hence, add 0.05 cm to each of the linear measurements used to calculate the volume having the greatest error.

- Measured volume:

$$20.0 \times 12.0 \times 5.0 = 1{,}200 \text{ cm}^3$$

- Volume with greatest error:

$$20.05 \times 12.05 \times 5.05 \approx 1{,}220.093 \text{ cm}^3$$

- Calculate the relative error:

$$\text{relative error} = \frac{\begin{pmatrix}\text{volume with}\\ \text{greatest error}\end{pmatrix} - \begin{pmatrix}\text{volume}\\ \text{measured}\end{pmatrix}}{\text{volume measured}}$$

$$= \frac{1{,}220.093 - 1{,}200}{1{,}200}$$

$$= \frac{20.093}{1{,}200}$$

$$\approx 0.0167 \text{ or } 1.67\%$$

The relative error in calculating the volume of the box, to the *nearest tenth of a percent*, is **1.7%**.

Example 6

The length and width of a rectangular garden are measured, to the *nearest foot*, as 12 feet by 18 feet. What is the least number of square feet in the area of the garden?

Solution:

- The smallest width that could round to 12 feet to the nearest foot is 11.5 feet.
- The smallest length that could round to 18 feet to the nearest foot is 17.5 feet.
- The least possible area of the garden is 11.5 × 17.5 = **201.25** square feet.

Check Your Understanding of Section 9.5

A. *Multiple Choice.*

1. If the diameter of a cylinder is tripled, then the volume of the cylinder is multiplied by
(1) 9 (2) 6 (3) 3 (4) 8

2. If the circumference of the circular base of a cylinder is doubled and the height is also doubled, then the volume of the cylinder is multiplied by
(1) 9 (2) 6 (3) 3 (4) 8

3. Which expression represents the surface area of a cylinder open at the top with radius r and height h?
(1) $2\pi r(r+h)$ (2) $\pi r(2r+h)$ (3) $\pi r(r+2h)$ (4) $\pi rh(r+2)$

4. What is the volume of a cube whose surface area is 294 square inches?
(1) 216 in.^{3} (2) 252 in.^{3} (3) 343 in.^{3} (4) 512 in.^{3}

5. A storage container in the shape of a right circular cylinder is shown in the accompanying diagram. What is the number of cubic inches in the volume of this container, to the *nearest hundredth*?
(1) 56.55 (3) 251.33
(2) 125.66 (4) 502.65

8 in.
10 in.

6. Vicki weighs exactly 115 pounds but weighs 111 pounds on a defective scale. What is the relative error in measurement of the defective scale to the *nearest tenth* of a percent?
(1) 3.4% (2) 3.5% (3) 3.6% (4) 4%

7. Ariel estimates the measurement of the volume of a container to be 282 cubic inches. The manufacturer of the container specifies the volume to be 289 cubic inches. What is the relative error of Ariel's measurement to the *nearest thousandth*?
(1) 0.024 (2) 0.025 (3) 0.096 (4) 1.025

8. The dimensions of a cube are measured with a ruler that is precise to 0.2 cm. Using this ruler, the edge length is measured as exactly 10 cm. What is the greatest relative error in calculating the volume of the cube correct to the *nearest tenth of a percent*?
(1) 2.8% (2) 2.9% (3) 3.0% (4) 3.1%

9. The lengths of the sides of a rectangular box are in error by as much as 10% of their measured value. If the measured values of the dimensions of this box are 10 cm by 15 cm by 20 cm, what is the greatest possible volume of the box?
(1) 2,187 cm^3 (2) 3,000 cm^3 (3) 3,300 cm^3 (4) 3,993 cm^3

B. *Show how you arrived at your answer.*

10. A cylindrical pitcher with a radius of 4 inches and a height of 20 inches is filled to the top with water. What is the greatest number of cylindrical cups with a radius of 1.5 inches and a height of 4 inches that can be filled to the top with water poured from the pitcher?

11. A cylindrical container is manufactured so that its radius is one-half of its height. The container must have a volume of at least 250 cubic inches, and the height must be a whole number. What is the least number of inches the height can be?

12. The radius and height of a cylinder are equal in length. If the surface area of the cylinder is 169π square inches, what is the volume of the cylinder in terms of π?

13. A rectangular box that is closed at the top is constructed so that its length, width, and height are in the ratio of $5 : 4 : 2$, respectively. If the surface area of the box is 931 square centimeters, what is the number of cubic centimeters in the volume of the box?

14. Sarah needs to know the area of her rectangular bedroom window when she purchases a new shade. She measured the window to be 36 inches by 42 inches. The actual measurements of the window are 36.5 inches by 42.5 inches. Determine, to the *nearest thousandth*, the relative error in calculating the area of the window using Sarah's measurements.

15. The sides of a rectangular prism are measured to be 6.0 ft by 5.1 ft by 5.1 ft. Find the relative error, to the *nearest thousandth*, of the surface area of the prism if the true measurements are 6.0 ft by 5.0 ft by 5.0 ft.

16. Spencer used a ruler to measure the sides of a rectangular prism. He found the sides to be 5 cm by 8 cm by 4 cm. The true measurements are 5.3 cm by 8.2 cm by 4.1 cm. Find Spencer's relative error in calculating the volume of the prism, to the *nearest thousandth*.

17. An oil company distributes oil in a metal can shaped like a cylinder that has an actual radius of 5.1 cm and an actual height of 15.1 cm. A worker incorrectly measured the radius as 5 cm and the height as 15 cm. Determine the relative error in calculating the surface area, to the *nearest thousandth*.

18. A rectangle is measured using a ruler that is accurate to 0.1 cm. The rectangle measures 18.2 cm by 11.4 cm. What is the relative error, to the *nearest tenth of a percent,* in calculating the area of the rectangle?

19. A rectangular box is measured with a ruler which is precise to $\frac{1}{8}$ inch. The dimensions of the box were recorded as 10 in., 8 in., and $2\frac{3}{8}$ in. Find the relative error, correct to the *nearest tenth of a percent*, in calculating the volume of the box.

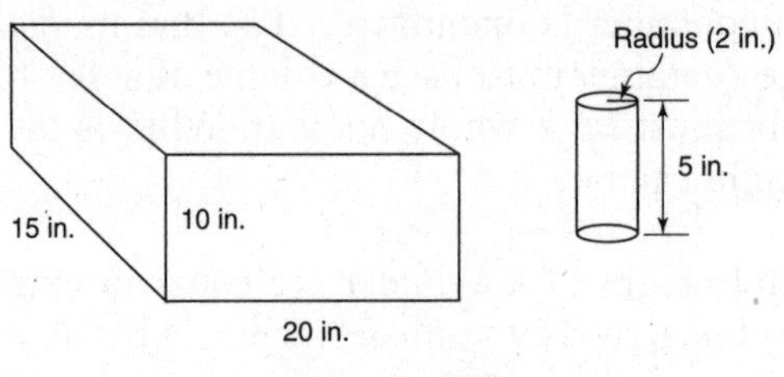

Exercise 20

20. In the accompanying diagram, a rectangular container with the dimensions 10 inches by 15 inches by 20 inches is to be filled with water, using a cylindrical cup whose radius is 2 inches and whose height is 5 inches. What is the maximum number of full cups of water that can be placed into the container without the water overflowing the container?

21. A rectangular piece of cardboard is to be formed into an uncovered box. The piece of cardboard is 2 centimeters longer than it is wide. A square that measures 3 centimeters on a side is cut from each corner. When the sides are turned up to form the box, its volume is 765 cubic centimeters.

a. Find the dimensions, in centimeters, of the original piece of cardboard.
b. Find the number of square centimeters in the surface area of the box.

CHAPTER 10
GRAPHING AND WRITING EQUATIONS OF LINES

10.1 SLOPE OF A LINE

KEY IDEAS

If you think of a line as a hill, then the **slope** of the line is a number that measures its steepness. If two lines have positive slopes, then the steeper line has the greater slope. It is customary to represent the slope of a line by the letter *m*.

Finding the Slope of a Line from Its Graph

When moving along a line from one point to another, the change in vertical distance is the *rise* and the change in horizontal distance is the *run*. The slope of a line is the ratio of the rise to the run for any two points on the line. In Figure 10.1, the rise in moving along the line from (1,2) to (4,6) is 4 and the run is 3 so the slope of the line is $\frac{4}{3}$.

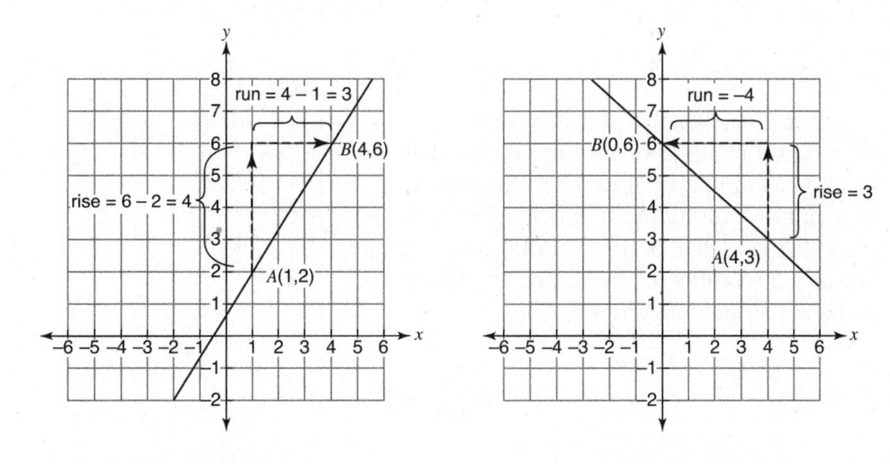

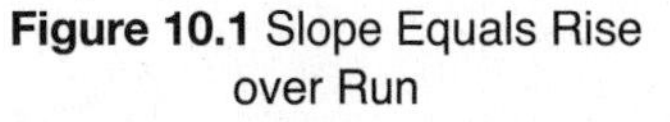
Figure 10.1 Slope Equals Rise over Run

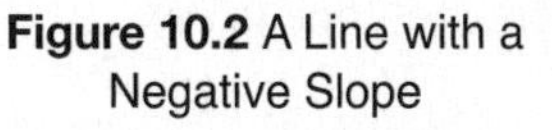
Figure 10.2 A Line with a Negative Slope

Negative Slope

As you move along the line in Figure 10.2 from (4,3) to (0,6), the rise is 3 and the run (from right to left) is –4 so the slope of the line is $\frac{3}{-4}$ or, equivalently, $-\frac{3}{4}$.

MATH FACTS

In the accompanying figure, m represents the slope of the line.

- A *positive* slope indicates that as x increases, the line *rises*.
- A *negative* slope means that as x increases, the line *falls*.

Example 1

What is the slope of line ℓ shown in the accompanying diagram?

(1) $-\frac{5}{3}$ (3) $\frac{5}{3}$

(2) $-\frac{3}{5}$ (4) $\frac{3}{5}$

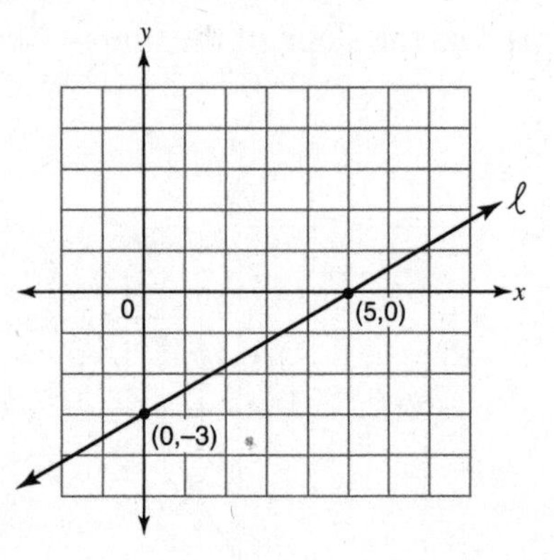

Solution: Because the line rises as x increases, the slope of the line is positive so choices (1) and (2) can be eliminated. As you move along the line from (0,–3) to (5,0), the rise is + 3 (up 3 units) and the run is + 5 (5 units to the right), as shown in the accompanying figure. Thus,

$$m = \frac{\text{rise}}{\text{run}} = \frac{3}{5}$$

The correct choice is **(4)**.

Slope Formula

The slope, m, of a line is the average rate of change between any two points $A(x_A,y_A)$ and $B(x_B,y_B)$ on the line. Thus,

$$m = \frac{y_B - y_A}{x_B - x_A}$$

As shown in Figure 10.3, Δy (read "delta y") and Δx (read "delta x") represent the difference or change in the corresponding coordinates of two points. The symbol Δ is the Greek letter delta. Using delta notation, $\frac{\Delta y}{\Delta x}$ represents the slope of a line.

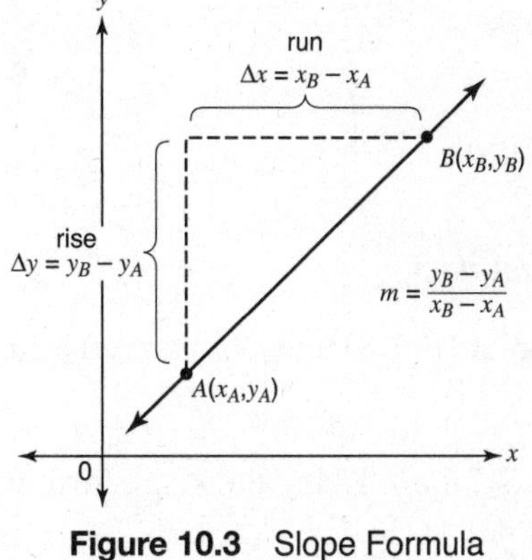

Figure 10.3 Slope Formula

MATH FACTS

The slope, m, of the line through the points $A(x_A,y_A)$ and $B(x_B,y_B)$ can be calculated without graphing the line by using the formula:

$$m = \frac{\Delta y}{\Delta x} = \frac{y_B - y_A}{x_B - x_A}$$

- The order in which the y-coordinates are subtracted in the slope formula must be the same as the order in which the corresponding x-coordinates are subtracted. It would *not* be correct to do the subtractions in the order indicated by writing $y_B - y_A$ over $x_A - x_B$.
- The slopes calculated using different pairs of points on the same line must be equal.

Example 2

Determine the slope of the line through the points $A(-2,3)$ and $B(1,7)$.

Solution: Use the slope formula where $A(x_A,y_A) = (-2,3)$ and $B(x_B,y_B) = (1,7)$:

$$m=\frac{y_B-y_A}{x_B-x_A}=\frac{7-3}{1-(-2)}=\frac{4}{3}$$

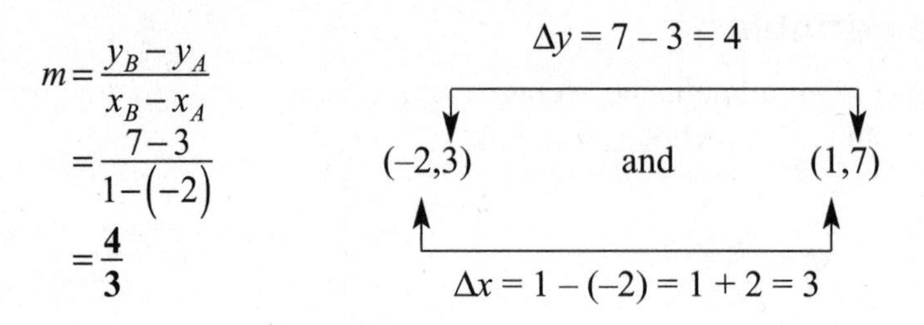

Example 3

If points (–1,8) and (9,k) are on a line whose slope is $-\frac{3}{4}$, what is the value of k?

Solution: Using the slope formula, represent the slope of the line in terms of k:

$$m=\frac{y_B-y_A}{x_B-x_A}=\frac{k-8}{10}$$

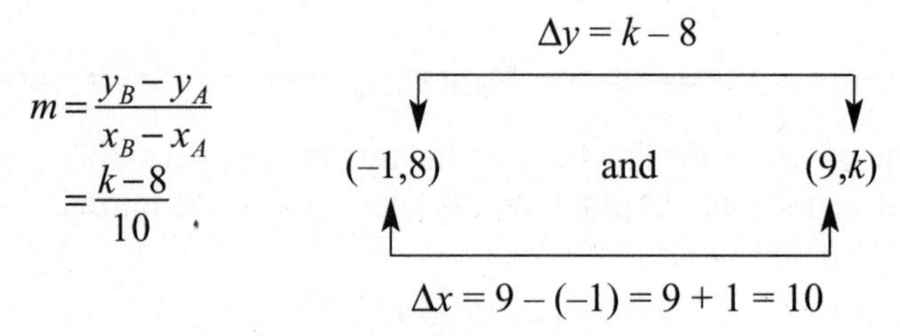

Set the slope equal to $-\frac{3}{4}$ and solve for k:

$$\begin{aligned}\frac{k-8}{10}&=\frac{-3}{4}\\4(k-8)&=-3(10)\\4k-32&=-30\\4k&=-30+32\\\frac{4k}{4}&=\frac{2}{4}\\k&=\frac{1}{2}\end{aligned}$$

Example 4

Determine if $C(-69,53)$ is on the same line as $A(3,5)$ and $B(1,8)$.

Solution: The slope of a line must be constant for any two points on that line. Hence, point C is on the same line as points A and B if the slope of $\overleftrightarrow{AB}$ is the same as the slope of $\overleftrightarrow{AC}$ $\left(or\ \overleftrightarrow{BC}\right)$.

- If $A(x_A, y_A) = (3,5)$ and $B(x_B, y_B) = (1,8)$, then the slope of $\overleftrightarrow{AB}$ is

$$m = \frac{y_B - y_A}{x_B - x_A} = \frac{8-5}{1-3} = -\frac{3}{2}$$

- To find the slope of $\overleftrightarrow{AC}$, use the slope formula with $A(x_A, y_A) = (3,5)$ and $C(x_C, y_C) = (-69,53)$:

$$m = \frac{y_C - y_A}{x_C - x_A} = \frac{53-5}{-69-3} = -\frac{48}{72} = -\frac{2}{3}$$

- Compare slopes. The slope of line $\overleftrightarrow{AB}$ is $-\frac{3}{2}$, while the slope of $\overleftrightarrow{AC}$ is $-\frac{2}{3}$. Because the two slopes of the two lines are not equal, $C(-69,53)$ does ***not*** lie on $\overleftrightarrow{AB}$.

Slopes of Special Lines

The slope of a slanted line is some nonzero real number.

- The slope of a *horizontal* line is 0, as shown in Figure 10.4. All pairs of points on a horizontal have the same *y*-coordinate so $m = \frac{\Delta y}{\Delta x} = \frac{0}{\Delta x} = 0$.
- The slope of a *vertical* line is undefined, as shown in Figure 10.5. All pairs of points on a vertical line have the same *x*-coordinate so $m = \frac{\Delta y}{\Delta x} = \frac{\Delta y}{0}$, which is not defined.

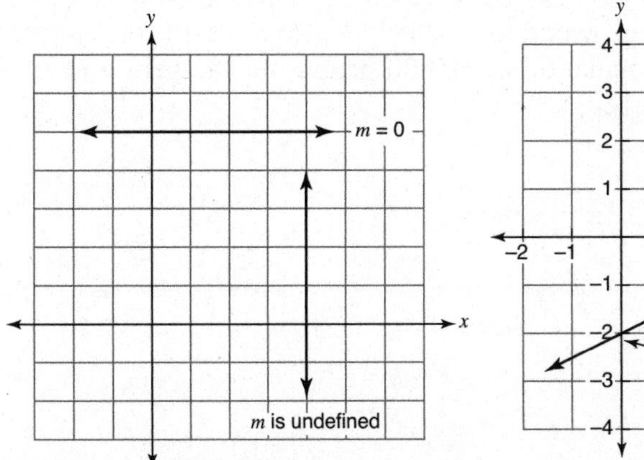

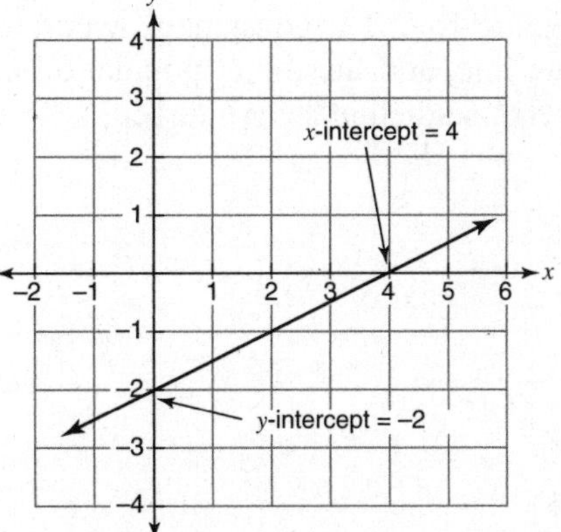

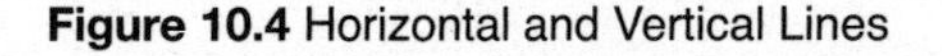

Figure 10.4 Horizontal and Vertical Lines

Figure 10.5 Intercepts of a Graph

Intercepts of a Graph

An **intercept** of a graph is a point at which the graph intersects either the x-axis or the y-axis. In Figure 10.5:

- the line crosses the x-axis at (4,0) so the x-intercept is 4.
- the line crosses the y-axis at (0,–2) so the y-intercept is –2.

MATH FACTS

An x-intercept of a graph is an x-value where the graph intersects the x-axis. At each such point, $y = 0$. Similarly, a ***y*-intercept** of a graph is a y-value where the graph intersects the y-axis. At each such point, $x = 0$.

- A horizontal line has no x-intercept, and a vertical line has no y-intercept.
- If you know an equation of a line, you can find either intercept by setting the other variable equal to zero and solving the resulting equation. For example, to find the x-intercept of $4x + 3y = 12$, set $y = 0$ and solve for x: $4x + 3(0) = 12$ and $x = \frac{12}{4} = 3$, so the line crosses the x-axis at (3,0).

Slope As a Rate of Change

The slope of a line represents the constant rate at which y changes over equal intervals of x. Suppose that the population of a town was 50,000 in the year 2006 and is expected to increase at a constant rate of 4000 people per year for the next 5 years. This situation can be represented graphically by the line in Figure 10.6. The constant growth rate of 4000 people per year is the slope of the line, and the initial population of 50,000 people for the period of time being studied is the y-intercept.

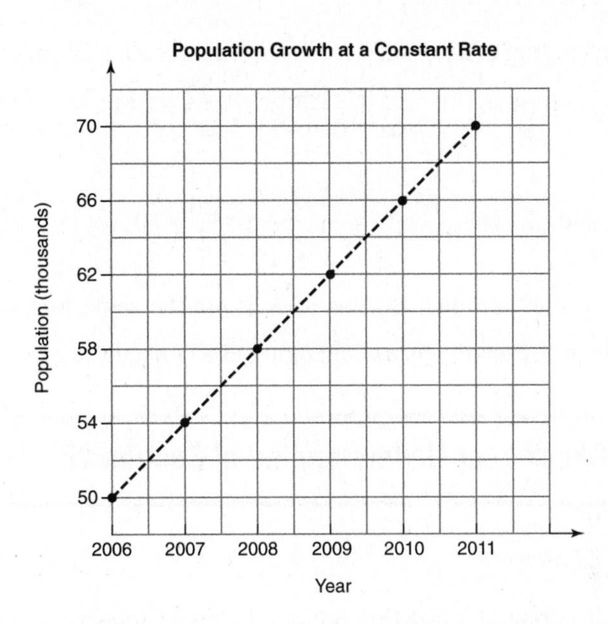

Figure 10.6 Illustrating a Constant Rate of Population Growth

Example 5

Using the data in each table, determine if y is changing at a constant rate.

a.

x	y
–2	–8
–1	–5
0	–2
1	1

b.

x	y
–2	15
0	–7
2	13
4	–9

Solution: Variable y is changing at a constant rate if $\dfrac{\Delta y}{\Delta x}$ is the same for each pair of points in the table.

a. (–2,–8) and (–1,–5): $\dfrac{\Delta y}{\Delta x}=\dfrac{-5-(-8)}{-1-(-2)}=\dfrac{-5+8}{-1+2}=3$

(–1,–5) and (0,–2): $\dfrac{\Delta y}{\Delta x}=\dfrac{-2-(-5)}{0-(-1)}=\dfrac{-2+5}{+1}=3$ ✓

(0,–2) and (1,1): $\dfrac{\Delta y}{\Delta x}=\dfrac{-2-(-5)}{0-(-1)}=\dfrac{-2+5}{+1}=3$ ✓

Variable y is changing at a constant rate of 3 units for each 1 unit change in x.

b. $(-2,15)$ and $(0,-7)$: $\frac{\Delta y}{\Delta x}=\frac{-7-15}{0-(-2)}=\frac{-22}{2}=-11$

$(0,-7)$ and $(2,13)$: $\frac{\Delta y}{\Delta x}=\frac{13-(-7)}{2-0}=\frac{20}{2}=10$

There is no need to continue. Because $\frac{\Delta y}{\Delta x}$ is not the same for two different pairs of points in the table, y is ***not*** changing at a constant rate.

Check Your Understanding of Section 10.1

A. Multiple Choice.

1. What is the slope of a line through points $(-4,2)$ and $(6,8)$?

(1) $-\frac{3}{5}$ (2) $\frac{3}{5}$ (3) $\frac{5}{3}$ (4) $-\frac{5}{3}$

2. If points $(3,2)$ and $(c,-5)$ are on a line whose slope is $-\frac{7}{2}$, what is the value of c?

(1) 5 (2) 6 (3) $\frac{15}{7}$ (4) 4

3. In the accompanying diagram, what is the slope of line ℓ?

(1) 0 (3) $\frac{1}{2}$

(2) -2 (4) $-\frac{1}{2}$

4. The point whose coordinates are $(4,-2)$ is on a line whose slope is $\frac{3}{2}$. The coordinates of another point on this line could be

(1) $(1,0)$ (2) $(2,1)$ (3) $(6,1)$ (4) $(7,0)$

5. What is the slope of the line whose y-intercept is 2 and which contains the point $(5,-8)$?

(1) $\frac{1}{2}$ (2) 2 (3) $-\frac{1}{2}$ (4) -2

6. What is the positive value of n if the slope of the line joining $(6,n)$ and $(7,n^2)$ is 20?
(1) 49 (2) 13 (3) 5 (4) 4

7. In the accompanying diagram, line ℓ intersects the horizontal line at P. If the slope of line ℓ is $\frac{1}{4}$, what are the coordinates (a,b) of point P?

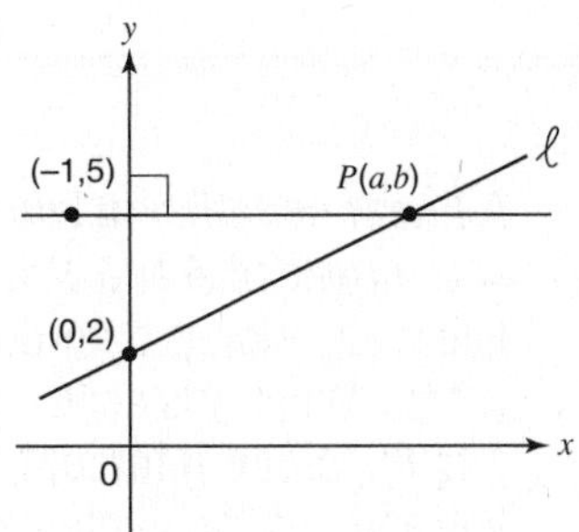

(1) (12,5) (3) (8,5)
(2) (16,5) (4) (5,–12)

B. *Show how you arrived at your answer.*

8. What is the slope of a line through points (–2,–5) and (6,–9)?

9. What is the slope of a line whose x-intercept is –3 and y-intercept is 5?

10. The slope of the line that passes through the points (–1,4) and $(6,y)$ is $\frac{5}{7}$. Find y.

11. What is the slope of the line whose y-intercept is –5 and which contains the point (–9,1)?

12. What is the slope of the line whose x-intercept is –1 and which contains the point (7,–6)?

13. If $C(k,14)$ is on the same line as $A(6,-1)$, and $B(2,5)$, what is the value of k?

14. Using the data in the accompanying table, determine if y is changing at a constant rate with respect to x. Give a reason for your answer.

x	–2	0	2	4
y	–15	–8	–1	6

15. The line that contains the points $(3,k)$ and (4,–1) has the same slope as the line that contains the points (–2,3) and $(k,1)$. What are the possible values of k?

16. A car purchased in 2005 for \$24,000 depreciates at a constant rate of \$1500 per year for the next 5 years. Represent this situation graphically.

17. A line contains the points $(a^2 + b, 2a)$ and $(b + b^2, a + b)$. Express the slope of the line in terms of a and b. Write your answer in simplest form.

10.2 SLOPE-INTERCEPT FORM OF A LINEAR EQUATION

KEY IDEAS

A **linear equation in two variables** is an equation of the form $Ax + By = C$, provided A and B are not both zero. A **solution** of a two-variable linear equation is an ordered number pair (x,y) for which the equation is true. When y is solved for in terms of x, the equation is transformed into the **slope-intercept** form $y = mx + b$ where m is the slope of the line and b is the y-intercept.

Solutions of Two-Variable Linear Equations

One solution of the equation $2x + y = 5$ is (1,3) since substituting 1 for x and 3 for y results in a true statement:

$$2x+y=2\cdot 1+3=5 \checkmark$$

Although it is not possible to list *all* solutions of the equation, the solution set can be represented graphically by a line. Each point (x,y) on the line in Figure 10.7 represents a solution of $2x + y = 5$. It is always true that:

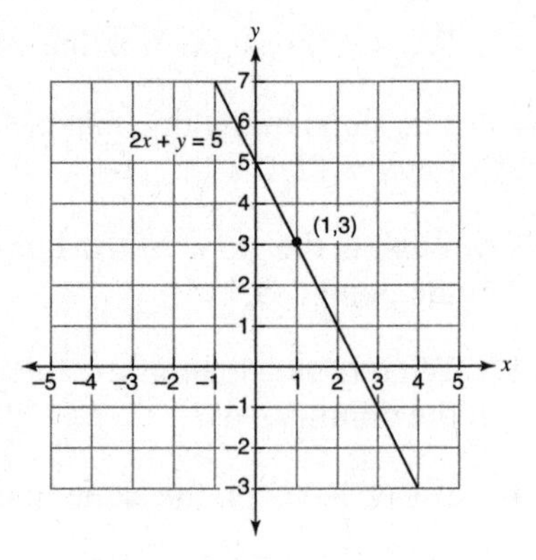

Figure 10.7 Graph of $2x + y = 5$

- The coordinates of each point on a line is a solution of the equation of that line.
- If the coordinates of a point do *not* satisfy an equation of a line, then the point is *not* on that line.

Example 1

The line whose equation is $y = 3x - 1$ contains the point $A(2,k)$. What is the value of k?

Solution: Because the point $A(2,k)$ lies on the line, its coordinates must satisfy its equation. To find k, replace x by 2 and y by k in $y = 3x - 1$, which makes $k = 3(2) - 1 = \mathbf{5}$.

Example 2

The line whose equation is $kx + 3y = 13$ passes through the point (8,–1). What is the value of k?

Solution: Since the line contains the point (8,–1), its coordinates must satisfy the given equation. Replacing x with 8 and y with –1 in $kx + 3y = 13$ gives $8k + 3(-1) = 13$ so $8k = 16$ and $k = \frac{16}{8} = \mathbf{2}$.

Slope-Intercept Form: $y = mx + b$

You can easily verify that (–2,3) and (4,6) satisfy the equation $y = \frac{1}{2}x + 4$. Figure 10.8 shows the result of graphing these two points and then drawing the line that passes through them. You can tell from the graph that the slope of the line is

$$m = \frac{\Delta y}{\Delta x} = \frac{3}{6} = \frac{1}{2}$$

and its y-intercept is 4. Thus, the coefficient of x in $y = \frac{1}{2}x + 4$ is the **slope** of the line and the constant term is the ***y*–intercept**. The equation $y = \frac{1}{2}x + 4$ is of the form $y = mx + b$, where m is the slope of the line and b is the y-intercept.

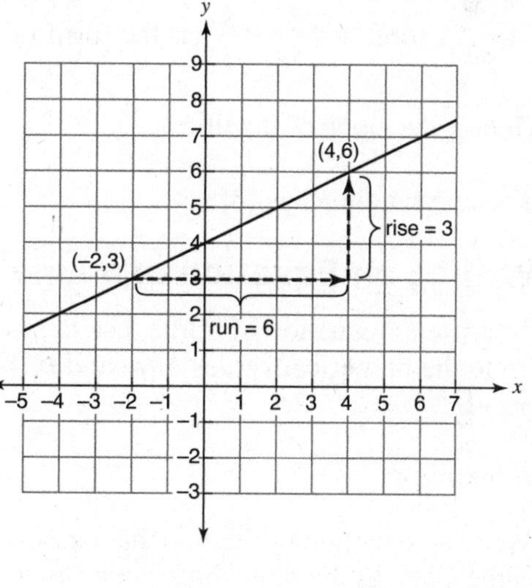

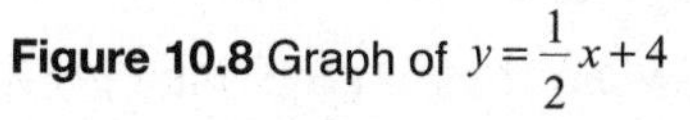

Figure 10.8 Graph of $y = \frac{1}{2}x + 4$

MATH FACT

An equation of a line is in **slope-intercept form** when it is written in the form $y = mx + b$, where m is the slope of the line and b is the y-intercept.

To read the slope and y-intercept of a line from its equation, it may be necessary to rewrite it in $y = mx + b$ form.

Example 3

What is the slope of the line whose equation is $4x - 3y - 8 = 0$?

(1) $\frac{3}{4}$ (2) $-\frac{3}{4}$ (3) $\frac{4}{3}$ (4) $-\frac{4}{3}$

Solution: Write the given equation in slope-intercept form by solving for y in terms of x:

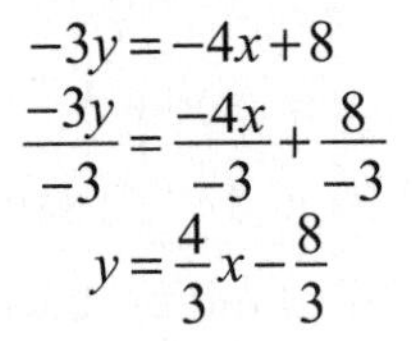

$$-3y = -4x + 8$$
$$\frac{-3y}{-3} = \frac{-4x}{-3} + \frac{8}{-3}$$
$$y = \frac{4}{3}x - \frac{8}{3}$$

The equation $y = \frac{4}{3}x - \frac{8}{3}$ has the form $y = mx + b$, where $m = \frac{4}{3}$ and $b = -\frac{8}{3}$.

Hence, the slope of the line is $\frac{4}{3}$.

The correct choice is **(3)**.

Writing an Equation of a Line

To write an equation of a line, use the information that is provided to determine the numerical values of m and b. Then substitute these values into $y = mx + b$.

Example 4

Write an equation of the line that passes through the point $(-1,3)$ and has the same slope as the line whose equation is $y - 2x = 3$.

Solution : If $y - 2x = 3$, then $y = 2x + 3$ so the slope of this line is 2.

- Because $m = 2$, the equation of the required line has the form $y = 2x + b$.
- It is also given that $(-1,3)$ is a point on the line. Find b by substituting $x = -1$ and $y = 3$ in $y = 2x + b$, which gives $3 = 2(-1) + b$ so $b = 5$.
- Since $m = 2$ and $b = 5$, an equation of the required line is $\mathbf{y = 2x + 5}$.

Example 5

A ball is thrown into the air with a velocity of v feet per second. The data in the accompanying table show the velocity t seconds after the ball has been thrown. A positive velocity indicates that the ball is rising, and a negative velocity means it is falling. If the velocity of the ball is changing at a constant rate with respect to time, find a linear equation that expresses v in terms of t.

t (sec)	1	3	5	6
v (ft/sec)	92	28	−36	−68

Solution: Since velocity of the ball is changing at a constant rate with respect to time, the required equation has the form $v = mt + b$ where m is the rate of change of v. Choose any two points from the table. Use the slope formula to find m, and then find b by substituting the coordinates of either point into $v = mt + b$.

- For (1,92) and (3,28):

$$m=\frac{\Delta v}{\Delta t}=\frac{28-92}{3-1}=\frac{-64}{2}=-32$$

- Replace m with −32 in $v = mt + b$:

$$v = 32t + b$$

- Use (1,92) to find b:

$$92=-32(1)+b$$
$$\text{so } b=92+32=124$$

The required equation is $\boldsymbol{v = -32t + 124}$.

Example 6

What is an equation of the line in the accompanying diagram?

(1) $y=-\frac{3}{2}x+2$

(2) $y=\frac{3}{2}x-3$

(3) $y=\frac{2}{3}x+2$

(4) $y=\frac{2}{3}x-3$

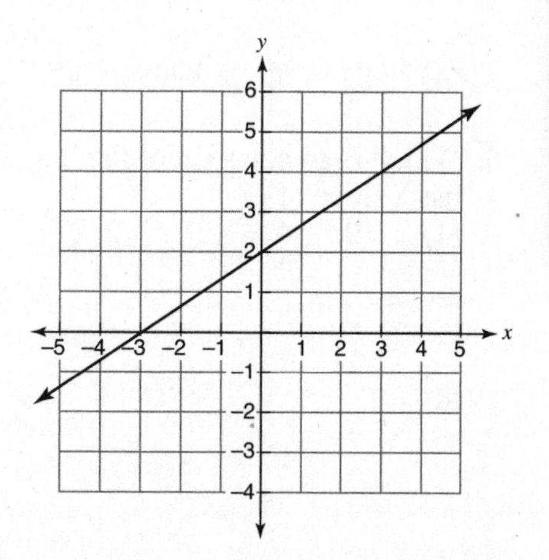

Solution: You can read from the graph that the y-intercept is 2. To find the slope of the line, calculate the rise over the run for any two convenient points along the line. For example, as you move from the y-intercept of (0,2) to (3,4), the run is 3 and the rise is 2 so the slope is $\frac{2}{3}$. Because $b = 2$ and $m = \frac{2}{3}$, an equation of the line is $y = \frac{2}{3}x + 2$.

The correct choice is **(3)**.

Check Your Understanding of Section 10.2

A. *Multiple Choice.*

1. Which point does not lie on the graph of $3x - y = 7$?
(1) (2,–1) (2) (3,2) (3) (–1,4) (4) (1,–4)

2. If the point $(k,2)$ is on the line whose equation is $2x + 3y = 4$, what is the value of k?
(1) 1 (2) 0 (3) –1 (4) $\frac{1}{2}$

3. Point $(k, -3)$ lies on the line whose equation is $x - 2y = -2$. What is the value of k?
(1) –8 (2) –6 (3) 6 (4) 8

4. The line whose equation is $3x - 2y = 12$ has
(1) slope $= \frac{3}{2}$; y-intercept $= -6$
(2) slope $= -\frac{3}{2}$; y-intercept $= 6$
(3) slope $= 3$; y-intercept $= -2$
(4) slope $= -3$; y-intercept $= -6$

5. Which is an equation of the line that passes through the point (0, 2) and has a slope of 4?
(1) $x = 2y - 4$ (2) $y = 2x + 4$ (3) $4x + y = 2$ (4) $y - 2 = 4x$

6. Which diagram could represent the graph of the equation $y + 1 = 2x$?

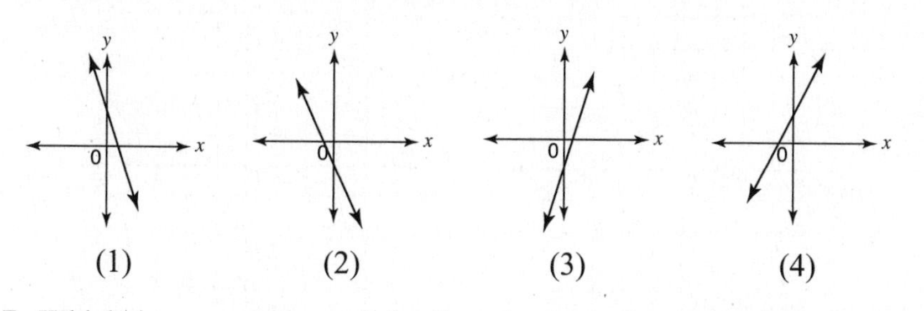

7. Which is an equation of the line that contains the points (1,3) and (−1,1)?

(1) $x = 1$ (2) $y = 2x + 1$ (3) $y = x + 2$ (4) $y = 3$

8. The accompanying diagram shows the graph of line m. Which equation represents this line?

(1) $y = 2x + 1$ (2) $y = \frac{1}{2}x + 2$ (3) $y = -2x + 1$ (4) $y = -\frac{1}{2}x + 2$

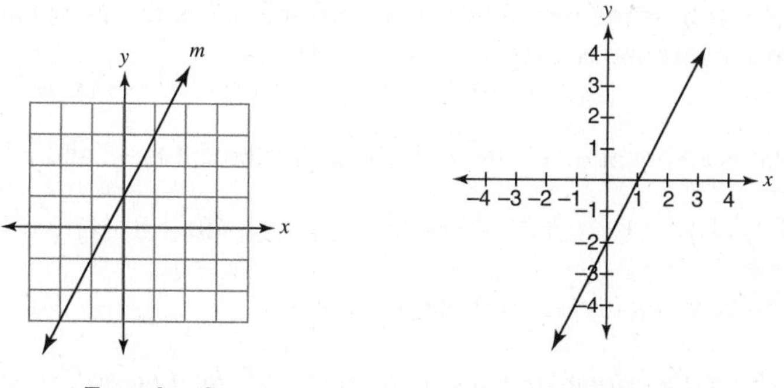

Exercise 8 **Exercise 9**

9. What is an equation of the line in the accompanying diagram?

(1) $2y = x - 2$ (2) $y = \frac{1}{2}x + 1$ (3) $y = -2x - 2$ (4) $y = 2x - 2$

10. What is an equation of line ℓ in the accompanying diagram?

(1) $y = \frac{3}{2}x + 2$ (2) $y = -\frac{3}{2}x + 3$ (3) $y = -\frac{2}{3}x + 2$ (4) $y = -\frac{3}{2}x + 2$

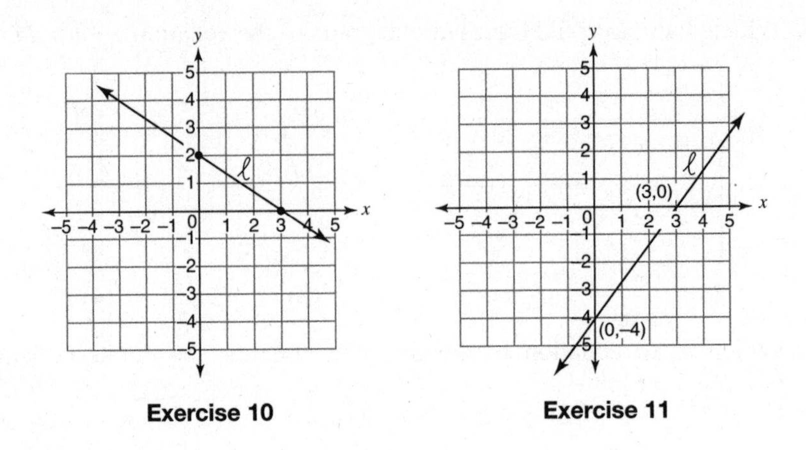

11. What is an equation of line ℓ in the accompanying diagram?

(1) $y=\frac{4}{3}x+3$ (2) $y=-\frac{4}{3}x+3$ (3) $y=\frac{4}{3}x-4$ (4) $y=-\frac{4}{3}x-4$

12. The graph of the equation $x + 3y = 6$ intersects the y-axis at the point whose coordinates are
(1) (0,2) (2) (0,6) (3) (0,18) (4) (6,0)

13. What is an equation of the line whose x-intercept is −3 and y-intercept is 4?
(1) $3y - 4x = 12$ (2) $4y - 3x = 12$ (3) $2y = -6x + 8$ (4) $y = 4x - 3$

B. *Show how you arrived at your answer.*

14–16. *Find the slope and the y-intercept of the line whose equation is given.*

14. $2y + x = 10$ **15.** $3y - 2x = 15$ **16.** $2x - 3y = 8$

17. Write an equation of the line that contains the point (2,−1) and has the same y-intercept as the line whose equation is $y + 7 = 5x$.

18. Write an equation of the line that contains the point (4,−3) and has the same slope as the line whose equation is $y + 3x = 2$.

19–21. *Write an equation of the line that contains each pair of points.*

19. (1,−7) and (3,5) **20.** (2,−9) and (−3,11) **21.** (−8,5) and (−4,−1)

22. The accompanying table shows the cost of CDs at a music store. If the relationship between the number of CDs purchased and the total cost were represented as a line, what would be an equation of the line?

Number of CDs, x	Total Cost, y
1	9
2	15
3	21

23. A ball is thrown into the air with a velocity of v feet per second. The data in the accompanying table shows the velocity t seconds after the ball has been thrown.

a. If the velocity is changing at a constant rate with respect to time, find an equation that expresses v in terms of t.

t (sec)	1	3	5	7
v (ft/sec)	80	16	−48	−112

b. How many seconds after the ball is thrown is the velocity of the ball $0 \frac{\text{ft}}{\text{sec}}$?

24. a. For the accompanying graph, write the equations of lines ℓ and m.

b. Using your answers from part a, confirm algebraically that (−3,5) is a point on line m but not on line ℓ.

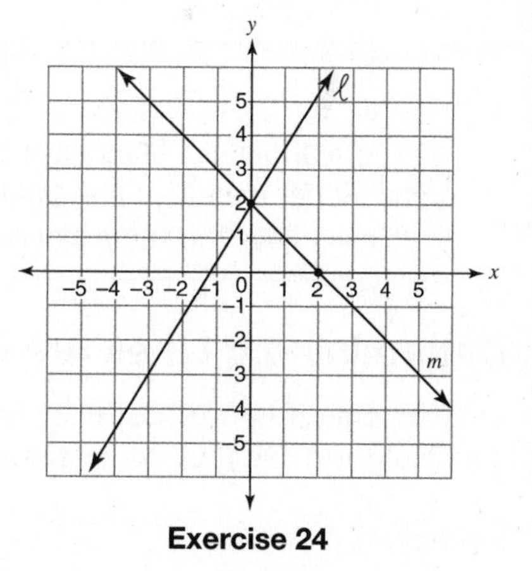

Exercise 24

25. a. Write an equation of the line whose x-intercept is 3 and y-intercept is −2.

b. Determine whether the point (−6,−6) is on the line.

26. Charles wants to compare Celsius and Fahrenheit temperatures by drawing a conversion graph. He knows that −40°C = −40°F and that 20°C = 68°F.

a. Copy the accompanying grid. Construct the conversion graph and, using the graph, find the Celsius equivalent of 50°F.

b. Determine an equation of the conversion graph.

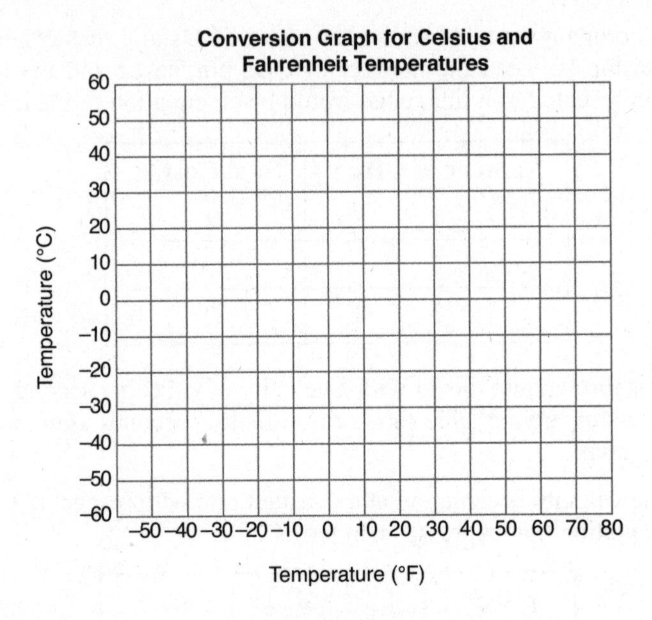

Exercise 26

10.3 SLOPES OF PARALLEL LINES

KEY IDEAS

Nonvertical parallel lines have the same slope. If two lines have the same slope, then they are parallel. Because the lines $y = 2x + 4$ and $y = 2x - 3$ have the same slope, they are parallel.

Determining If Lines Are Parallel

To determine if nonvertical lines are parallel, write their equations in slope-intercept form and then compare their slopes.

Example 1

An equation of line ℓ is $2y + 3x = 6$. Which equation represents a line that is parallel to line ℓ?

(1) $y=\frac{2}{3}x+3$ (2) $y=-\frac{2}{3}x+3$ (3) $y=\frac{3}{2}x+2$ (4) $y=-\frac{3}{2}x+2$

Solution: Compare the slopes of the two lines after the equation of line ℓ is expressed in slope-intercept form.

- Find the slope of line ℓ. If $2y + 3x = 6$, then solving for y gives $y = -\frac{3}{2}x + 3$. Hence, the slope of line ℓ is $-\frac{3}{2}$.

- The slope of the required line must also be $-\frac{3}{2}$ since the lines are parallel.

- Because the slope of the line in answer choice (4) is $-\frac{3}{2}$, this line is parallel to line ℓ.

The correct choice is **(4)**.

Example 2

Kim graphed the line represented by the equation $y + 3x = 5$.

a. Write an equation of a different line that is parallel to the line Kim graphed.
b. Write an equation of a line that is parallel to the line Kim graphed and that contains the point (1,4).
c. Write an equation of a line whose graph coincides with the line Kim graphed but has different numerical coefficients.

Solution: If $y + 3x = 5$, then $y = -3x + 5$. The slope of the line Kim graphed is -3, and its y-intercept is 5.

a. A different line that is parallel to $y + 3x = 5$ has the same slope but a different y-intercept, such as $\mathbf{y = -3x + 1}$.
b. As the required line is parallel to the line Kim graphed, its slope is also -3 so its equation has the form $y = -3x + b$.

- Because (1,4) is a point on this line, find b by substituting $x = 1$ and $y = 4$ into $y = -3x + b$, which gives $4 = -3(1) + b$ so $b = 7$.
- As $m = -3$ and $b = 7$, an equation of the required line is $\mathbf{y = -3x + 7}$.

c. Multiplying the equation $y + 3x = 5$ by any nonzero number produces another linear equation whose graph coincides with the line Kim graphed. One such equation is $\mathbf{2y + 6x = 10}$. You should verify that the slope of this line is -3 and the y-intercept is 5.

Lines Parallel to a Coordinate Axis

As shown in Figure 10.9, $x = 3$ is an equation of a line parallel to the y-axis. The x-coordinate of each point on the line is 3. Similarly, $y = 5$ is an equation of a line parallel to the x-axis. The y-coordinate of each point on the line is 5. In general,

- The graph of $x = h$ is a vertical line parallel to the y-axis, where h is the x-coordinate of each point on the line.
- The graph of $y = k$ is a horizontal line parallel to the x-axis, where k is the y-coordinate of each point on the line.

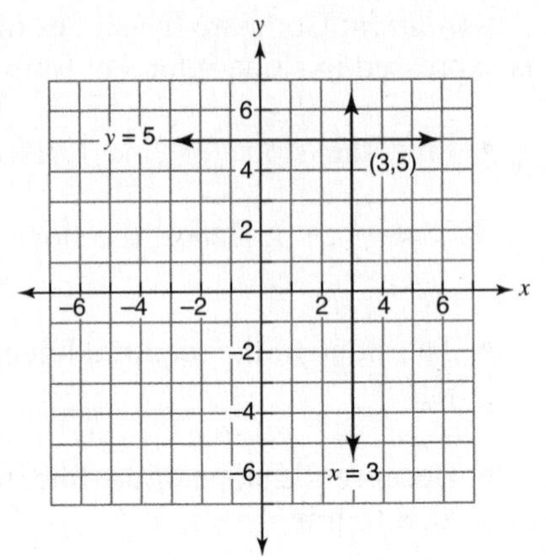

Figure 10.9 Equations of Vertical and Horizontal Lines

Example 3

Write an equation of the line that satisfies each condition:

a. Parallel to the y-axis and a distance of 2 units to the left of it.
b. Parallel to the x-axis and a distance of 1 unit below it.
c. Parallel to the x-axis and contains the point (−3,2).

Solutions:

a. $\boldsymbol{x = -2}$ b. $\boldsymbol{y = -1}$ c. $\boldsymbol{y = 2}$

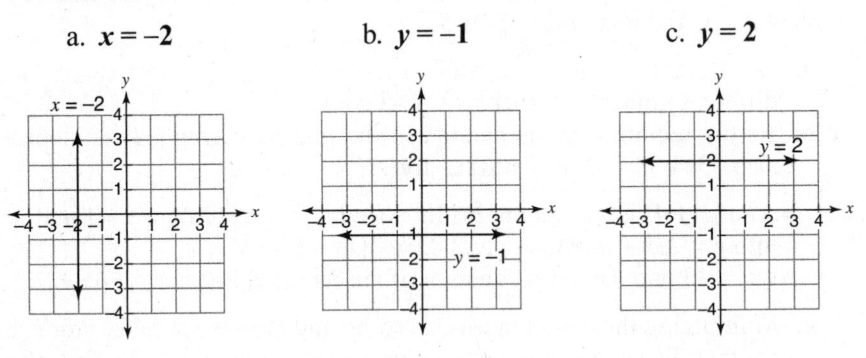

Check Your Understanding of Section 10.3

A. *Multiple Choice.*

1. What is an equation of a line that is parallel to the y-axis and contains the point (–3,1)?
(1) $x = -3$ (2) $x = 1$ (3) $y = -3$ (4) $y = 1$

2. Which pair of points determine a line that is parallel to the y-axis?
(1) (1,1) and (2,3)
(2) (1,1) and (3,3)
(3) (2,3) and (2,5)
(4) (2,5) and (4,5)

3. Which pair of points determine a line that is parallel to the x-axis?
(1) (1,3) and (–2,3)
(2) (1,–1) and (–1,1)
(3) (1,3) and (1,–1)
(4) (1,1) and (–3,–3)

4. What is an equation of a line that is parallel to the x-axis and contains the point (4,–2)?
(1) $x = 4$ (2) $x = -2$ (3) $y = 2$ (4) $y = -2$

5. Which is an equation of a line that is parallel to the y-axis and 2 units to its right?
(1) $x = 2$ (2) $x = -2$ (3) $y = 2$ (4) $y = -2$

6. Which is an equation of the line that is parallel to $y = 2x - 8$ and passes through the point (0,–3)?
(1) $y = 2x + 3$ (2) $y = 2x - 3$ (3) $y = -\frac{1}{2}x + 3$ (4) $y = -\frac{1}{2}x - 3$

7. The graph of the equation $x - 3y = 6$ is parallel to the graph of which equation?
(1) $y = -3x + 7$ (2) $y = -\frac{1}{3}x + 5$ (3) $y = 3x - 8$ (4) $y = \frac{1}{3}x + 8$

8. Which is an equation of the line that is parallel to $y = 3x - 5$ and has the same y-intercept as $y = -2x + 7$?
(1) $y = 3x - 2$ (2) $y = -2x - 5$ (3) $y = 3x + 7$ (4) $y = -2x + 7$

9. An equation of line ℓ is $3y - 4x = 8$. Which equation represents a line that is parallel to line ℓ?
(1) $y = \frac{4}{3}x + 2$ (2) $y = -\frac{4}{3}x + 2$ (3) $y = \frac{3}{4}x + 3$ (4) $y = -\frac{3}{4}x + 3$

10. Which pair of equations represents parallel lines?

(1) $2y-x=7$
$x+2y=4$

(2) $x=3y+1$
$2x-6y=5$

(3) $y=\frac{1}{2}x-1$
$y=2x+1$

(4) $2y-3x=0$
$4x+12y=8$

B. *Show how you arrived at your answer.*

11. Find the area of the rectangle formed by the intersection of the lines $x = 2$, $x = 9$, $y = -1$, and $y = 7$.

12. Given the points $A(4,6)$ and $B(0,-2)$. Write an equation of the line that is parallel to $\overleftrightarrow{AB}$ and that contains the point $(2,-5)$.

13. The equations for two lines are $3y - 2x = 6$ and $3x + ky = -7$. For what value of k are the two lines parallel?

14. Write an equation of a line that is parallel to the line whose equation is $3y - 12x = 6$ and that intersects the line whose equation is $2y + x = 10$ at its y-intercept.

15. The vertices of $\triangle PQR$ are $P(1,2)$, $Q(-3,6)$, and $R(4,8)$. A line through Q is parallel to $\overline{PR}$ and passes through $(k,14)$. What is the value of k?

16. a. Find the area of the triangle whose vertices are $A(0,6)$, $B(0,0)$, and $C(-8,0)$.

b. Write an equation of the line that passes through B and is parallel to $\overline{AC}$.

10.4 GRAPHING LINEAR EQUATIONS

KEY IDEAS

When graphing a linear equation in two variables using graph paper, follow these general guidelines:

- Draw and label the coordinate axes.
- Plot at least two points that satisfy the equation of the line. Although two points determine a line, plotting a third point serves as a check since the required line must contain all three points.
- Draw a line through the plotted points and label it with its equation.

Three-Point Table Method

To graph the equation $3x - y = 1$, first write the equation in slope-intercept form as $y = 3x - 1$. Then prepare a table of values that includes three points whose coordinates satisfy the equation of the line:

- Pick three easy values for x and list them in the first column of a table of values. Include negative and positive values of x as well as 0:

x	$y = 3x - 1$	**Solutions**
-1	$y = 3(-1) - 1 = -4$	$(-1,-4)$
0	$y = 3(0) - 1 = -1$	$(0,-1)$
1	$y = 3(1) - 1 = 2$	$(1,2)$

- On each row in the second column of the table, substitute the x-value into the equation of the line to obtain the corresponding y-value.
- In the third column of the table, list the ordered number pairs that satisfy the equation. Plot the three solution points, as shown in Figure 10.10. Then connect these points with a straight line. The third point serves as a check. If it is not possible to draw a line that contains all three points, then you made a mistake either in calculating the coordinates of at least one of the points or in plotting them.

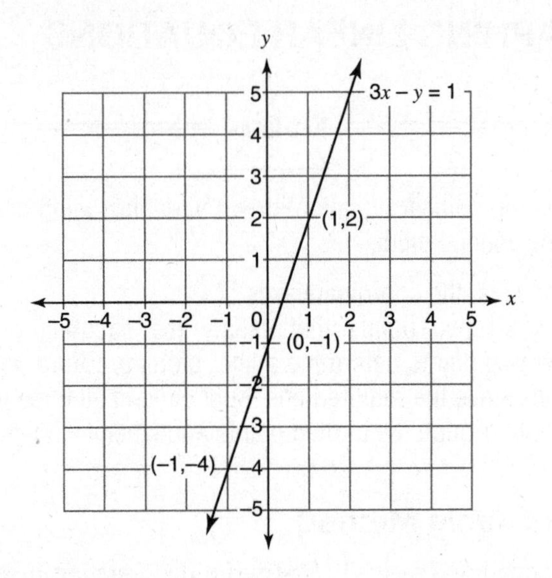

Figure 10.10 Graph of $3x - y = 1$ Using a Table of Values

Slope-Intercept Method

A line can be graphed from the slope-intercept form of its equation by starting at its y-intercept and then using the slope of the line to locate additional points on the line. To graph the equation $4y - 3x = 12$ using this method:

- Write the equation in slope-intercept form as $y = \frac{3}{4}x + 3$. The slope of the line is $\frac{3\,(= \text{rise})}{4\,(= \text{run})}$ and the y-intercept is 3.
- Plot the y-intercept at (0,3). Locate a second point on the line by moving 4 units to the right of the y-intercept and 3 units up. The coordinates of this point are (4,6), as shown in Figure 10.11.
- Locate a third point on the line by starting at (4,6) and repeating the "rise-run" procedure, which gives (8,9) as another point on the line.
- Draw a line through (0,3), (4,6), and (8,9). Label the line with its equation.

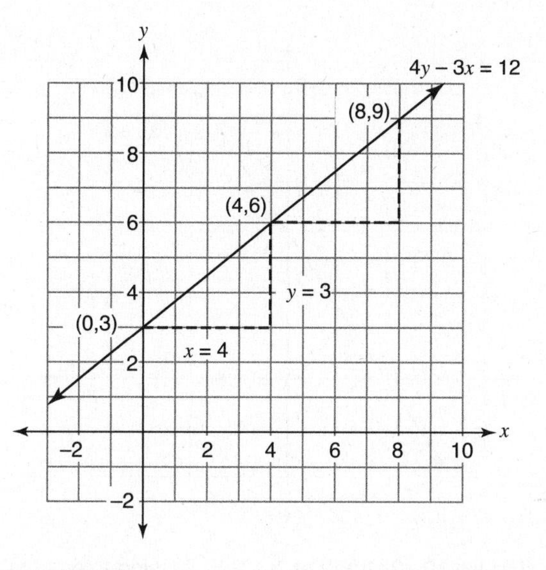

Figure 10.11 Graphing $4y - 3x = 12$ Using the Slope-Intercept Method

Graphing Using Intercepts

Some oblique (slanted) lines can be graphed quickly by using its two *intercepts*. To find the intercepts of a graph from its equation, set each variable, in turn, equal to 0 and solve for the other variable. To graph $2y - 3x = 3$ using its intercepts:

- Find the y-intercept by setting $x = 0$, then $2y = 3$ so $y = \frac{3}{2}$. The line intersects the y-axis at $\left(0, \frac{3}{2}\right)$.
- Find the x-intercept by setting $y = 0$, which makes $-3x = 3$ so $x = \frac{3}{-3} = -1$. The line intersects the x-axis at $(-1,0)$.
- To avoid estimating when plotting the y-intercept, scale the coordinate axes so that every 2 square boxes corresponds to one unit, as shown in the accompanying figure. Then draw a line through the two intercepts, as shown in Figure 10.12.

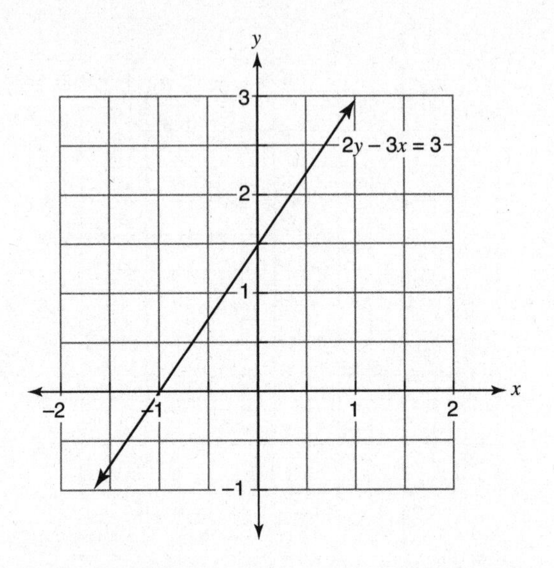

Figure 10.12 Graph of $2y - 3x = 3$ Using Intercepts

The intercepts method of graphing a line has a couple of disadvantages:

- Only two points are used to graph the line. Because a third "check" point is not used, a careless error may go unnoticed.
- The method does not work well when the intercepts are difficult fractions.

Graphing Using Technology

Here is a quick review of the steps to follow when graphing a linear equation such as $3x - y = 1$ using a calculator:

- Write the equation in slope-intercept form as $y = 3x - 1$.
- Press $\boxed{Y=}$ and set Y_1 equal to $3x - 1$ by pressing

  $\boxed{3}\ \boxed{x, T, \theta, n}\ \boxed{-}\ \boxed{1}$.

- Press $\boxed{\text{GRAPH}}$. Adjust the size of the viewing rectangle, if necessary, as described in Section 3.2. Figure 10.13 shows the graph in a $[-4.7, 4.7] \times [-6, 6]$window.

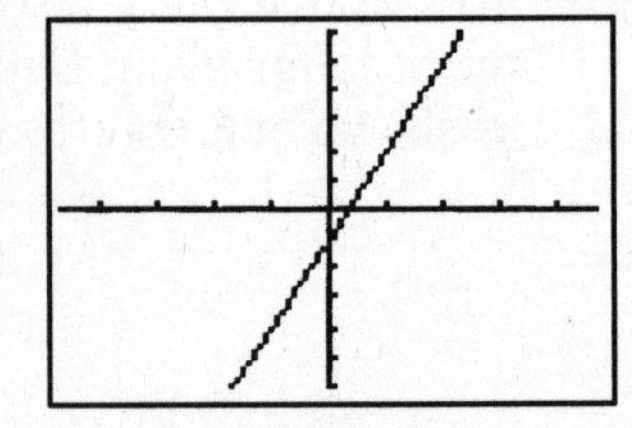

Figure 10.13 Graph of $3x - y = 1$

Choosing a Graphing Method

If you are not directed to use a particular method of graphing a line, use the method that is most comfortable for you. The intercepts method is a quick way to graph an equation that has the form $Ax + By = C$, provided the two intercepts are easy to graph. When the equation is in $y = mx + b$ form, or can be easily put in this form, the slope-intercept method can be used, provided the y-intercept is easy to graph. Otherwise, draw the graph based on a table of values that includes at least three convenient points that satisfy the equation of the line. It may be helpful to check your work by comparing a line drawn on graph paper with the line obtained using a graphing calculator.

Check Your Understanding of Section 10.4

Show how you arrived at your answer.

1–3. *Use a table of values to graph each equation.*

1. $y = 2x - 1$ **2.** $y = \frac{1}{2}x + 3$ **3.** $y = -2x + 1$

4. Graph $y = -3x - 4$ using $\{-3, -2, -1, 0, 1, 2\}$ as the replacement set for x.

5–10. *Graph each equation using the intercepts method.*

5. $y = 2x - 4$ **7.** $2x - y = 6$ **9.** $3x - 2y = 12$

6. $x + 2y = 8$ **8.** $3x + 2y = 6$ **10.** $5x - 2y = 5$

11–13. *Graph each equation using the slope-intercept method.*

11. $y = 2x + 3$ **12.** $2y - x = 6$ **13.** $3y + x = 12$

14. The senior class is sponsoring a dance at a gym. The cost of a student disk jockey is \$45 and the cost of renting the gym is \$255.

a. If tickets sell for \$4 each, write a linear equation that represents the relationship between the number of tickets sold and the profit from the dance.

b. Graph the equation determined in part a. From the graph, determine how many tickets must be sold to break even.

10.5 DIRECT VARIATION

KEY IDEAS

If the number of cups of flour in a cake recipe are doubled, then the number of cups of milk that the recipe requires must also be doubled. The number of cups of flour *varies directly* as the number of cups of milk because their ratio, called the **constant of variation**, must remain the same. When *y* **varies directly** as *x*, the ratio of *y* to *x* is the same for all ordered number pairs (x,y) that satisfy this relation. The graph of a direct variation is a line through the origin whose slope is the constant of variation.

Representing a Direct Variation

A direct variation between two quantities may be represented algebraically or graphically.

- The equation $\frac{y}{x} = k$ describes a direct variation between x and y, where k is a nonzero number called the **constant of variation**. When y varies directly as x, multiplying the value of one variable by some nonzero number produces a corresponding change in the other variable so that the ratio of the two variables remains the same. For example, if y is doubled, x is also doubled.
- The direct variation equation $\frac{y}{x} = k$ may be rewritten as $y = kx$, which represents a line through the origin. The slope of the line is k, the constant of variation. The perimeter, p, of a square varies directly as its side length, s, according to the equation $p = 4s$. Figure 10.14 shows p plotted against s. The slope of the line is 4, which represents the constant of variation. At each point on the line, the ratio of p to s is 4.

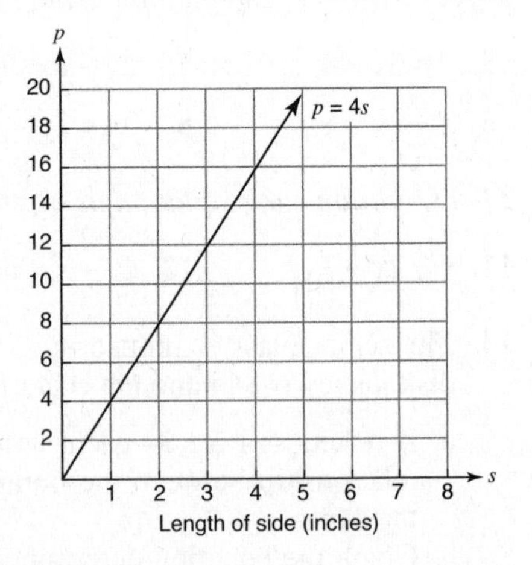

Figure 10.14 Graphing a Direct Variation

Solving Direct Variation Problems

To solve a problem in direct variation, write a proportion that states the ratios of the two quantities in direct variation, taken in the same order, are equal.

Example 1

If y varies directly as x and $y = 36$ when $x = 16$, what is the value of y when $x = 24$?

Solution: Two ordered pairs that satisfy this direct variation are (16,36) and(24,y). Because the ratio of y to x is constant:

$$\frac{y\text{-value}}{x\text{-value}}=\frac{36}{16}$$
$$\frac{y}{24}=\frac{36}{16}$$
$$16y=36\times 24$$
$$\frac{16y}{16}=\frac{864}{16}$$
$$y=\mathbf{54}$$

Example 2

The weight of a set of identical coins varies directly as the number of coins. If 20 of the identical coins weigh 42 grams, how many of these coins weigh 105 grams?

Solution: If x represents the number of identical coins that weigh 105 grams, then the ordered pairs that satisfy this direct variation are (20,42) and (x,105):

$$\frac{\text{weight of the coins}}{\text{number of coins}}=\frac{42}{20}$$
$$\frac{105}{x}=\frac{42}{20}$$
$$42x=20\cdot 105$$
$$\frac{42x}{42}=\frac{2100}{42}$$
$$x=50$$

50 coins weigh 105 grams.

Example 3

Anita's wages vary directly as the number of hours she works. If she earns \$29.80 for working 4 hours, how much will she earn when she works 30 hours?

Solution: If w represents Anita's wages for working 30 hours, then the ordered pairs that satisfy this direct variation are (4,\$29.80) and (30,w):

$$\frac{\text{wages}}{\text{hours worked}} = \frac{\$29.90}{4}$$
$$\frac{w}{30} = \frac{\$29.90}{4}$$
$$4w = 30(\$29.90)$$
$$\frac{4w}{4} = \frac{\$897}{4}$$
$$w = \$224.25$$

Anita will earn **\$224.25** for working 30 hours.

Check Your Understanding of Section 10.5

A. *Multiple Choice.*

1. If y varies directly as x and $y = 5.4$ when $x = 3$, what is the value of y when $x = 5$?
(1) 3.24 (2) 9 (3) 7.4 (4) 15

2. A cake recipe calls for 1.5 cups of milk and 3 cups of flour. Ben made a mistake and used 4 cups of flour. How many cups of milk should be used so that the milk and flour are combined in the correct proportions?
(1) 1.75 (2) 2.0 (3) 2.25 (4) 2.5

3. When rice is prepared, the amount of rice varies directly as the amount of water required. If 2 cups of rice require 4.5 cups of water, what is the total number of cups of water needed to prepare 5 cups of rice?
(1) 9 (2) 10 (3) 11.25 (4) 22.5

B. *Show how you arrived at your answer.*

4–6. *In each case, y varies directly as x.*

4. If $y = 39$ when $x = 26$, find x when $y = 21$.

5. If $y = 7.2$ when $x = 1.8$, find y when $x = 0.25$.

6. If $y = \frac{9}{8}$ when $x = \frac{1}{4}$, find x when $y = 6$.

7. If b varies directly as p and $b = 260$ when $p = 13$, what is the value of b when $p = 17$?

8. If x varies directly as y and $x = 18$ when $y = 4$, find x when $y = 10$.

10.6 POINT-SLOPE FORM OF A LINEAR EQUATION

KEY IDEAS

When the slope and the coordinates of a point on a line are given, an equation of the line can be written quickly using the **point-slope form** $y - k = m(x - h)$, where m is the slope of the line and (h,k) are the coordinates of any point on the line.

Point-Slope Form: $y - k = m(x - h)$

To write an equation of the line that contains the point (4,3) and whose slope is –2, use the point-slope form of the equation of a line $y - k = m(x - h)$, where $h = 4$, $k = 3$, and $m = -2$:

$$y - k = m\,(x - h)$$
$$\downarrow \quad \downarrow \quad \downarrow$$
$$y - 3 = -2\,(x - 4)$$

If necessary, the equation $y - 3 = -2(x - 4)$ can be written in $y - mx + b$ form by removing the parentheses on the right side of the equation and isolating y:

$$y - 3 = -2(x - 4)$$
$$y - 3 = -2x + 8$$
$$y = -2x + 11$$

Example 1

Write an equation of the line that contains (1,–7) and (3,5).

Solution: Find the slope of the line. Then substitute the coordinates of either point into the point-slope form of the equation of the line.

- Find the slope:

$$\begin{aligned} m = \frac{\Delta y}{\Delta x} &= \frac{5-(-7)}{3-1} \\ &= \frac{12}{2} \\ &= 6 \end{aligned}$$

- Read the values of h and k from either of the two given points. Using the point (1,–7), substitute $m = 6$, $h = 1$, and $k = -7$ into the point-slope form of the equation of the line:

$$\begin{aligned} y-k &= 6(x-h) \\ y-(-7) &= 6(x-1) \\ \boldsymbol{y+7} &= \boldsymbol{6(x-1)} \end{aligned}$$

If required, the equation can also be written in $y = mx + b$ form:

$$\begin{aligned} y+7 &= 6(x-1) \\ y+7 &= 6x-6 \\ y &= 6x-13 \end{aligned}$$

Using Technology: Finding an Equation of a Line

To use a graphing calculator to find an equation of the line that contains (1,–7) and (3,5), open the statistics editor by pressing STAT ENTER.

- Enter the x-coordinates of the two points in the first column labeled L1 and enter the corresponding y-coordinates in the column to the right labeled L2, as shown in Figure 10.15.

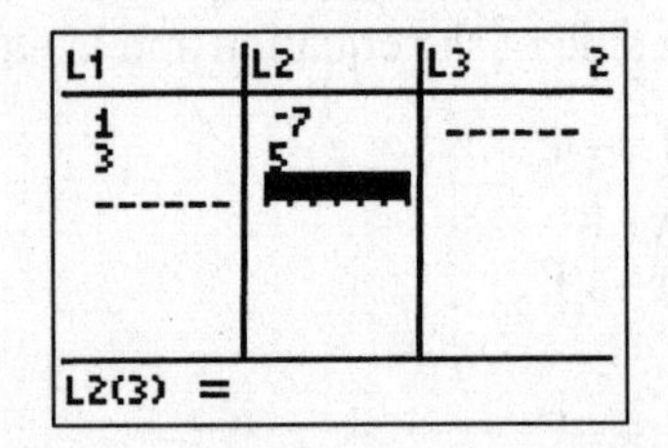

Figure 10.15 Entering Coordinates

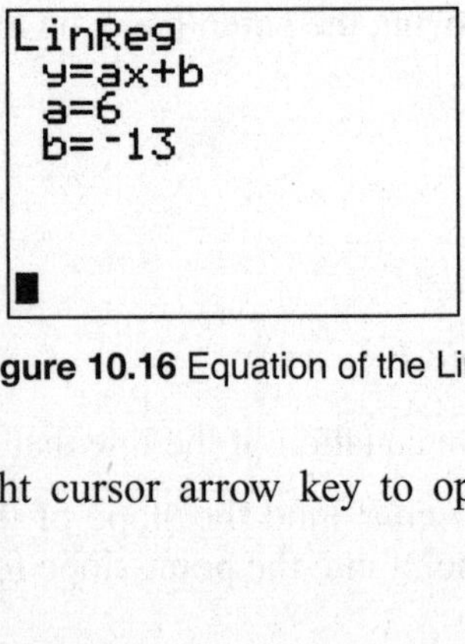

LinReg
y=ax+b
a=6
b=-13

Figure 10.16 Equation of the Line

- Press STAT followed by the right cursor arrow key to open the CALC menu.

- Select the **Lin**ear**Reg**ression option by pressing [4] [ENTER]. The display now shows the values of constants a and b in the $y - ax + b$ form of the equation of the line, as shown in Figure 10.16. Because $a = 6$ and $b = -13$, an equation of the line through the two given points is $y = 6x - 13$.

A similar procedure is used when it is necessary to find an equation of the line that "best fits" a collection of data points, which is discussed in Section 11.5.

Check Your Understanding of Section 10.6

Show how you arrived at your answer.

1–3. *Write an equation in point-slope form of the line that contains each pair of points.*

1. (8,2) and (3,–8) **2.** (–5,–1) and (9,6) **3.** (7,–3,) and (–1,1)

4. Express the equation obtained in Exercise 2 in slope-intercept form. Then check your answer by comparing this equation with the equation obtained using the statistics features of your graphing calculator.

5. Express the equation obtained in Exercise 3 in slope-intercept form. Then check your answer by comparing this equation with the equation obtained using the statistics features of your graphing calculator.

6. Write an equation of a line that is parallel to $y - 4x = 1$ and that passes through the point (7,–5).

7. An equation of line p is $y - 3 = k(x - 1)$. Line p intersects the line $y - 9 = 2x$ at its y-intercept. What is the numerical value of the slope of line p?

8. Write an equation of the line that bisects the segment whose endpoints are (4,–4) and (–2,8) and whose y-intercept is 3.

9. Molly charges a fixed amount for babysitting plus an additional charge for each hour that she works. Molly's total fee for babysitting for 3 hours one evening was $39, and her total fee for babysitting 5 hours on another evening was $57.

a. Molly's next babysitting job is for h hours. Find an equation that represents the number of dollars, D, expressed in terms of h, that Molly earns babysitting for h hours.
b. What is Molly's hourly rate for babysitting?

CHAPTER 11

FUNCTIONS, GRAPHS, AND MODELS

11.1 FUNCTION CONCEPTS

KEY IDEAS

If you assume wisdom depends on age, then wisdom *is a function of* age. The term *function* in mathematics also implies a dependent relationship between variables. The area, A, of a circle depends on its radius, r. The equation $A = \pi r^2$ expresses A as *a function of* r. When $r = 3$, $A = 9\pi$, which is described by the ordered pair $(3, 9\pi)$. Allowing r to take on different positive values produces a set of ordered pairs of the form (r, A) in which no two ordered pairs have the same first member, but different second members. Any set of ordered pairs with this property is a **function**.

Informal Definition of a Function

You can think of a function as a machine in which an x-value goes in, one or more operations are performed on it, and then exactly one y-value comes out. For the machine to be a function, if the same x-value is put in again, the same y-value as before must come out. The set of all possible x-values that the machine can accept is called the **domain** of the function. The set of corresponding y-values that comes out is called the **range**.

MATH FACT

A function is a rule that tells how to assign each element of one set, called the domain, to an element of another set, called the range, such that each element of the domain is assigned to exactly one element of the range. A function may pair different elements in the domain with the same element in the range. However, for each x-value there is exactly one y-value.

Representing a Function as a Set of Ordered Pairs

Any set of ordered pairs of the form (x,y) represents a function provided no two ordered pairs have the same x-values but different y-values.

- The set of ordered pairs {(1,3), (2,2), (3,4)} represents a function.
- The set of ordered pairs {(5,–1), (8,3), (5,4)} is not a function since the x-value of 5 is paired with two different y-values, –1 and 4.
- A function may assign different elements in the domain to the same element in the range. For example, {(2,–3), (1,7), (4,–3)} is a function that assigns two different x-values to the same y-value of –3.

Example 1

Which set of ordered pairs does *not* represent a function?

(1) {(3,−2), (−2,3), (4,−1), (−1,4)}
(2) {(1,−2), (2,1), (4,−1), (4,−3)}
(3) {(3,−2), (4,−3), (−3,4), (6,3)}
(4) {(3,−2), (5,−2), (4,−2), (−1,−2)}

Solution: Examine each of the answer choices in turn until you find the one that pairs the same x-value with two different y-values. In choice (2), the same x-value of 4 is paired with both –1 and –3, so the set of ordered pairs is not a function.

The correct choice is **(2)**.

Function Notation

A function is usually referred to by a lowercase letter such as f. If the input for a function f is 3, the corresponding output may be represented as $f(3)$, which is read as "f of 3." To find $f(3)$ when $f(x) = x^2 - 1$, replace x with 3 and simplify:

$$\begin{aligned} f(3) &= 3^2 - 1 \\ &= 9 - 1 \\ &= 8 \end{aligned}$$

Since $f(3) = 8$ when $x = 3$, $y = 8$. So the ordered pair (3,8) belongs to function f. Here are some more facts about functions.

- Although the letter f is commonly used to name a function, any other single letter may be used to name a function.
- The graph of function f is the graph of the equation $y = f(x)$. The graph of function f consists of the set of all ordered pairs $(x, f(x))$ that results from x taking on each of its possible values.
- Because $y = f(x)$, the terms "function value" and "y-value" are sometimes used interchangeably.
- A function is sometimes described by the type of equation that defines it. For example, a function that is defined by a first-degree equation such as $f(x) = 3x - 2$ is called a **linear function**. A function that is defined by a second-degree equation such as $f(x) = 3x^2 - 2x + 5$ is called a **quadratic** function.

Example 2

If $f(x) = \sqrt{x} + 2x$, what is the value of $f(9)$?

Solution: To find the value of $f(9)$, replace x with 9 and simplify:

$$\begin{aligned} f(9) &= \sqrt{9} + 2 \cdot 9 \\ &= 3 + 18 \\ &= 21 \end{aligned}$$

Example 3

If $g(x) = 2x + c$, for what value of c is $g(5) = 9$?

Solution: Replace x with 5 and $g(5)$ with 9. Then solve for c:

$$\begin{aligned} g(5) &= 2 \cdot 5 + c \\ 9 &= 10 + c \\ 9 - 10 &= c \\ c &= -1 \end{aligned}$$

Example 4

For what values of x is the function $f(x) = \dfrac{3x+1}{x^3 - 9x}$ undefined?

Solution: A fraction with a variable denominator is undefined for any value of the variable that makes the denominator evaluate to 0. To find the values of x that make function f undefined, set $x^3 - 9x = 0$. Solve for x by factoring the left side of the equation:

$$\begin{aligned} x^3 - 9x &= 0 \\ x(x^2 - 9) &= 0 \\ x(x+3)(x-3) &= 0 \end{aligned}$$

$$x = 0 \quad or \quad x + 3 = 0 \quad or \quad x - 3 = 0$$

$$x = 0 \quad or \quad x = -3 \quad or \quad x = 3$$

Function f is undefined when $x = -3$, 0, or 3.

Zeros of a Function

If f is a function, then those values of x that make $f(x) = 0$ are called the **zeros** of the function. The zeros of function f are the x-intercepts of the graph of $y = f(x)$. To find the zeros of $f(x) = (x - 5)(x^2 - 4)$, set each factor equal to 0 and solve for x:

- If $x - 5 = 0$, then $x = 5$.
- If $x^2 - 4 = 0$, then $x = -2$ *or* $x = +2$.

The zeros of f are -2, 2, and 5.

Representing A Function In Different Ways

A function can be represented in different ways: algebraically as an equation, numerically as a table of values, visually as a graph, or simply as a rule expressed in words. Suppose the possible side lengths of a certain square are 1, 3, and 5. The function that pairs the possible side lengths of the square to the corresponding perimeters is $\{(1,4), (3,12), (5,20)\}$.

In this example, the domain is $\{1,3,5\}$ and the range is $\{4,12,20\}$. Figure 11.1 represents this function in different ways where s is the side length of the square and p is the corresponding perimeter.

Algebraic Representation	Numerical Representation	Graphical Representation
$p = 4s$ where the domain of s is $\{1,3,5\}$.	s / p: 1 / 4; 3 / 12; 5 / 20	Graph of points (1,4), (3,12), (5,20); y-axis: Perimeter (4, 8, 12, 16, 20, 24); x-axis: Side (1, 2, 3, 4, 5, 6)

Figure 11.1 Representing the Same Function in Different Ways

You can also express this same function as a verbal rule by stating, "If I give you the side length of a square, you give me the perimeter of that square."

Example 5

Determine whether each relation is a function. Give a reason for your answer.

a. $g = \{(-3,3), (7,4), (2,2), (1,4)\}$

b.

c. $x = y^2$

d.

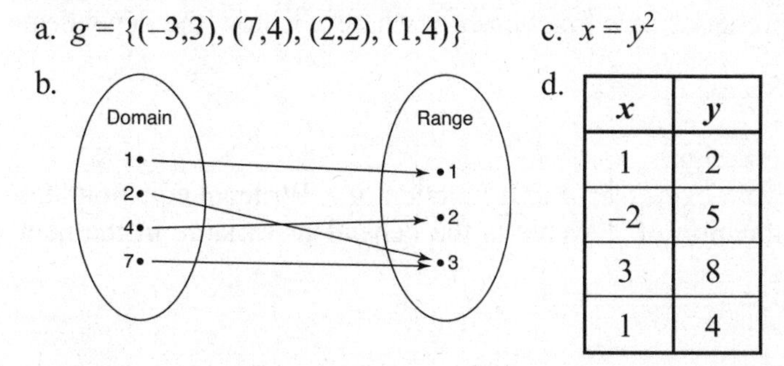

x	y
1	2
–2	5
3	8
1	4

Solution:

a. The set of ordered pairs represents a function because no two ordered pairs have the same x-value paired with different y-values. It does not matter that different ordered pairs, namely (7,4) and (1,4), have the same y-values paired with different x-values.
b. The mapping diagram represents a function that consists of a set of ordered pairs, $\{(1,1), (2,3), (4,2), (7,3)\}$, in which no x-value is paired with different y-values.
c. Not all equations describe functions. The equation $x = y^2$ produces ordered pairs with the same x-value but different y-values such as (4,2) and (4,–2). Hence, this equation does *not* represent a function.
d. Each row of the table corresponds to an ordered pair. Two of these ordered pairs, ($\underline{1}$,2) and ($\underline{1}$,4), have the same x-value but different y-values. Hence, this relation is ***not*** a function.

Independent Versus Dependent Variables

In the equation $p = 4s$, variable s is the **independent variable** since it can be assigned any value from its domain. Variable p is the **dependent variable** because its value depends on what value is chosen for s.

- The independent variable is always the first member of each ordered pair of a function:

$$f = \{(\underbrace{\text{independent variable}}_{x\text{-value}}, \underbrace{\text{dependent variable}}_{y\text{-value}})\}$$

- It is customary to write a function equation in a form in which the dependent variable, say y, is solved for in terms of the independent variable, say x, as in $y = 2x - 1$. When written in this form, the dependent variable y is said to be *a function of* the independent variable x.
- Solving $y = 2x - 1$ for x in terms of y gives $x = \frac{y+1}{2}$. The transformed equation expresses x as *a function of* y where x is now the dependent variable.

Example 6

The equation $A = \pi r^2$ expresses A as a function of r. Write an equation that expresses r as a function of A. What is the dependent variable in the new equation?

Solution: To express r as a function of A, solve for r in terms of A:

$$A = \pi r^2$$
$$r^2 = \frac{A}{\pi}$$
$$\boldsymbol{r} = \sqrt{\frac{A}{\pi}}$$

In the new equation, $\boldsymbol{r}$ is the dependent variable.

Example *7*

If $2p+3q=8$, express p as a function of q. In the transformed equation, what are the independent and dependent variables?

Solution: To express p as a function of q, solve the equation for p in terms of q.

$$2p+3q=8$$
$$\frac{2p}{2}=\frac{-3q}{2}+\frac{8}{2}$$
$$\boldsymbol{p}=\frac{-3q}{2}+4$$

In this equation, $\boldsymbol{q}$ is the independent variable, and $\boldsymbol{p}$ is the dependent variable.

Restricting the Domain

If the domain of a function is not explicitly stated, assume it is the set of real numbers or the largest possible subset of it. The domain may need to be restricted by excluding any x-value that does not produce a real value for y.

- The equation $y=\frac{x}{x+2}$ represents a function. The domain is understood to include all real numbers *except* -2 since when $x=-2$, the fraction is undefined.
- The equation $y=\sqrt{x-1}$ represents a function. The domain includes only those real numbers greater than or equal to 1 because when $x<1$, the corresponding y-values are not real numbers.

Vertical Line Test

Although a function may take the form of a graph, not all graphs describe functions. A graph represents a function if *no* vertical line can be drawn that intersects the graph in more than one point. Because the graph in Figure 11.2 passes the vertical line test, it represents a function.

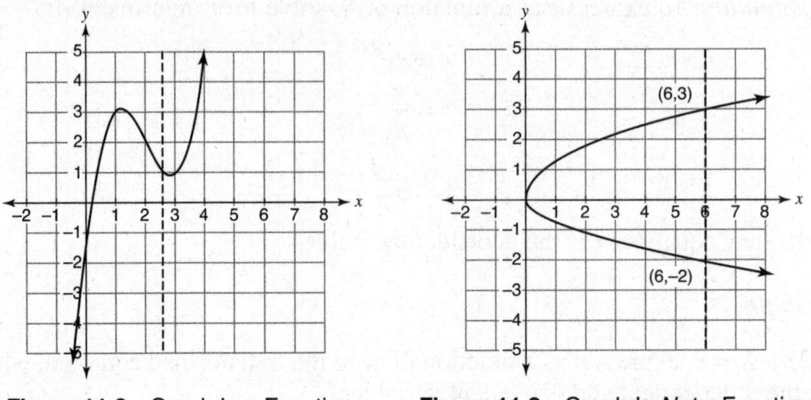

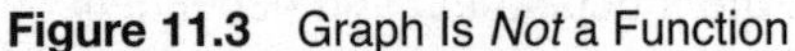

Figure 11.2 Graph Is a Function **Figure 11.3** Graph Is *Not* a Function

In Figure 11.3, it is possible to draw a vertical line that intersects the graph in two different points indicating that the graph includes two ordered pairs with the same x-value but different y-values. Because the graph fails the vertical line test, it does *not* represent a function.

Example 8

Which of the accompanying diagrams does *not* describe a function?

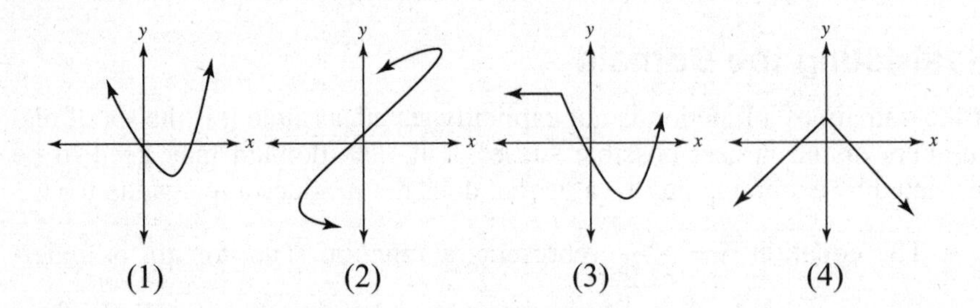

Solution: Each graph passes the vertical line test except the graph in choice (2), as shown in the accompanying figure. Thus, the graph in **choice (2)** does not represent a function.

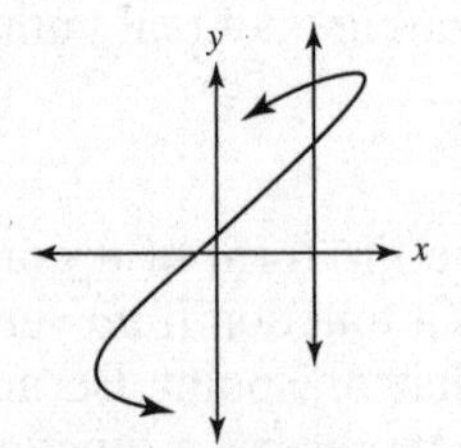

Check Your Understanding of Section 11.1

A. Multiple Choice.

1. Each graph below represents a possible relationship between temperature and pressure. Which graph does *not* represent a function?

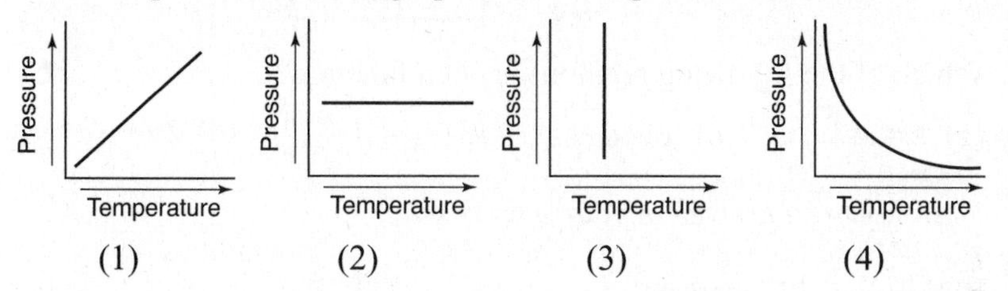

2. Which set of ordered pairs is *not* a function?

(1) {(3,1), (2,1), (1,2), (3,2)}
(2) {(4,1), (5,1), (6,1), (7,1)}
(3) {(1,2), (3,4), (4,5), (5,6)}
(4) {(0,0), (1,1), (2,2), (3,3)}

3. Which of the accompanying diagrams describes a function?

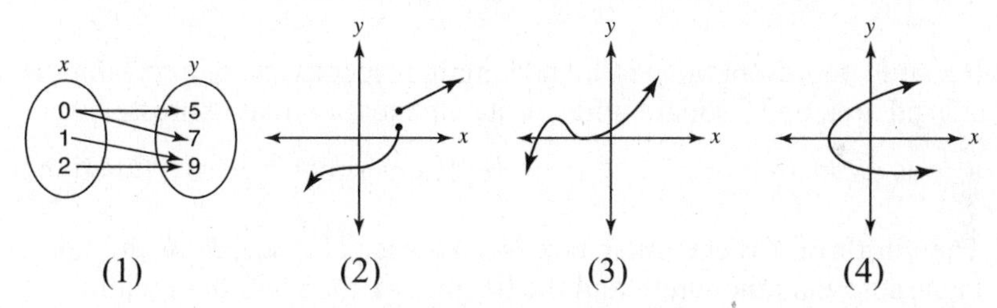

4. Which of the accompanying diagrams does not describe a function?

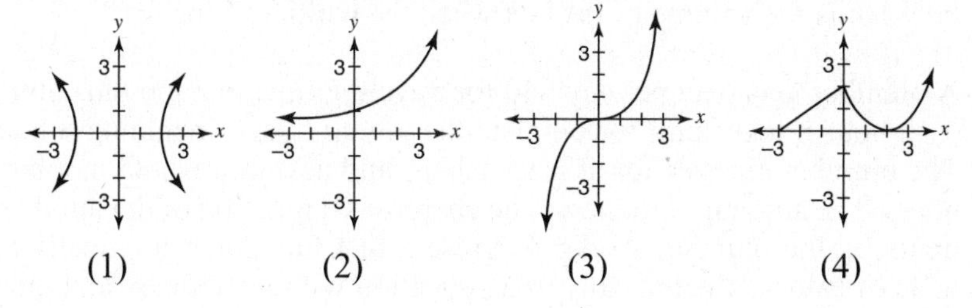

5. Which linear function represents the data in the accompanying table?
(1) $d = 1.50c$
(2) $d = 1.50c + 20.00$
(3) $d = 20.00c + 1.50$
(4) $d = 21.50c$

c	d
0	20.00
1	21.50
2	23.00
3	24.50

6. Which of the following relations is not a function?

(1) $x = |y|$ (2) $y = \sqrt{x}$ (3) $x = \sqrt{y}$ (4) $2x + 3y = 6$

B. *Show how you arrived at your answer.*

7. If $4x + 3y = 12$, express:

a. y as a function of x

b. x as a function of y

8. If x represents the length of a rectangle with a fixed perimeter of 20, write an equation that expresses the area, A, of the rectangle as a function of x.

9. If x and y represent the width and length, respectively, of a rectangle with a fixed area of 36 square units, write an equation that expresses:

a. y as a function of x

b. The perimeter, p, as a function of x

10. The width of a rectangular box is x inches. The length of the box is 1 inch more than the width and the height is 1 inch less than width.

a. Express the volume, V, of the box as a function of x.
b. What is the volume of the box when the width is 4 inches?

11. A plumber gets paid per day \$40 for traveling time to a job plus \$60 for each hour it takes him to complete the job up to and including 8 hours. The plumber charges for at least 1 hour and always a whole number of hours. The amount of money, y, he charges is a function of the number of hours, x, the plumber works. Represent this function numerically as a table of values, algebraically as an equation with its domain, and graphically as a set of points.

11.2 FUNCTION GRAPHS AS MODELS

KEY IDEAS

Function graphs are often used to describe the relationship between two real-world quantities. When a real-world function is graphed, the independent "x-variable" is measured along the horizontal axis, and the dependent "y-variable" is measured along the vertical axis. The slope of a line represents the *rate of change* between the quantities measured along the coordinate axes.

Graphs as Models

A graph can model a situation by giving a visual picture of how one real-world quantity depends on another.

Example 1

Janet decided as part of a school project to study the drinking habits of her dog Rover and to summarize her findings in a graph. The accompanying graph shows the amount of water left in Rover's water dish over a period of time. How many seconds did Rover wait from the end of his first drink to the start of his second drink of water?

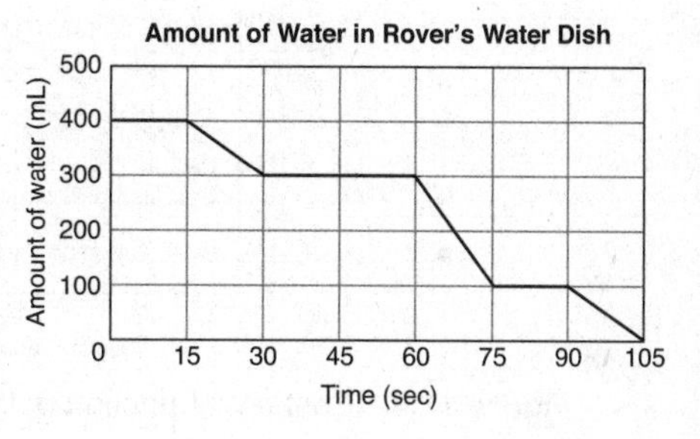

Solution: Since the graph is flat from 0 to 15 seconds, the water level did not change so Rover did not drink any water.

- From 15 to 30 seconds, Rover took his first drink of water which ended at the 30-second mark.
- From 30 to 60 seconds, the water level did not change so Rover did not drink. At the 60-second mark, Rover began his second drink of water.
- Rover waited $60 - 30 =$ **30** seconds from the end of his first drink to the start of his second drink of water.

Linear Functions

A **linear function** is a function whose graph is a nonvertical line so its equation has the form $y = ax + b$. Because the slope of a line is constant, linear functions are characterized by a constant rate of change.

- The linear function $F = 1.8C + 32$ tells how to convert a given Celsius temperature to an equivalent Fahrenheit temperature. Because the slope of the corresponding line is 1.8, each 1 degree change in the Celsius temperature produces a change of 1.8 degrees in the equivalent Fahrenheit temperature.
- The cost *C*, in dollars, of manufacturing *x* units of a certain product increases at a constant rate as measured by the slope of the line in Figure 11.4, which shows the overall pattern of how cost depends on the number of units manufactured.

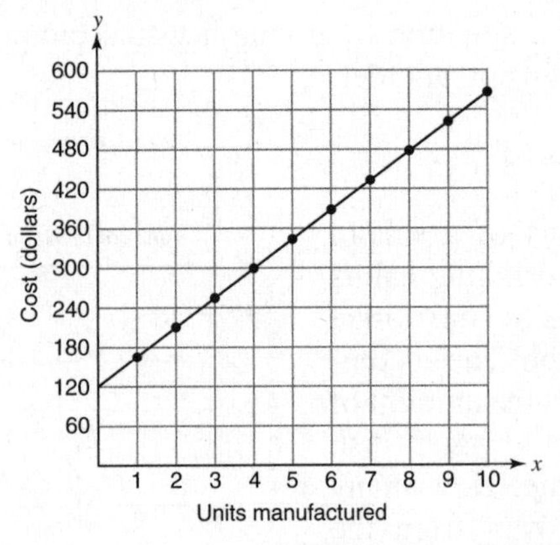

Figure 11.4 Cost as a Function of the Number of Units Manufactured

To find the rate at which the cost changes for each additional unit manufactured, determine the slope of the line by choosing any two convenient points with integer coordinates such as (4,300) and (8,480):

$$\text{slope} = \frac{\Delta y}{\Delta x} = \frac{480 - 300}{8 - 4} = \frac{180}{4} = 45\ \frac{\text{dollars}}{\text{unit}}$$

Thus, the cost of manufacturing each additional unit is **\$45**.

Example 2

Find an equation for the linear function in Figure 11.4.

Solution: The line is the graph of a linear function so its equation has the general form $C = ax + b$, where a is the slope of the line and b is the y-intercept. Because $a = 45$ and $b = 120$, $\boldsymbol{C = 45x + 120}$.

Example 3

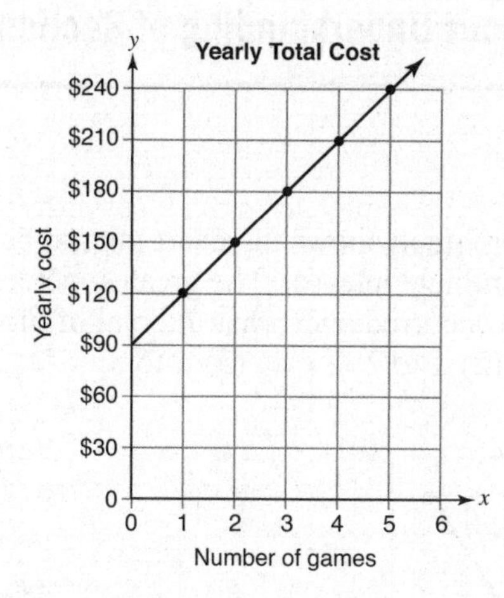

The accompanying graph represents the yearly cost of playing 0 to 5 games of golf at the Shadybrook Golf Course, which includes a yearly membership fee.

a. What is the cost of playing one game of golf?
b. Write a linear function that expresses the total cost in dollars, C, of joining the club and playing n games during the year?
c. Using your answer in part a, find the cost of playing 10 games during the year.

Solution:

a. The slope of the line represents the cost per game. Pick any two points on the line with integer coordinates such as (1,120) and (2,150) and find the slope of the line:

$$\text{slope} = \frac{\Delta y}{\Delta x} = \frac{150-120}{2-1} = 30\frac{\text{dollars}}{\text{game}}$$

The cost of one game is **$30**.

b. When the number of games is 0, the cost is $90. Hence, the y-intercept of the line represents the yearly cost of membership. Because the slope of the line is 30 and its y-intercept is 90, an equation of the line is $\boldsymbol{C = 30n + 90}$.
c. To find the total cost of playing 10 games during the year, substitute 10 for n in $C = 30n + 90$, which gives $y = 30 \cdot 10 + 90 = 390$. The total cost of joining the club and playing 10 games during the year is **$390**.

Check Your Understanding of Section 11.2

Multiple Choice.

1. The accompanying graph shows the heart rate, in beats per minute, of a jogger during a 4-minute interval. The greatest percent of increase in the jogger's heart rate occurred over what interval of time, in minutes?
(1) 0 to 1 (2) 1 to 2 (3) 2 to 3 (4) 3 to 4

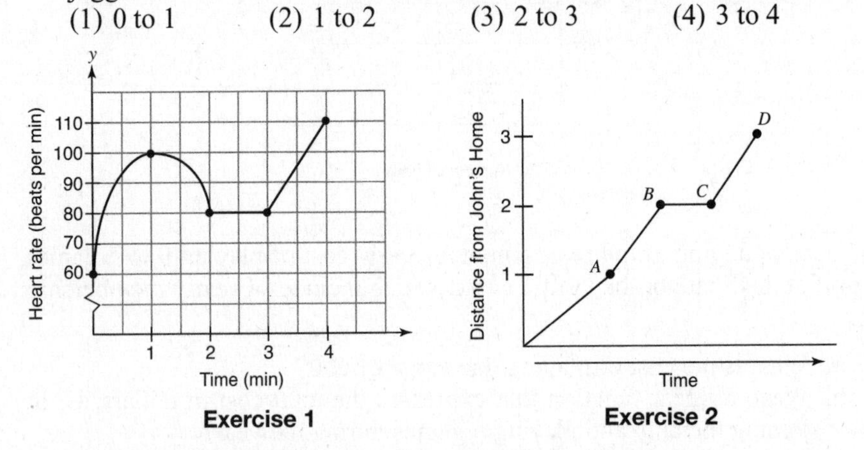

2. John left his home and walked 3 miles to his friend's house, as shown in the accompanying diagram. Between which two points was John's average rate of speed the greatest?
(1) *O* to *A* (2) *A* to *B* (3) *B* to *C* (4) *C* to *D*

3. The cost C, in dollars, of manufacturing x units of a certain product is given by the linear function $C = ax + b$, as shown in the accompanying graph, where a and b are positive constants. Which of the following functions, h, represent the *average* cost per unit, in dollars, of manufacturing x units?

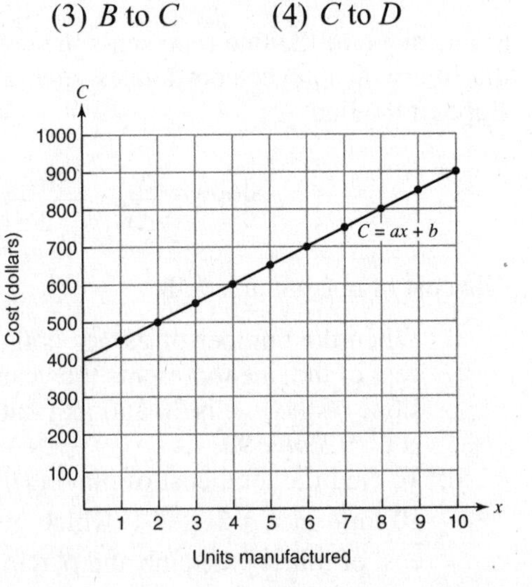

(1) $h = 50x + 400$
(2) $h = 400x + 50$
(3) $h = \frac{400}{x} + 50$
(4) $h = \frac{50}{x} + 400$

4–5. *The accompanying graph shows the speed in miles per hour at which Brad drove from 9:00 to 11:30.*

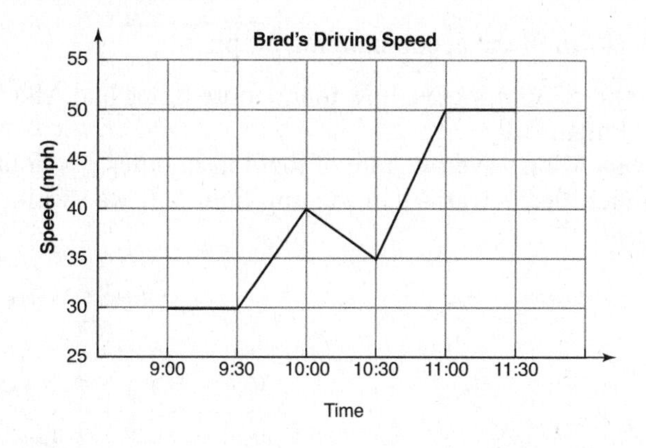

4. During which interval of time did Brad's driving speed increase at the greatest rate?

(1) 9:30 to 10:00 (2) 10:00 to 10:30
(3) 10:30 to 11:00 (4) 11:00 to 11:30

5. What was Brad's average driving speed, in miles per hour, from 9:00 to 10:00?

(1) 32.5 (2) $33\frac{1}{3}$ (3) 35 (4) 37.5

B. *Show how you arrived at your answer.*

6. A certain rectangle has a fixed length of 5 inches but its width can vary.

a. If x represents the number of inches in the width of the rectangle and P represents the perimeter of the rectangle, write an equation that expresses P as a function of x.

b. Graph the equation in part a. Using the graph, find the perimeter of the rectangle when the width is 8 inches.

7. When Carol works as a babysitter, she charges a fixed amount of $10 plus an additional $8 per hour for each hour that she works.

a. If y represents the total amount of money Carol earns for working x hours during one babysitting job, write an equation that expresses y as a function of x.

b. Graph the equation written in part a. From the graph, determine how much Carol earns when she babysits for $4\frac{1}{2}$ hours.

8. During a gym class, Allan jogged and Bob walked at constant rates around a circular $\frac{1}{4}$-mile track. Their times in minutes and distances in miles are shown in the accompanying graph.

a. At the end of 30 minutes, how many more times had Allan completed the track than Bill?
b. What was Allan's average rate of jogging in miles per hour?
c. How much faster was Allan jogging than Bill was walking in miles per hour?

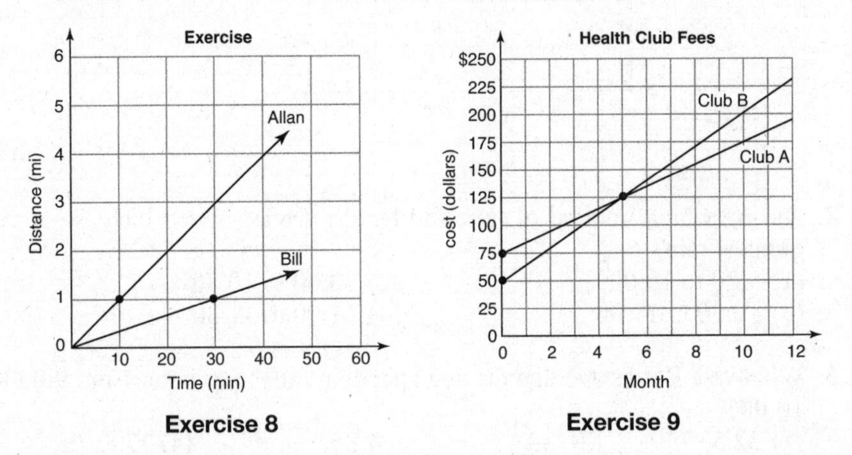

Exercise 8 **Exercise 9**

9. Two health clubs offer different membership plans. The accompanying graph represents the total cost of belonging to Club A and Club B for one year. The yearly cost for each club includes an initial membership fee plus a monthly charge.

a. By what amount does the membership fee for Club B exceed the membership fee for club A?
b. If Jack's membership in Club A begins in January and Jill's membership in Club B begins in the same month, in what month will the total cost in club membership be the same for Jack and Jill?
c. What is the *monthly* charge for Club B?
d. Write a linear function that expresses the total charges, y, for membership in Club A for a period of x months.

10. The accompanying figure represents the distances traveled by car A and car B at the end of 6 hours.

a. Car A is traveling how many miles per hour faster than car B?

b. Represent the distance each car travels as a function of time.

c. If the cars continue to travel at the same rates of speed, how many more miles has car A traveled at the end of 8 hours than car B?

11. Two heating systems for a new apartment complex, electric and oil, are being considered. The accompanying graph compares the total costs using electric and oil systems to heat the apartment complex.

a. For which heating system does the cost increase at the greatest rate? Explain your answer.

b. After 10 years, which system is projected to be more expensive and by how many thousands of dollars?

c. Write an equation that represents the total cost of oil heat, C, in Y years where C is expressed in terms of Y.

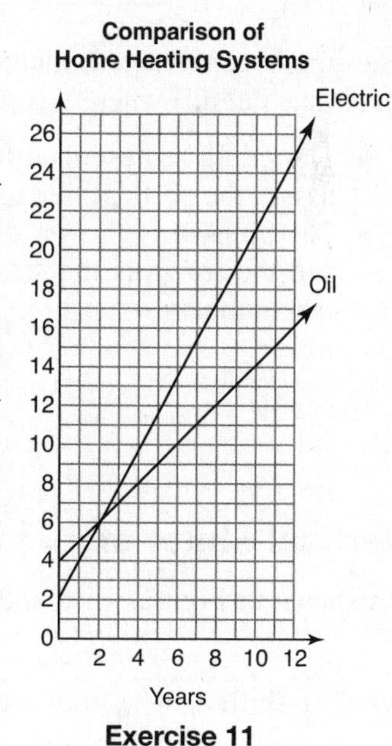

Exercise 11

11.3 THE ABSOLUTE VALUE FUNCTION

KEY IDEAS

The equation $y=|x|$ represents the **absolute value function.** Because the absolute value of a real number is always nonnegative, $y=|x|=x$ when $x \geq 0$ and $y=|x|=-x$ when $x < 0$. Hence, the graph of the absolute value function includes the origin and the parts of the lines $y = x$ and $y = -x$ that lie above the x–axis. The graph of $y=|x|$ can be shifted up or down by combining y with a number.

Using Technology: Graphing $y = |x|$

To graph the absolute value function $y = |x|$ using your graphing calculator, open the $Y =$ editor and press

The graph is **V**-shaped, includes the origin, and is otherwise restricted to quadrants I and II where y is positive, as shown in Figure 11.5.

- The graph represents a function since it passes the vertical line test.
- The *domain* is the set of real numbers, and the *range* is the set of nonnegative real numbers.

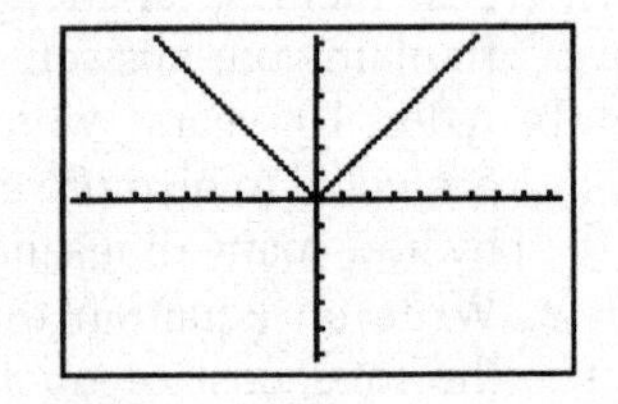

Figure 11.5 Graph of $y = |x|$

Vertical Shifts of $y = |x|$

As shown in Figures 11.6 and 11.7, the graph of $y=|x|+2$ is the graph of $y = |x|$ shifted up 2 units, and the graph of $y=|x|-2$ is the graph of $y = |x|$ shifted down 2 units.

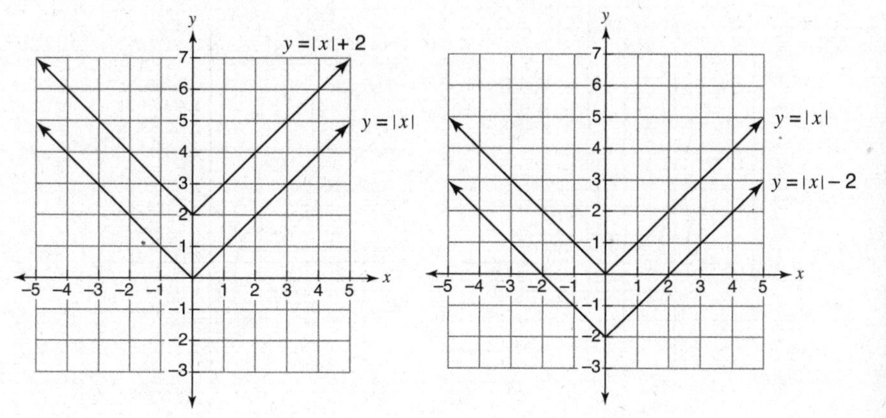

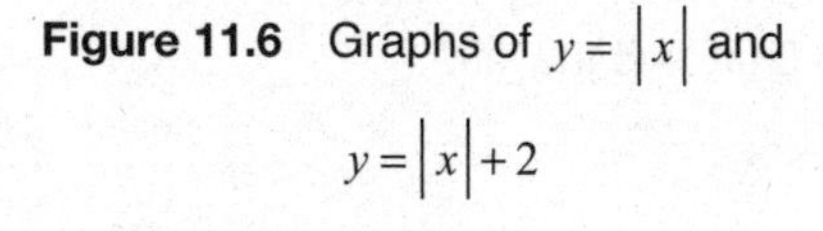

Figure 11.6 Graphs of $y = |x|$ and $y = |x| + 2$

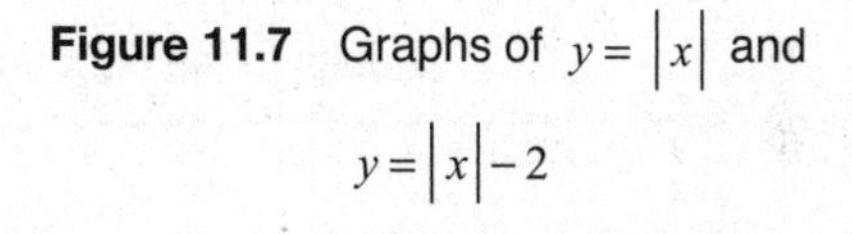

Figure 11.7 Graphs of $y = |x|$ and $y = |x| - 2$

Based on these figures, you can generalize that if k is a positive number,

- The graph of $y = |x + k|$ is the graph of $y = |x|$ shifted **up** k units, as in Figure 11.6.
- The graph of $y = |x - k|$ is the graph of $y = |x|$ shifted **down** k units, as in Figure 11.7.

These generalizations hold true for the graph of any function when a positive number k is added to the function or subtracted from it. For example, the graph of $y = 2x + 3$ is the graph of $y = 2x$ shifted *up* 3 units.

Horizontal Shifts of $y = |x|$

Adding a number inside the absolute value sign of $y = |x|$ shifts its graph sideways, as illustrated in Figure 11.8. If k stands for a positive number.

- The graph of $y = |x - k|$ is the graph of $y = |x|$ shifted horizontally k units to the *right*.
- The graph of $y = |x + k|$ is the graph of $y = |x|$ shifted horizontally k units to the *left*.

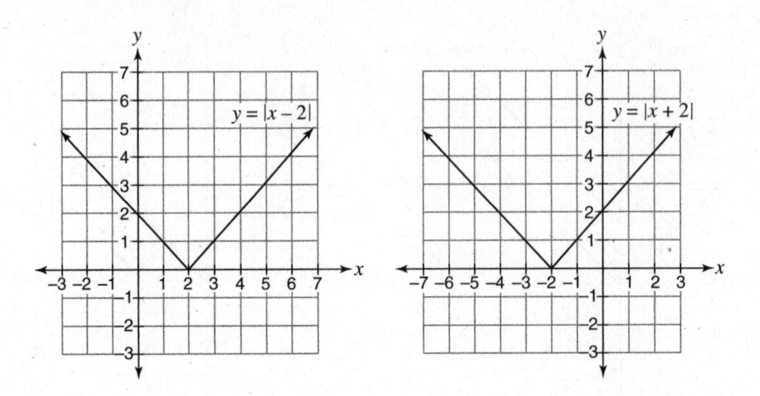

Figure 11.8 Shifting the graph of $y=|x|$ horizontally 2 units to the right and 2 units to the left.

Graph of $y = |ax|$

The coefficient of x in $y=|ax|$ affects how narrow the graph is compared with the graph of $y=|x|$. Figure 11.9 compares the graphs of $y=|ax|$ for $a=\frac{1}{2}$, $a=1$, and $a=2$.

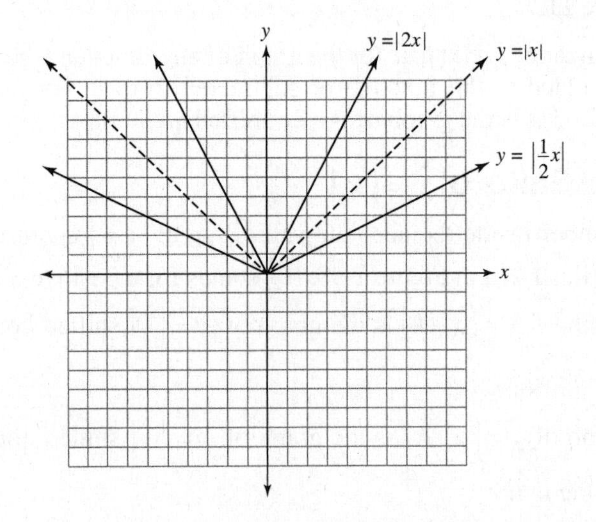

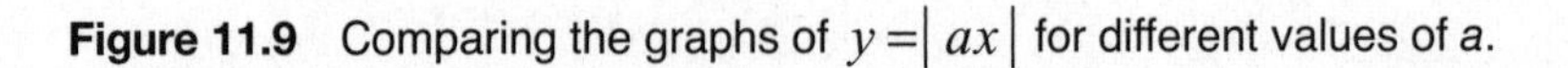

Figure 11.9 Comparing the graphs of $y=|ax|$ for different values of a.

You should know the following facts about the graph of $y = |ax|$:

- The value of the constant a in $y = |ax|$ affects the steepness of the rays of the graph. For example, as a increases from $\frac{1}{2}$ to 1 to 2, the rays of the graph of $y = |ax|$ become steeper and closer to the y-axis.
- The sign of a in $y = |ax|$ has no effect on the shape of the graph since a is multiplied by x inside the absolute value sign. For example, you should verify that the graphs of $y = |2x|$, $y = |-2x|$, $y = 2|x|$ are the same.

Graph of $y = a|x|$

You should know the following facts about the graph of $y = a|x|$:

- When graphing a function of the form $y = a|x|$ where $a \neq 0$, the sign of the constant a matters. The graph of $y = -2|x|$ is the graph of $y = 2|x|$ reflected over the x-axis, as shown in Figure 11.10.
- The value of the constant a in $y = a|x|$ affects the steepness of the rays of the graphs. As the value of the constant a increases in the positive direction, the rays of the graph become steeper. When the constant a decreases in the negative direction, the rays of the graph also become steeper.

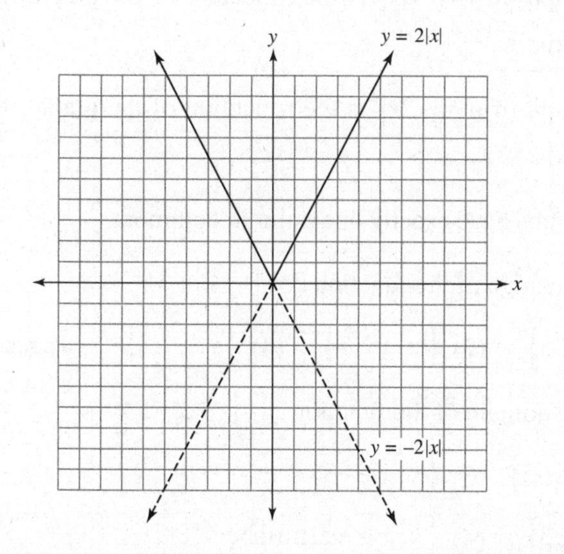

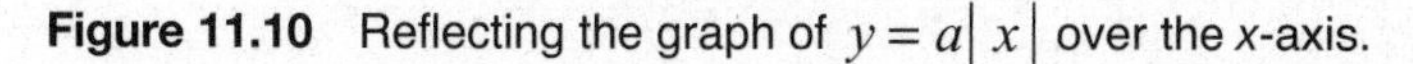

Figure 11.10 Reflecting the graph of $y = a|x|$ over the x-axis.

Check Your Understanding of Section 11.3

A. Multiple Choice.

1. If $y=|x|-x^3$, what is the value of y when $x=-2$?
(1) −10 (2) 10 (3) −6 (4) 4

2. When graphed on the same set of axes, in how many points do the graphs of $y=|x|$ and $y=3$ intersect?
(1) 1 (2) 2 (3) 0 (4) 4

3. When graphed on the same set of axes, in how many points do the graphs of $y=|x|-1$ and $y=-2$ intersect?
(1) 1 (2) 2 (3) 0 (4) 4

4. For which function is the domain $\{x\,|\,x\geq -1\}$?

(1) $y=|x+1|$ (2) $y=\sqrt{x+1}$ (3) $y=(x+1)^2$ (4) $y=\dfrac{1}{x+1}$

5. Which statement is true about the graphs of $y=|2x|$ and $y=|-2x|$?
(1) The graphs coincide.
(2) The graph of $y=|-2x|$ is the reflection of the graph of $y=|2x|$ in the x-axis.
(3) The graph of $y=|-2x|$ is the reflection of the graph of $y=|2x|$ in the y-axis.
(4) The graphs have exactly one point in common.

6. What is the range of the function $y=|x-2|+3$?

(1) $\{y: y\geq 3\}$ (2) $\{y: y\geq 2\}$ (3) $\{x: x\geq 3\}$ (4) $\{x: x\geq 2\}$

7. What is the domain of the function $y=|x+3|-5$?

(1) $\{y: y\geq -5\}$ (3) $\{x: x\geq -3\}$
(2) $\{y: y\geq 3\}$ (4) $\{x: x\in \text{real numbers}\}$

B. Show how you arrived at your answer.

8. Use your graphing calculator to solve the equation $|x+1|=4$ by graphing $y=|x+1|$ and $y=4$ on the same set of axes.

9. Graph $x=|y|$ on graph paper. Does the graph represent a function? Explain your answer.

10. What is the domain and range of the function $y=\frac{x}{|x|}$?

11. a. Using your graphing calculator, graph $y=|x|, y=|x+3|$, and $y=|x-3|$ on the same set of axes.

b. Based on part a, describe how the graph of $y=|x+c|$ compares to the graph of $y=|x|$ when c is positive and when c is negative.

11.4 CREATING A SCATTER PLOT

KEY IDEAS

A **scatter plot** is a graph of data points whose x-coordinates are from one set of measurements and whose y-coordinates are the corresponding values from another set of measurements. A scatter plot may show at a glance whether the relationship between the two sets of measurements is approximately linear.

Univariate Versus Bivariate Data

An **experimental** or **statistical variable** is a quantity that can be measured or observed. A set of measurements for a single experimental variable is called **univariate** data. Univariate data might consist of the pulse rates of a group of men. Statistical studies typically involve the relationships between two or more statistical variables. **Bivariate** data are a set of paired measurements for *two* related experimental variables such as the pulse rates of a group of men and their body weights.

Unbiased Versus Biased Samples

Suppose you wanted to conduct a survey to find out how students in your school feel about a school policy that affects the entire student body. It may not be practical or even possible to interview every student. In such cases, you can select to interview a smaller, representative group of students using some random selection process. For example, you could interview one out of every five students entering the school cafeteria on a particular day. In this example, the entire school population represents the **target population** and the smaller subset of students selected to be interviewed is a **sample**. A sample may be biased or unbiased.

- A sample is **unbiased** if each member of the target population has an equal chance of being included in the sample.
- If the selection process is not random and favors a particular group, the sample is **biased**.

Because the set of students selected to participate in this survey is chosen at random from the entire student body, the sample in our example is *unbiased*. If only senior boys who are members of the basketball team were selected to participate in the survey, the sample would most likely be *biased*. If a survey is biased, any general conclusions drawn from that survey may not be valid.

Qualitative Versus Quantitative Data

Data items that can be measured numerically or can be arranged on a numerical scale are referred to as **quantitative data**. Data collected based on the answers to questions that involve key words such as "how many," "what amount," or "how frequently" are typically quantitative data. Quantitative data can be analyzed using statistical or graphical methods. Nonnumerical data are sometimes referred to as **qualitative data**. Suppose a group of college graduates are interviewed.

- If their ages are collected, the type of data collected is quantitative.
- If the names of the colleges they graduated from are collected, the type of data collected is qualitative.

Scatter Plots for Bivariate Data

Suppose height and weight measurements of a group of people are collected. If the data are plotted as a set of ordered pairs of the form (height, weight), then the graph that results is called a **scatter plot**, as shown in Figure 11.11. Although it is not possible to draw a single line through all the data points, the data appear to be clustered about a line. Whenever this is the case, the relationship between the two sets of measurements is said to be *approximately* linear.

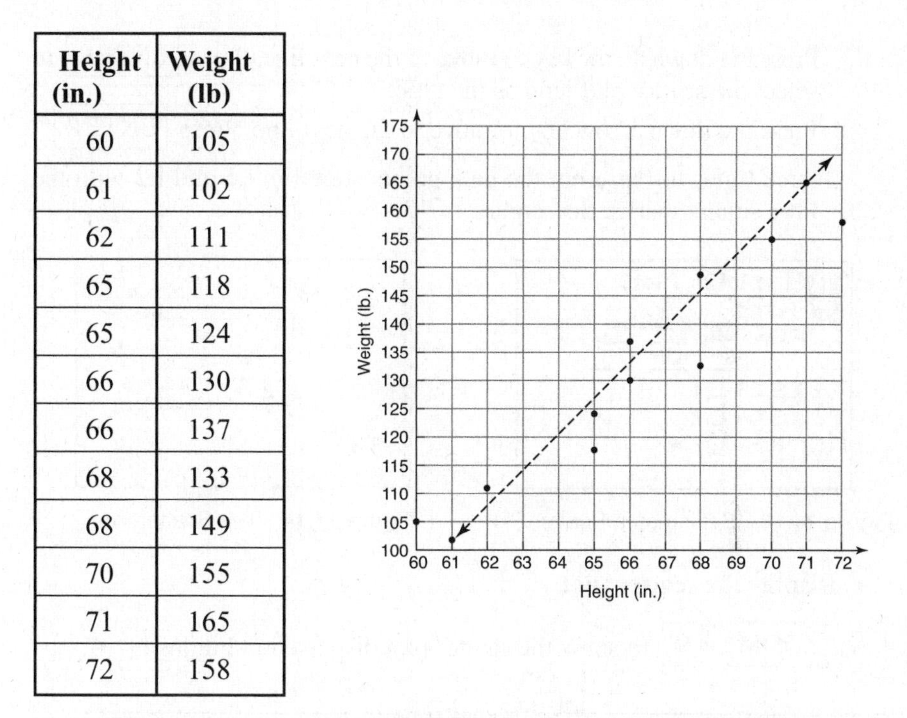

Height (in.)	Weight (lb)
60	105
61	102
62	111
65	118
65	124
66	130
66	137
68	133
68	149
70	155
71	165
72	158

Figure 11.11 A Scatter Plot for the (height, weight) Measurements Given in the Table

Using Technology: Creating a Scatter Plot

You can draw a scatter plot of the data points in Figure 11.11 either using graph paper or a graphing calculator. To use a graphing calculator, work as follows:

- **Store the data.**

  Press STAT ENTER. Enter the x-values (heights) in list L1 and the corresponding y-values (weights) in list L2, as shown in Figure 11.12.

L1	L2	L3 2
60	105	------
61	102	
62	111	
65	118	
65	124	
66	130	
66	137	

L2(7) =137

Figure 11.12 Storing the Data as Lists

- **Set up the scatter plot.**

1. Press 2nd [STAT PLOT] 1 ENTER to activate **Plot1**.

2. Press the down arrow key to move to the next line. Press ENTER to select the scatter plot icon as the type.
3. Press the down arrow key to move to the next line. Press ENTER three times to represent the data points stored in L1 and L2 with the little square marks. See Figure 11.13.

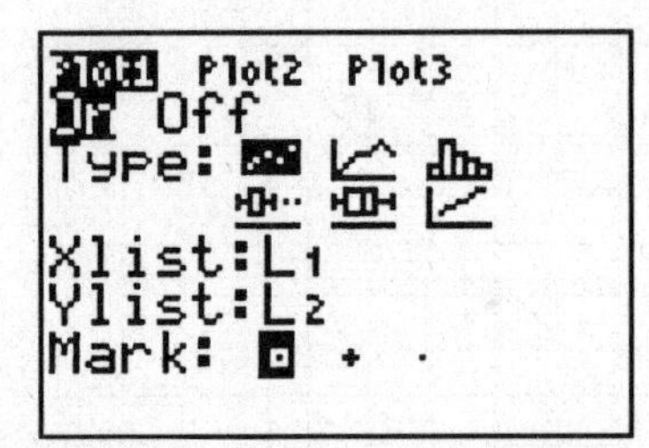

Figure 11.13 Setting up a Scatter Plot

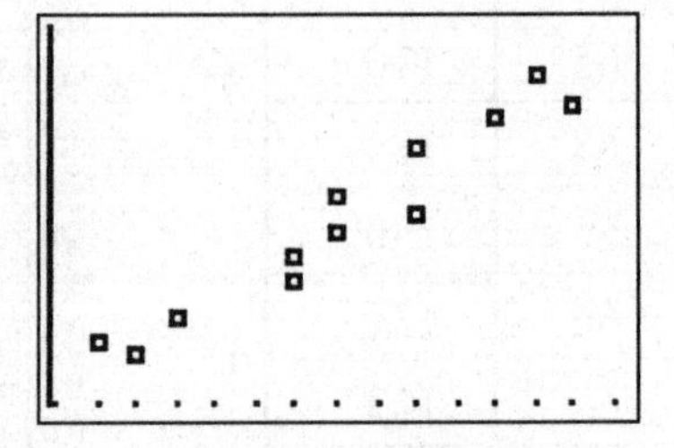

Figure 11.14 The Scatter Plot

- **Display the scatter plot.**

Press ZOOM 9 to show the scatter plot displayed in Figure 11.14.

MATH FACT

Data can be entered into a list from the home screen by enclosing the list of data values in braces and storing it in a list. For example, to store the numbers 1, 3, and 5 in list L1, press

[{] 1 , 3 , 5 [}] STO→ [L1] ENTER

Significance of a Scatter Plot

A scatter plot may suggest how x and y are related. If a single line can be drawn through all of the data points, there is a perfect linear relationship between x and y. This is rarely the case with experimental data. If all of the data points do not fall on the same line, but are clustered about a line, then x and y may still be linearly related. Errors in the collection or measurement of the data could help explain why a line does not contain all of the scatter plot points. If two experimental variables are linearly related, then an equation of the form $y = ax + b$ can be used to approximate their relationship.

Check Your Understanding of Section 11.4

1. The SAT exam is a college entrance exam that consists of a mathematics section and a verbal section. Draw a scatter plot using the data in the accompanying table of SAT scores for 10 students.

Math SAT	500	540	590	600	620	640	640	650	680	700
Verbal SAT	510	540	550	640	580	600	670	600	650	720

2. The accompanying table compares the number of hours that 12 students in Mr. Euclid's math class studied before taking their midterm exam with their test grades. Using this data, draw a scatter plot.

Hours, x	0	1	1.5	2	2.5	3	4.5	5	5.5	6	6.5	7
Grades, y	60	55	60	70	75	70	70	96	90	92	95	94

11.5 FINDING A LINE OF BEST FIT

If a scatter plot suggests that data are linearly related, then a *reasonable* line of best fit for the data can be approximated and drawn. Alternatively, a graphing calculator can be used to determine an equation of the line of best fit.

Approximating a Line of Best Fit

The line that is "closest" to the data points in a scatter plot is called the **line of best fit**. To find an equation of a reasonable line of best fit for the data in Figure 11.11 on page 311:

- Use a ruler to draw the line that appears to your eye to best fit the data points, such as the line drawn in Figure 11.15.

- Find the slope using any two points on the line with integer coordinates, such as (66,130) and (70,155):

$$m=\frac{\Delta y}{\Delta x}=\frac{155-130}{70-66}=6.25$$

- Find the y-intercept using either of the two points, say (66,130):

$$\begin{aligned} y&=6.25x+b \\ 130&=(6.25\times 66)+b \\ b&=130-412.5 \\ &=-282.5 \end{aligned}$$

An equation of the best-fit line that has been drawn in Figure 11.15 is $y=6.25x-282.5$.

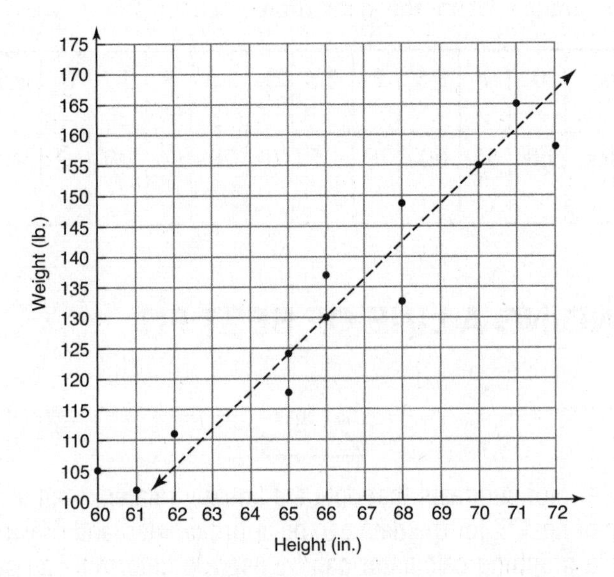

Figure 11.15 A Scatter Plot for the (height, weight) Measurements Given in the Table

Using Technology: Finding the Best-Fit Line

Although different lines may appear to the eye to fit the data points equally well, there is only one line of *best* fit whose equation can be determined using a graphing calculator. Because of the way an equation of the best-fit line is calculated, it is sometimes referred to as the **regression equation** and its graph as the **regression line**. Assuming the data from Figure 11.11 have already been entered in your calculator and the scatter plot settings have been stored (see page 311), work as follows:

- *Calculate the regression equation.* Press STAT, the right arrow key, and 4 to select the LinReg(ax + b) menu option. Then press VARS, the right arrow key, and 1 1 to store the calculated regression equation as Y_1.
- *View the regression equation.* Press ENTER to get the display in Figure 11.16 where a is the slope of the regression line and b is the y-intercept. The equation of the regression line is approximately $y = 5.29x - 218.06$.

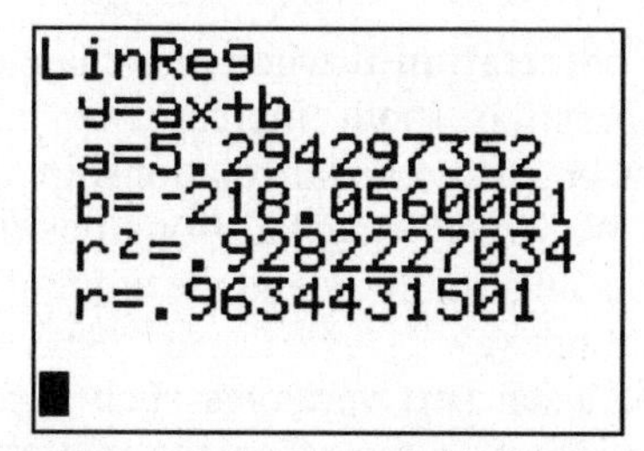

Figure 11.16 Regression Equation

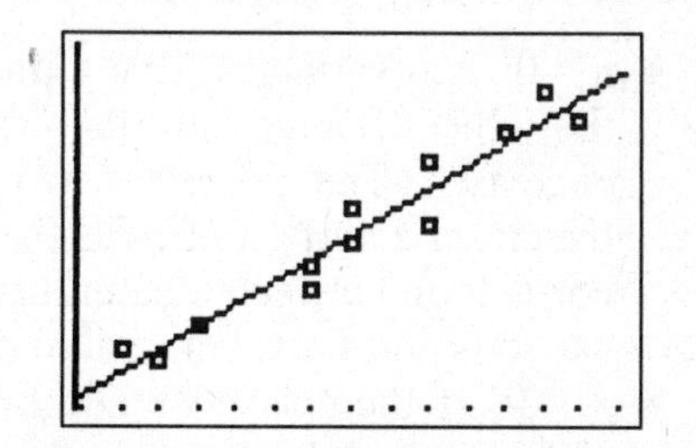

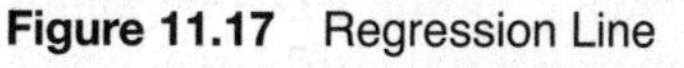

Figure 11.17 Regression Line

- *Graph the regression line on a scatter plot.* Press ZOOM 9 to get the display in Figure 11.17.

MATH FACTS

A regression line always contains the point $(\bar{x}, \bar{y})$ where $\bar{x}$ is the average (mean) of the x-data values and $\bar{y}$ is the average (mean) of the y-data values.

Coefficient of Linear Correlation

The r-value that appears in Figure 11.16 is the **coefficient of linear correlation** and measures how close the regression line comes to passing through all the actual data points. The value of r can range from -1 to $+1$. The closer the absolute value of r is to 1, the closer the regression line fits the data. Because the r-value is approximately 0.97, the calculated regression line fits the data points very closely, as can be seen in Figure 11.17.

If the r value does not appear with the regression equation, you can activate this feature by pressing

- 2nd [CATALOG].
- x^{-1} to scroll down to the catalog entries that begin with **D**.
- The down arrow cursor key until **DiagnosticsOn** is highlighted. Then press ENTER two times.

The next time the regression line is calculated, the r value will also be displayed.

Comparing Coefficients of Correlation

Figure 11.18 suggests that there are three cases to consider:

- $r > 0$. Two variables have a **positive correlation** if when one changes, the other changes in the same direction (both increase or both decrease). The closer r is to 1, the stronger the linear relationship and the closer a best-fit line fits the data. We would expect a strong positive correlation between the number of cars on a highway and the amount of car emission fumes measured on that highway.
- $r \approx 0$. If there is no relationship between two variables such that a change in one has a negligible or no effect on the other, the two variables are not correlated. The closer r is to 0, the weaker the linear relationship between two variables. Because there is little or no relationship between men's shoe sizes and the model year of the car that the men drive, their coefficient of correlation would be very close to 0. A scatter plot showing their relationship would contain widely scattered data points showing no pattern.
- $r < 0$. Two variables have a **negative correlation** if when one changes, the other changes in the opposite direction (when one increases, the other decreases). The closer r is to -1, the stronger the linear relationship and the closer a falling best-fit line fits the data. We would expect a strong negative correlation between the speed at which a student bicycles from his home to a park and the time the student takes to arrive at the park.

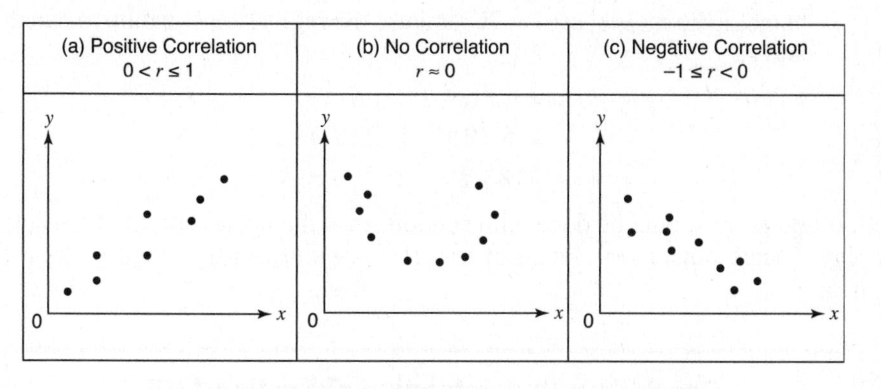

Figure 11.18 Comparing Coefficients of Correlation

Causation Versus Correlation

A strong linear correlation between x and y does not imply that x *causes* y. You would expect to find a high positive correlation between shoe size and the reading level of children. This does not mean that larger feet causes a higher reading level. Rather, reading level and foot size both increase as the age of a child increases.

Although a correlation coefficient close to 1 or −1 implies a strong *linear* relationship between the paired data, a correlation coefficient of approximately 0 does *not* imply the data are unrelated. It simply means that no *linear* relationship exists between the paired data.

Interpolation and Extrapolation

When there is strong linear correlation $\left(r \approx \pm 1\right)$, the regression equation becomes a useful model for predicting y using a value of x not included in the original set of observations.

- Estimating *within* the range of observed measurements for x is called **interpolation**. To estimate or *interpolate* the weight of an individual whose height is 69 inches, set $x = 69$, and use the regression equation to calculate y:

$$\begin{aligned} y &= 5.29x - 218.06 \\ &= (5.29)(69) - 218.06 \\ &= 146.95 \text{ pounds} \end{aligned}$$

- **Extrapolation** is estimation *outside* the range of observed measurements for x. To predict or *extrapolate* the weight of an individual whose height is 76 inches, set $x = 76$, and use the regression equation to calculate y:

$$\begin{aligned} y &= 5.29x - 218.06 \\ &= (5.29)(76) - 218.06 \\ &= 183.98 \end{aligned}$$

Extrapolation should be done with caution since the linear pattern of the data may change outside the range of observed x-values in the original sample data.

Check Your Understanding of Section 11.5

A. *Multiple Choice.*

1. Which scatter plot shows the strongest positive correlation?

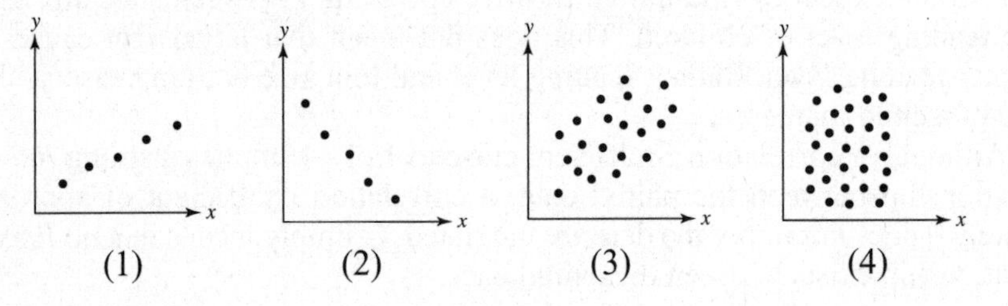

(1) (2) (3) (4)

2. Which scatter plot represents data that would have a correlation coefficient closest to -1?

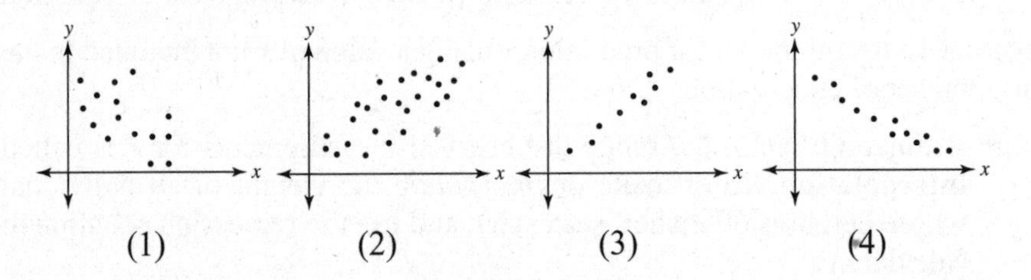

(1) (2) (3) (4)

3. The relationship of a woman's shoe size and length of a woman's foot, in inches, is given in the accompanying table.

Woman's Shoe Size	5	6	7	8
Foot Length (in.)	9.00	9.25	9.50	9.75

The linear coefficient of correlation for this relationship is
(1) 1 (2) -1 (3) 0.5 (4) 0

4. A best-fit linear equation between a student's attendance and the degree of success in school is $h = 0.5x + 68.5$. The correlation coefficient, r, for these data would be
(1) $0 < r < 1$ (2) $-1 < r < 0$ (3) $r = 0$ (4) $r = -1$

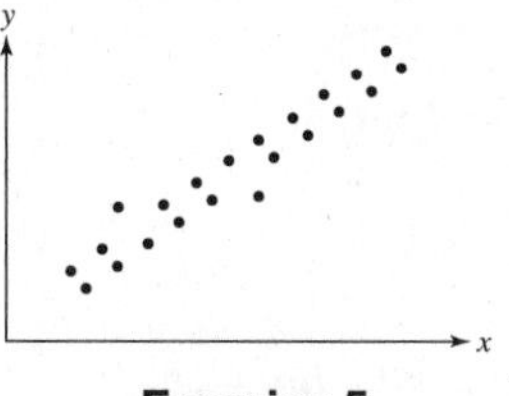

Exercise 5

5. What could be the approximate value of the correlation coefficient for the accompanying scatter plot?
(1) −0.85 (2) −0.16 (3) 0.21 (4) 0.90

6. Which phrase best describes the relationship between the number of incorrect answers on a test and the test score?
(1) causal, but not correlated (3) both correlated and causal
(2) correlated, but not causal (4) neither correlated nor causal

7. Which situation describes a correlation that is *not* a causal relationship?
(1) the length of the edge of a cube and the volume of the cube
(2) the number of hours Kathy works babysitting and the amount of money Kathy earns babysitting
(3) heights of students and their scores on the Integrated Algebra Regents exam
(4) the time of day and the amount of harmful car fumes emitted on a highway

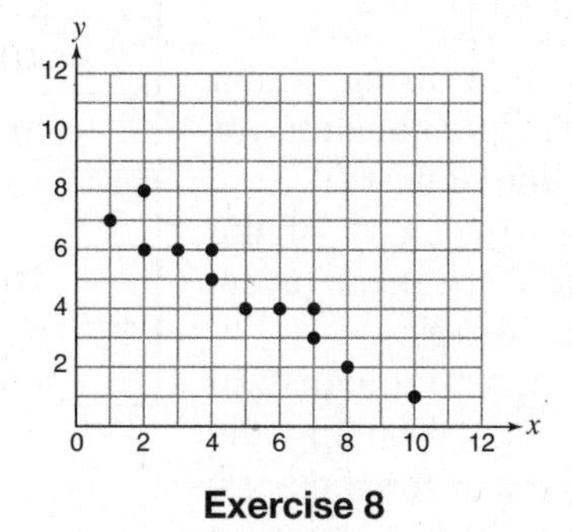

Exercise 8

8. Which equation most closely represents the line of best fit for the accompanying scatter plot of data points?

(1) $y = -\frac{2}{3}x + 8$ (2) $y = \frac{2}{3}x + 8$ (3) $y = \frac{3}{2}x + 10$ (4) $y = -\frac{3}{2}x + 10$

9. Which equation most closely represents the line of best fit for the scatter plot of data points?

(1) $y = \frac{3}{2}x + 3$ (3) $y = \frac{1}{2}x + 2$

(2) $y = \frac{3}{2}x + 1$ (4) $y = \frac{1}{2}x + 4$

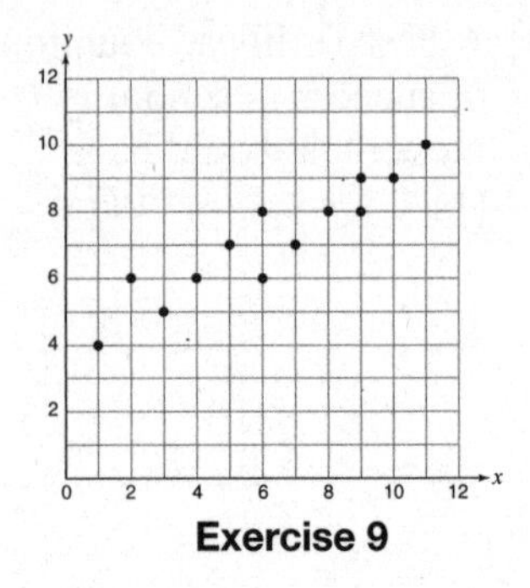

Exercise 9

B. Show how you arrived at your answer.

10. The data table to the right shows water temperatures at various depths in an ocean.

a. Construct a scatter plot for this data, and draw a line of best fit.
b. Write an equation of the line of best fit drawn in part a.
c. Using this equation, predict the temperature (°C), to the *nearest integer*, at a water depth of 275 meters.

Water Depth (x) (meters)	Temperature (y) (°C)
50	20
75	16
100	10
150	4
200	2

Exercise 10

11. The accompanying table shows the results of an experiment that relates the height at which a ball is dropped, x, to the height of its first bounce, y.

a. Draw a scatter plot using graph paper and then find an equation for an *approximate* line of best fit.
b. Find $\bar{x}$, the mean (average) of the drop heights, and $\bar{y}$, the mean (average) of the bounce heights.
c. Find the linear regression equation for this set of data, rounding the coefficients to three decimal places. Then show that $(\bar{x}, \bar{y})$ is a point on the regression line.

Drop Height, x (cm)	Bounce Height, y (cm)
100	26
90	23
80	21
70	18
60	16

Exercise 11

CHAPTER 12

SYSTEMS OF LINEAR EQUATIONS AND INEQUALITIES

12.1 SOLVING LINEAR SYSTEMS GRAPHICALLY

KEY IDEAS

A **system of equations** is a set of equations involving the same variables. When the two equations

$$x + y = 5$$
$$y + 4 = 2x$$

are considered together, they form a system of linear equations or, more simply, a **linear system**. A **solution of a system of linear equations** in two variables is an ordered number pair (x,y) that satisfies each equation in the system. Linear systems can be solved by finding the point at which their graphs intersect.

Solving a Linear System By Graphing

To solve the linear system

$$x + y = 5$$
$$y + 4 = 2x$$

graph each equation in the same coordinate plane, as shown in Figure 12.1. Because the solution must satisfy each equation in the system, the solution must be a point on both lines. The only point that is on both lines is the point at which two lines intersect, (3,2). To confirm algebraically that (3,2) is the solution, substitute $x = 3$ and $y = 2$ in each of the original equations and verify that the resulting statements are both true:

Equation 1	**Equation 2**
$x + y = 5$ $3 + 2 = 5$ ✓	$y + 4 = 2x$ $2 + 4 \boxed{?} 2(3)$ $6 = 6$ ✓

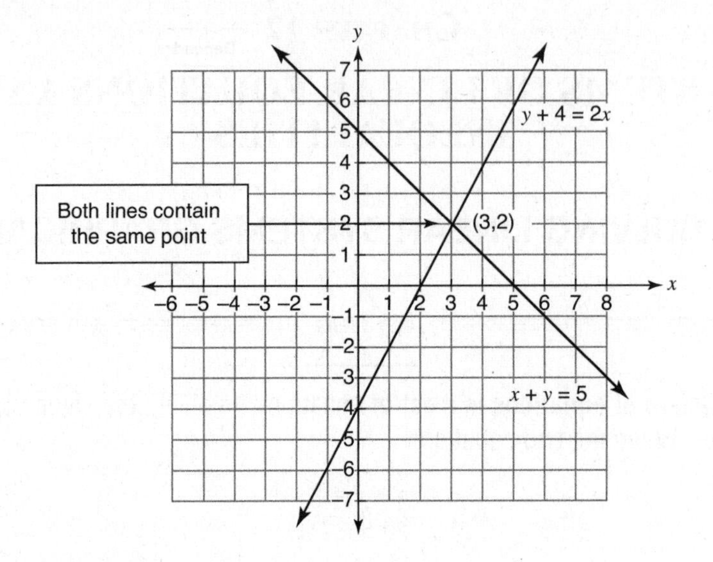

Figure 12.1 Solving $x + y = 5$ and $y + 4 = 2x$ Graphically

Example 1

At which point do the graphs of the equations $x - 2y = 6$ and $2x + y = 7$ intersect?

(1) (0,–3) (2) (2,3) (3) (4,–1) (4) (8,1)

Solution: The point of intersection can be obtained by graphing the two equations on the same set of axes. Since this is a multiple-choice question, a simpler method is to test the ordered pairs in the answer choices, in turn, until you find the ordered pair that satisfies both equations. You should confirm that choices (1), (2), and (4) do not work. For choice (3), $x = 4$ and $y = -1$:

- Check in first equation: $4 - 2(-1) = 4 + 2 = 6$ ✓
- Check in second equation: $2(4) + (-1) = 8 - 1 = 7$ ✓

The correct choice is **(3)**.

Inconsistent and Dependent Systems

Although the graphs of two linear equations generally intersect at a point, this is not always the case, as illustrated in Figure 12.2.

- The graphs of the equations $y = 2x + 4$ and $y = 2x - 3$ are parallel lines since the lines have the same slope but different y-intercepts. Because the two equations have no ordered pair in common, the linear system is **inconsistent**.

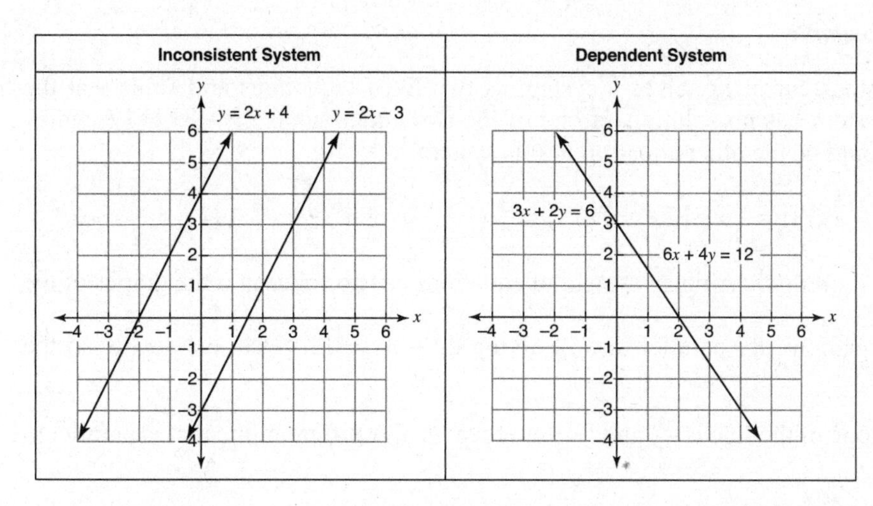

Figure 12.2 Inconsistent and Dependent Systems of Equations

- The equation $6x + 4y = 12$ is a multiple of the equation $3x + 2y = 6$, so their graphs coincide. Because the two equations have infinitely many ordered pairs in common, the linear system is **dependent**.

MATH FACTS

System	Equations and Their Graphs	Number of Solutions
Inconsistent	• Equations: Same slope but different y-intercepts. • Graph: Parallel lines.	None
Dependent	• Equations: Same slope and the same y-intercept. • Graph: Lines that coincide.	Infinitely many

To determine without graphing whether a linear system is inconsistent or dependent, compare the slope-intercept forms of their equations.

Example 2

James correctly solves a system of two linear equations and finds that the system has no solution. If one of the two equations is $2y - 3x = 12$, which could be the other equation in this system?

(1) $y=\frac{2}{3}x+12$ (2) $y=\frac{3}{2}x$ (3) $y=-\frac{3}{2}x$ (4) $y=\frac{3}{2}x+6$

Solution: A linear system of equations has no solution if the graphs of the equations are parallel lines. Solving $2y - 3x = 12$ gives $y=\frac{3}{2}x+6$ so the slope of the line is $\frac{3}{2}$ and its y-ntercept is 6. A different line that is parallel to $y=\frac{3}{2}x+6$ has the same slope but a different y-intercept, such as $y=\frac{3}{2}x$.

The correct choice is **(2)**.

Using Technology: Solving a Linear System

To use a graphing calculator to solve a system of linear equations such as $2x + y = 5$ and $y - x = -4$, write each equation in slope-intercept form. In the $Y=$ editor, set $Y_1 = -2x + 5$ and $Y_2 = x - 4$.

- To solve the system graphically, press ZOOM 4 to draw the graphs in a decimal window. Use either the TRACE feature or the intersect feature from the CALCULATE menu to locate the point of intersection at (3,–1). See Figure 12.3.

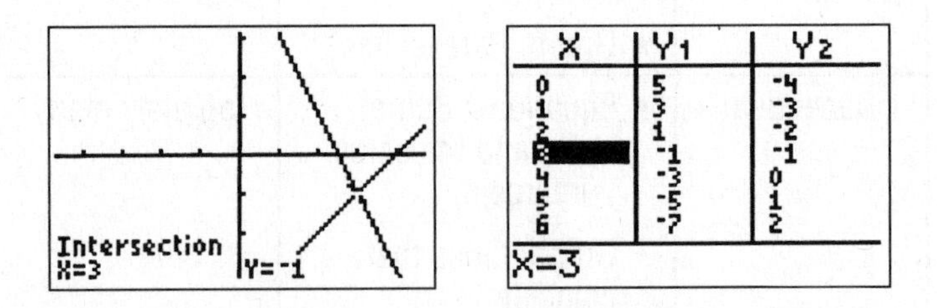

Figure 12.3 Graphical Solution **Figure 12.4** Numerical Solution

- To solve the system numerically, create a table by pressing 2nd [GRAPH]. Scroll up or down until you find the row on which $Y_1 = Y_2$, as shown in Figure 12.4. The solution may be written as either **(3,–1)** or $\boldsymbol{x = 3, y = -1}$. Be aware that if the point of intersection of two lines does not have integer coordinates, you will need to adjust the value of ΔTbl in the table setup or use a different approach.

Area of a Triangular Region

When lines are graphed on the same set of axes, the lines may enclose a triangular region whose base and altitude are parallel to the coordinates axes.

Example 3

Find the number of square units in the area of the triangular region formed in the first quadrant by the intersection of the lines $y = x$, $x + y = 6$, and the x-axis.

Solution: Using any convenient method, graph the lines $y = x$ and $x + y = 6$ on the same set of axes thereby forming $\triangle ABC$, as shown in the accompanying figure.

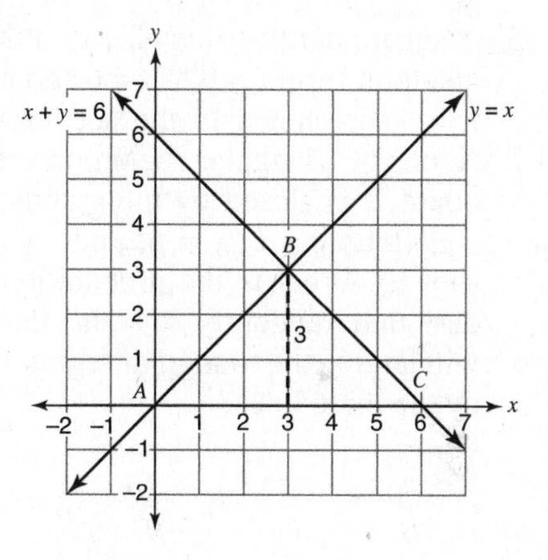

- The base of $\triangle ABC$ is the length of $\overline{AC}$, which measures 6 units.
- The height of $\triangle ABC$ is the distance from point B to the x-axis which is 3 units.
- Hence, area of

$$\triangle ABC = \frac{1}{2} \times \text{base} \times \text{height}$$

$$= \frac{1}{2} \times 6 \times 3$$

$$= \mathbf{9} \text{ square units}$$

Check Your Understanding of Section 12.1

A. *Multiple Choice.*

1. Which statement is true about the graphs of the equations $y = x - 3$ and $x = y - 3$?

(1) They are identical.
(2) They intersect.
(3) They are parallel.
(4) They have the same y-intercept.

2. When $y - 3x = 1$ and $2y - 1 = 6x$ are graphed on the same set of axes, the number of common solutions is
(1) 1 (2) 2 (3) 0 (4) infinite

3. Which system of equations has exactly one solution?

(1) $y=-x+20$, $y= x-20$ (2) $y=0.5x+30$, $2y=x-60$ (3) $y=\frac{3}{5}x+12$, $y=0.6x-19$ (4) $y=-x+15$, $y=-x+25$

4. Vanessa correctly solves a system of two linear equations and finds that the system has an infinite number of solutions. If one of the two equations is $3y + 4x = 6$, which could be the other equation in this system?

(1) $y=\frac{3}{4}x+2$ (2) $y=-\frac{4}{3}x$ (3) $y=-\frac{4}{3}x+2$ (4) $y=-\frac{4}{3}x+6$

5. A square dartboard is placed in the first quadrant from $x = 0$ to $x = 6$ and $y = 0$ to $y = 6$, as shown in the accompanying figure. A triangular region on the dartboard is enclosed by the graphs of the equations $y = 2$, $x = 6$, and $y = x$ (not shown). What is the probability that a dart that randomly hits the dartboard will land in the triangular region formed by the three lines?

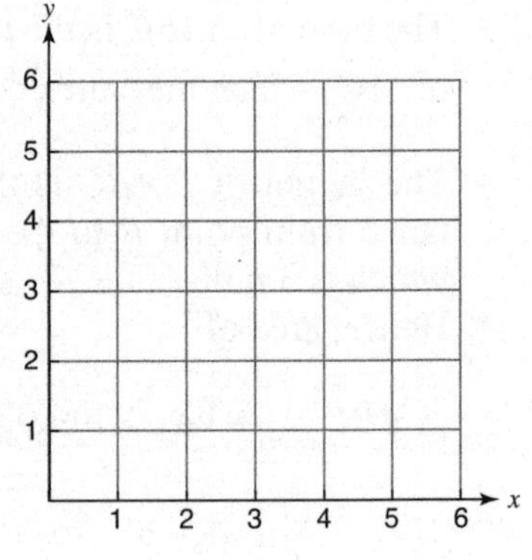

(1) $\frac{4}{9}$ (2) $\frac{2}{9}$

(3) $\frac{7}{36}$ (4) $\frac{15}{36}$

6. Which system of equations has *no* solution?
(1) $y = -2x + 1$ and $2y + 4x = 2$ (3) $y = 3x$ and $y=\frac{1}{3}x$
(2) $x + y = 3$ and $x + y = 7$ (4) $y = 2x$ and $4y + 8x = 12$

7. At which point do the graphs of the equations $-y = 2x$ and $x = 2y + 5$ intersect?
(1) (−2,4) (2) (1,−2) (3) (−1,2) (4) (−2,1)

B. *Show how you arrived at your answer.*

8–13. *Solve each system of equations graphically using graph paper. Check your answer algebraically.*

8. $2x = 8 - 5y$
$x + y = 1$

9. $3x = y + 4$
$x - y = 6$

10. $x + 2y = 4$
$y - 7 = 2x$

11. $2x + 2y = -10$
$3x - \frac{1}{2}y = 6$

12. $2y - 3x = -11$
$y - 5 = -2x$

13. $y = |x|$
$y + \frac{1}{3}x = 4$

14–16. *Solve each system of equations graphically using graph paper. Check your answer by solving the same system using a graphing calculator.*

14. $y + 5 = 2x$
$x + 3y = 6$

15. $y + 6 = \frac{2}{3}x$
$y + x = -1$

16. $y - 1 = \frac{x-2}{3}$
$5y - 2x = 7$

17. Solve the equation $1.5x - 5.7 = 3(1.6 - x) + x$ graphically by representing each side of the equation as a different line. Confirm your answer by solving the equation algebraically.

18. Determine the coordinates of the vertices of the triangle formed when the following three lines are graphed on the same set of axes:
(1) $y - 2x = 1$ (2) $3x + y = 6$ (3) $y = -3$

19-20. *Find the number of square units in the area of the triangular region formed in the second quadrant by the given set of lines.*

19. $y = -x$, $y = 2x + 12$, and x-axis

20. $x + y = 3$, $y + 7 = x$, and y-axis

21. Find the number of square units in the area of the triangle whose vertices are the points of intersection of the lines $y = 5$, $x = -4$, and $y = \frac{5}{4}x + 5$.

22. What is the number of square units in the area of the trapezoid formed in the first quadrant by the intersection of the lines $y = x$, $y = 4$, $x = 8$,and the x-axis.

23. a. On the same set of axes, graph the lines $y = -4$ and $2x + y = 6$.
b. Find the number of square units in the area of the trapezoid in the fourth quadrant bounded by the x-axis, the y-axis, and the lines graphed in part a.

24. Find the number of square units in the area of the parallelogram formed when the following four lines are graphed on the same set of axes:
(1) $y = 2$ (2) $y = 6$ (3) $y = 2x + 12$ (4) $y = 2x - 12$

25. Video store A charges $15 for membership and $2 for each video that is rented. Video store B does not charge a membership fee but charges $3.50 for each video that is rented. Let C represent the cost, in dollars, of renting n videos.

a. For each store, write an equation that expresses C as a function of n.
b. Determine graphically the number of video rentals for which the cost, including membership fees, is the same for the two clubs.

12.2 SOLVING LINEAR SYSTEMS BY SUBSTITUTION

KEY IDEAS

Solving a linear system graphically is not always the best approach. The equations may be difficult to graph, or the point of intersection of their graphs may be hard to read. Solving a linear system using algebraic methods overcomes the limitations of a graphical solution.

Eliminating a Variable by Substitution

To solve a linear system algebraically using the **substitution method**, solve one of the equations for either variable. Then substitute that solution into the second equation thereby eliminating that variable from it. To find the solution set of

$$y + 2x = 0$$
$$7x + 2y = 21$$

pick an equation in which it is easy to solve for one variable in terms of the other. Because the coefficient of y in the first equation is 1, solve it for y, which gives $y = -2x$.

- Eliminate y in the second equation by substituting $-2x$ for it. The second equation becomes $7x + 2(-2x) = 21$ so $3x = 21$ and $x = \frac{21}{3} = 7$.
- Find y by substituting $x = 7$ in either equation. Replacing x with 7 in the first equation makes $y + 2(7) - 0$ so $y = -14$.
- The solution set is **{(7,–14)}**. The check is left for you.

Example 1

Solve the following system of equations algebraically:

$$\begin{aligned} x+2y&=7 \\ y-1&=2x \end{aligned}$$

Solution 1: Solve the second equation for y which gives $y = 2x + 1$.

- Eliminate y in the first equation by replacing it with $2x + 1$:

$$\begin{aligned} x+2y&=7 \\ x+2(2x+1)&=7 \\ x+4x+2&=7 \\ \frac{5x}{5}&=\frac{5}{5} \\ x&=1 \end{aligned}$$

- Find y by substituting 1 for x in either equation:

$$\begin{aligned} y-1&=2x \\ y&=2(1)+1 \\ y&=3 \end{aligned}$$

Hence, the solution set is **{(1,3)}**.

- Check that the solution ($x = 1, y = 3$) works in *each* of the original equations:

Equation 1	Equation 2
$\underline{x + 2y = 7}$	$\underline{y - 1 = 2x}$
$1 + 2(3) \boxed{?} 7$	$3 - 1 \boxed{?} 2(1)$
$1 + 6 = 7$ ✓	$2 = 2$ ✓

Solution 2: Solve the first equation for x, which gives $x = 7 - 2y$. Eliminate x in the second equation by replacing it with $7 - 2y$:

$$\begin{aligned} y-1 &= 2x \\ y-1 &= 2(7-2y) \\ y-1 &= 14-4y \\ y+4y &= 14+1 \\ \frac{5y}{5} &= \frac{15}{5} \\ y &= 3 \end{aligned}$$

Find x by substituting 3 for y in either of the original equations.

Solving Word Problems

When a word problem involves two conditions, translate each condition into an equation using the same two variables.

Example 2

The sum of two positive numbers is 14. If the larger number exceeds the smaller number by 9, what are the two numbers?

Solution 1: Represent the smaller of the two numbers by x and the larger number by y. Translate each condition of the problem into an equation:

CONDITION 1: The sum of the smaller and larger number is 14: $x + y = 14$
CONDITION 2: The larger number is 9 more than the smaller number: $y = x + 9$

- The linear system is

$$\begin{aligned} x+y &= 14 \\ y &= x+9 \end{aligned}$$

- Solve the linear system by using the second equation to eliminate y in the first equation:

$$x+(x+9)=14$$

$$2x = 14 - 9$$

$$x=\frac{5}{2}$$

- Find y by substituting $\frac{5}{2}$ for x in the second equation:

$$y = x + 9 = \overbrace{\frac{5}{2}}^{x} + \overbrace{\frac{18}{2}}^{9} = \frac{23}{2}$$

The two numbers are $\mathbf{\frac{5}{2}}$ and $\mathbf{\frac{23}{2}}$.

Solution 2: Use one equation. Represent the two numbers by x and $14 - x$. Thus, $14 - x = x + 9$, $2x = 5$, and $x = \frac{5}{2}$ so $14 - x = 14 - \frac{5}{2} = \frac{23}{2}$.

Example 3

The Town Recreation Department ordered a total of 100 balls and bats for the summer baseball camp. Balls cost \$4.50 each, and bats cost \$20.00 each. The total purchase was \$822. How many of each item were ordered?

Solution 1: Represent the number of balls ordered by x and the number of bats ordered by y. Translate each condition of the problem into an equation:

CONDITION 1: The total number of balls and bats is 100: $x + y = 100$
CONDITION 2: The total purchase is \$822 when one ball costs \$4.50 and one bat costs \$20.00:

$$\overbrace{\$4.50x}^{\text{cost of } x \text{ balls}} + \overbrace{\$20.00y}^{\text{cost of } y \text{ bats}} = \$822$$

- The linear system is

$$\begin{aligned} x + y &= 100 \\ \$4.50x + \$20.00y &= \$822 \end{aligned}$$

- Solve the linear system. From the first equation, $y = 100 - x$. Eliminate y in the second equation by replacing it with $100 - x$:

$$\begin{aligned} \$4.50x + \$20(100 - x) &= \$822 \\ \$4.50x + \$2000 - \$20x &= \$822 \\ -\$15.50x &= \$822 - \$2000 \end{aligned}$$

$$x = \frac{-\$1178}{-\$15.50}$$
$$= 76$$

- Find y by substituting 76 for x in the first equation:

$$76 + y = 100 \text{ so } y = 100 - 76 = 24$$

Thus, **76 balls** and **24 bats** were ordered.

Solution 2: Use one variable. If x balls are ordered, then $100 - x$ represents the number of bats ordered so $\$4.50x + \$20(100 - x) = \$822$, which can be solved as before.

Example 4

An automobile repair shop wants to mix a solution that is 35% pure antifreeze with another solution that is 75% pure antifreeze. How many liters of each solution must be used in order to produce 80 liters of solution that is 50% pure antifreeze?

Solution 1: Assume x and y represent the number of liters of the 35% and 75% solutions, respectively, in the mixture.

CONDITION 1: The mixture will contain 80 liters of solution: $x + y = 80$
CONDITION 2: The sum of the number of liters of pure antifreeze in the two ingredients must be equal to the number of liters of pure antifreeze in the mixture:

$$0.35x + 0.75y = 0.50(80) = 40$$

- The linear system is

$$x + y = 80$$
$$0.35x + 0.75y = 40$$

- Solve the linear system. From the first equation, $y = 80 - x$. Eliminate y in the second equation:

$$0.35x + 0.75(80 - x) = 40$$
$$0.35x + 60 - 0.75x = 40$$
$$-0.40x = 40 - 60$$
$$\frac{-0.40x}{-0.40} = \frac{-20}{-0.40}$$
$$x = 50$$

- Find y by substituting 50 for x in the first equation:

$$50 + y = 80 \text{ so } y = 80 - 50 = 30$$

Thus, **50 liters** of the 35% solution and **30 liters** of the 75% solution must be used.

Solution 2: Use one variable where x and $80 - x$ represent the number of liters of the 35% and 75% solutions, respectively, so $0.35x + 0.75(80 - x) = 40$, which can be solved as before.

Check Your Understanding of Section 12.2

A. Multiple Choice.

1. What is the solution set for the following system of equations?

$$\begin{aligned} x + y &= 0 \\ x + 2y &= 6 \end{aligned}$$

(1) (–2,2) (2) (2,–2) (3) (6, –6) (4) (–6,6)

2. What is the solution for x in the following system of equations?

$$\begin{aligned} -y &= 2x - 3 \\ y &= -x + 1 \end{aligned}$$

(1) $\frac{2}{3}$ (2) 2 (3) $\frac{4}{3}$ (4) 4

3. What is the solution for y in the following system of equations?

$$\begin{aligned} \frac{3}{4} &= \frac{2}{x} \\ 6x + y &= 17 \end{aligned}$$

(1) 1 (2) –1 (3) 9 (4) –9

B. *Show how you arrived at your answer.*

4. Solve the following system of equations for y:

$$\begin{aligned} 2x - y &= 13 \\ 3x &= y \end{aligned}$$

5. Solve the following system of equations for s:

$$3s = t + 4$$
$$s - t = 6$$

6–11. *Solve each of the following systems of equations algebraically and check.*

6. $y + 5x = 0$
$3x - 2y = 26$

7. $\frac{y}{4} - x = 0$
$2y = 3x - 10$

8. $\frac{a-2}{3} = b$
$2a + 3b = -5$

9. $y - x = 1$
$7y - 11x = -5$

10. $0.7x + 0.4y = 16$
$x + y = 10$

11. $0.3y - 0.2x = 2.1$
$0.75y = 2.25x$

12. Five of the same type of pen cost the same as two of the same type of notebook. If one pen and two notebooks costs \$4.20, what is the cost of one pen?

13. Two angles are complementary. The difference between four times the measure of the smaller angle and the measure of the larger angle is 10. What is the measure of the larger angle?

14. The denominator of a fraction exceeds its numerator by 7. If the numerator is increased by 3 and the denominator is decreased by 2, the new fraction equals $\frac{4}{5}$. Find the original fraction.

15. A cellular telephone company has two plans. The cost, y, of Plan A is \$11 a month plus \$0.21 per minute. The cost, y, of Plan B is \$20 a month plus \$0.10 per minute. After how much time, to the *nearest minute*, will the cost of Plan A be equal to the cost of Plan B?

16. A chemical company has in storage a 15% solution and a 25% solution of a disinfectant. How many liters of each should be used to make 50 liters of a 22% solution?

17. In the accompanying diagram, lines AB and CD intersect at $E(x, y)$.

a. Determine the equations of lines AB and CD.

b. Using the equations obtained in part a, determine the coordinates of point E *algebraically*.

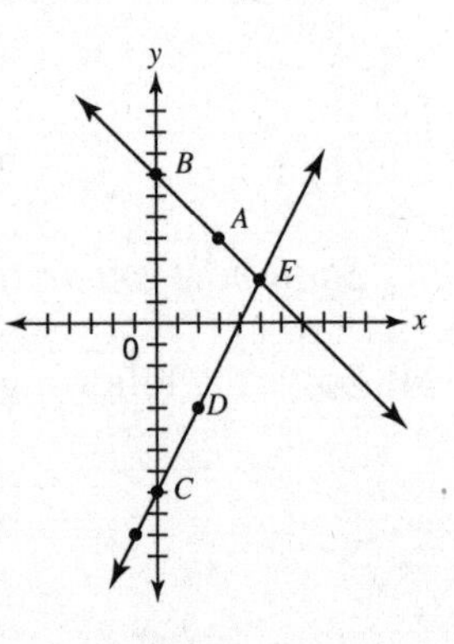

Exercise 17

18. A store charges a fixed amount to fax one page and another amount for faxing each additional page that is faxed to the same telephone number. If the cost of faxing 5 pages is \$3.05 and the cost of faxing 13 pages is \$6.65, what is the cost of faxing only one page? What is the cost of faxing each additional page after the first?

19. In $\triangle ABC$, the measure of $\angle B$ is twice the measure of $\angle A$. If the measure of $\angle A$ is subtracted from the measure of $\angle C$, the difference is 20. Find the measure of $\angle C$.

20. There were 100 more balcony tickets than main-floor tickets sold for a concert. The balcony tickets sold for \$4 and the main-floor tickets sold for \$12. The total amount of sales for both types of tickets was \$3056. Find the number of balcony tickets that were sold.

21. The senior class at Northwest High School needed to raise money for the yearbook. A local sporting goods store donated hats and T-shirts. The number of T-shirts was three times the number of hats. The seniors charged \$5 for each hat and \$8 for each T-shirt. If the seniors sold every item and raised \$435, what was the total number of hats and the total number of T-shirts that were sold?

12.3 SOLVING LINEAR SYSTEMS BY COMBINING EQUATIONS

KEY IDEAS

The linear system shown at the right is difficult to solve using the substitution method, but it can be easily solved by *adding* the two equations thereby eliminating y.

$$\begin{aligned} 2x - 9y &= 17 \\ 5x + 9y &= 11 \\ \hline 7x + 0y &= 28 \\ x &= \frac{28}{7} \\ &= 4 \end{aligned}$$

Combining Equations to Eliminate a Variable

To solve a linear system by combining equations, write each equation in the standard form $Ax + By = C$. If the two equations contain like variable terms with the same (or opposite) numerical coefficients, eliminate that variable by subtracting (or adding) the two equations.

Example 1

Find the solution set of $4y + 39 = 5x$
$3x = 17 - 4y$

Solution: Write each equation in the standard form $Ax + By = C$:

$$-5x + 4y = -39$$
$$3x + 4y = 17$$

- Since the y terms have the same numerical coefficient, eliminate y by *subtracting* the second equation from the first. Then solve the resulting equation for x:

$$\begin{array}{r} -5x+4y=-39 \\ \underline{-[3x+4y=17]} \end{array} \quad \Rightarrow \quad \begin{array}{r} -5x+4y=-39 \\ \underline{-3x-4y=-17} \\ -8x+0y=-56 \\ \frac{-8x}{-8}=\frac{-56}{-8} \\ x=7 \end{array}$$

- Find y by substituting 7 for x in either of the original equations:

$$4y+39=5x$$
$$4y+39=5(7)$$
$$4y=35-39$$
$$\frac{4y}{4}=\frac{-4}{4}$$
$$y=-1$$

The solution set is **{(7,–1)}**. The check is left for you.

Using Multipliers

It may be necessary to multiply one or both equations in a linear system by a number that makes the same variable in the two equations have opposite numerical coefficients.

Example 2

Find the solution set:

$$4y=3x+11$$
$$6x-5y=-16$$

Solution 1:

- Write the first equation in the standard form $Ax + By = C$:

$$\begin{aligned} -3x + 4y &= 11 \\ \underline{6x - 5y} &\underline{= -16} \end{aligned}$$

- Eliminate x. Change the coefficient of x in the first equation to -6 by multiplying the equation by 2. Then add the resulting equation to the second equation in the original system:

$$\begin{aligned} (2)[-3x+4y&=11] \\ 6x-5y&=-16 \end{aligned} \quad \Rightarrow \quad \begin{aligned} -6x+8y&=22 \\ 6x-5y&=-16 \\ \hline 0+3y&=6 \\ y&=\frac{6}{3}=2 \end{aligned}$$

- Find x by substituting 2 for y in either of the original equations:

$$\begin{aligned} 6x-5y&=-16 \\ 6x-5(2)&=-16 \\ 6x&=-16+10 \\ 6x&=-6 \\ \frac{6x}{6}&=\frac{-6}{6} \\ x&=-1 \end{aligned}$$

The solution set is **{(–1, 2)}**. The check is left for you.

Solution 2: Use the substitution method. From the first equation, $x = 3y - 1$. Eliminate x from the second equation by substituting $3y - 1$ for it. Solve $-2(3y-1)+5y=7$ for y and then use the solution, $y = -5$, to find x in the usual way.

Example 3

Find the solution set:

$$\begin{aligned} 3m+4p&=9 \\ 5m+6p&=21 \end{aligned}$$

Solution 1: Multiply *both* equations by suitable numbers that will make the same variable in the two equations have opposite numerical coefficients.

- Eliminate m. Change the coefficients of m in the two equations to $+15$ and -15 by multiplying the first equation by 5 and the second equation by -3. Then add the resulting equations:

$$\left.\begin{array}{r} 5[3m+4p=9] \\ -3[5m+6p=21] \end{array}\right\} \quad \Rightarrow \quad \begin{array}{r} +15m+20p=45 \\ -15m-18p=-63 \\ \hline \frac{2p}{2}=\frac{-18}{2} \\ p=-9 \end{array}$$

- Find m by substituting -9 for p in either of the original equations:

$$\begin{aligned} 3m+4p&=9 \\ 3m+4(-9)&=9 \\ 3m-36&=9 \\ \frac{3m}{3}&=\frac{45}{3} \\ m&=15 \end{aligned}$$

The solution set is **{(15,-9)}**. The check is left for you.

Solution 2: Eliminate p by multiplying the first equation by 3, the second equation by -2 , and then adding the resulting equations:

$$\begin{array}{r} 9m+12p=\ 27 \\ -10m-12p=-42 \\ \hline -m \qquad =-15 \end{array} \quad \text{so } m=15$$

Find p by substituting 15 for m in either of the original equations.

Example 4

Carla and Steve went shopping at Office Plus. Carla bought 3 boxes of computer disks and 2 notebooks for a total cost of \$15.00 excluding tax. Steve bought 2 boxes of the same computer disks and 5 of the same notebooks for a total cost of \$18.25 excluding tax. Find the cost of one box of computer disks.

Solution: Represent the price of one box of computer disks by x and the price of one notebook by y.

- Translate the conditions of the problem into a linear system:

$$\begin{aligned} \text{Carla's purchase: } & 3x+2y=\$15.00 \\ \text{Steve's purchase: } & \underline{2x+5y=\$18.25} \end{aligned}$$

- Eliminate y. Change the coefficients of y in the two equations to -10 and $+10$ by multiplying the first equation by -5 and the second equation by 2. Then add the resulting equations:

$$\begin{array}{r} -15x-10y=-\$75.00 \\ 4x+10y=\ \ \$36.50 \\ \hline -11x\qquad\ \ =-\$38.50 \end{array}$$

- Solve for x:

$$\frac{-11x}{-11}=\frac{-\$38.50}{-11}$$

$$x=\$3.50$$

One box of computer disks costs **\$3.50**.

Check Your Understanding of Section 12.3

A. Multiple Choice.

1. Which ordered pair is the solution to the following system of equations?

$$\begin{aligned} 2x-y&=10 \\ x+y&=2 \end{aligned}$$

(1) (4,–2) (2) (4, 2) (3) (2,–4) (4) (–4,2)

2. At which point do the graphs of the equations $2x + y = 8$ and $x - y = 4$ intersect?

(1) (0,4) (2) (4,0) (3) (–4,0) (4) (5,–2)

3. What is the solution for x in the following system of equations?

$$\begin{aligned} -3x+2y&=9 \\ 3(x+y)&=11 \end{aligned}$$

(1) $-\frac{1}{3}$ (2) 2 (3) $-\frac{4}{3}$ (4) 4

B. Show how you arrived at your answer.

4. Solve the following system of equations for n.

$$\begin{aligned} 2m+n&=10 \\ 3m-n&=15 \end{aligned}$$

5. Three shirts and two pairs of socks cost \$69. At the same prices, two shirts and three pairs of socks cost \$61. Find the cost of one shirt and one pair of socks.

6–14. *Solve each of the following systems of equations algebraically and check.*

6. $2x + y = 12$
$3y = 2x - 4$

7. $2x + 7y = 8$
$3x + 2y = -5$

8. $\frac{2}{3}s + t = 13$
$-s + 2t = 5$

9. $2x + 3y = -6$
$5x + 2y = 7$

10. $2x + 3y = 17$
$3x - 2y = -0.5$

11. $3u + v = 9$
$\frac{3u}{2} + \frac{4v}{5} = \frac{27}{5}$

12. $0.4a + 1.5b = -1$
$1.2a - b = 8$

13. $2b = 5c + 8$
$3b + 2c = 31$

14. $3x - 2y = 6$
$\frac{y-1}{x} = \frac{1}{2}$

15. The sophomore class at South High School raised \$860 from the sale of tickets to a concert. Tickets sold for \$2.50 if purchased in advance and for \$4.00 if purchased at the door. If a total of 275 tickets were sold, what was the number of tickets sold at the door?

16. Three bags of potatoes and four cases of corn cost \$40. Five bags of potatoes and two cases of corn cost \$34. Find the cost of one bag of potatoes and the cost of one case of corn.

17. Billy rented a sprayer and a generator. On his first job, he used each piece of equipment for 6 hours at a total cost of \$90. On his second job, he used the sprayer for 4 hours and the generator for 8 hours at a total cost of \$100. Find the hourly cost of the generator.

Food	Protein	Calories
Cereal	5g	90
Milk	8g	80

18. The accompanying table shows the number of grams of protein and the number of calories in a single serving of bran flakes cereal and milk. How many servings of each are needed to get a total of 35 grams of protein and 470 calories?

19. Cedric and Zelda went shopping at Price Buster. Cedric bought 2 jumbo rolls of aluminum foil and 3 packages of AA batteries for a total cost of \$21. Zelda bought 5 identical jumbo rolls of aluminum foil and 2 identical packages of AA batteries for a total cost of \$25. Find the cost of 1 jumbo roll of aluminum foil and the cost of 1 package of AA batteries.

20. At the local video rental store, Jose rents two movies and three games for a total of \$15.50. At the same time, Meg rents three movies and one game for a total of \$12.05. How much money does Chris need to rent a combination of one game and one movie from the same video rental store?

21. The cost of a long-distance telephone call is determined by a flat fee for the first 5 minutes and a fixed amount for each additional minute. Roseanne is charged \$3.25 for a long-distance telephone call that lasts 15 minutes and \$5.17 for a long-distance call that lasts 23 minutes. At the same rate, what is the cost of a long-distance telephone call that lasts 30 minutes?

12.4 GRAPHING SYSTEMS OF LINEAR INEQUALITIES

KEY IDEAS

A two-variable linear inequality can be graphed using graph paper or using a graphing calculator. The solution set of a system of two linear inequalities is the region where the individual solutions sets of the two inequalities overlap.

Graphing a Linear Inequality

To graph a linear inequality such as $y - 2x > 3$,

- Replace the inequality sign with an equal sign and graph the resulting equation as a broken line since points on the line do not make $y - 2x > 3$ a true statement, as shown in Figure 12.5. If the inequality relation is $\leq$ or $\geq$, draw a solid line to indicate that the solution set includes the points on the line.

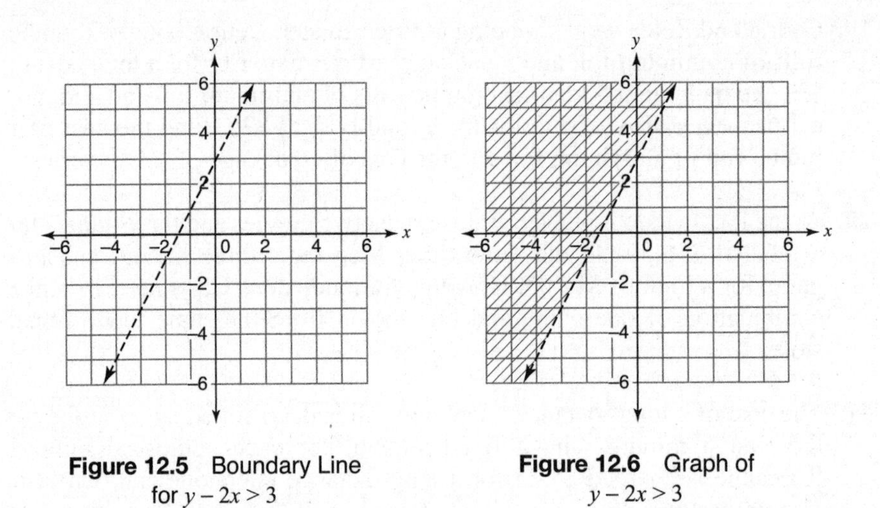

Figure 12.5 Boundary Line for $y - 2x > 3$

Figure 12.6 Graph of $y - 2x > 3$

- Decide which side of the boundary line, above or below it, represents the solution set. If you are not sure, pick a convenient test point not on the boundary line such as (0,0). Determine if (0,0) satisfies the inequality:

$$\begin{aligned} &y > 2x + 3 \\ &0 > 2(0) + 3 \ ? \\ &0 > 0 + 3 \quad ? \\ &0 > 3 \qquad \text{No!} \end{aligned}$$

The test point (0,0) does *not* satisfy the inequality so the solution set is the region *above* the boundary line since that region does *not* include (0,0).

- Shade in the region that represents the solution set, as shown in Figure 12.6.

Using Technology: Graphing Inequalities

To enter an inequality in a graphing calculator, open the $Y=$ editor and place the cursor over the diagonal that comes before Y_1.

- If the inequality relation is greater than, press ENTER until you get the solid right triangle, as shown in Figure 12.7. This symbol indicates that the area above the graph will be shaded. After entering the right side of the inequality, display the graph, as shown in Figure 12.8.

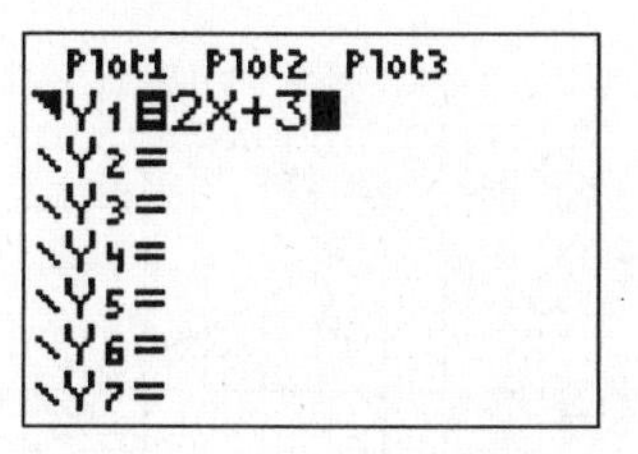

Figure 12.7 Entering $y > 2x + 3$

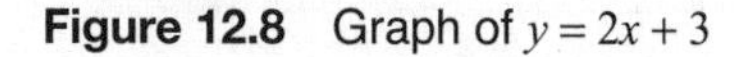

Figure 12.8 Graph of $y = 2x + 3$

- If the inequality relation is less than, press ENTER until you get the solid right triangle, as shown in Figure 12.9. This symbol indicates that the region below the line will be shaded. Display the graph, as shown in Figure 12.10.

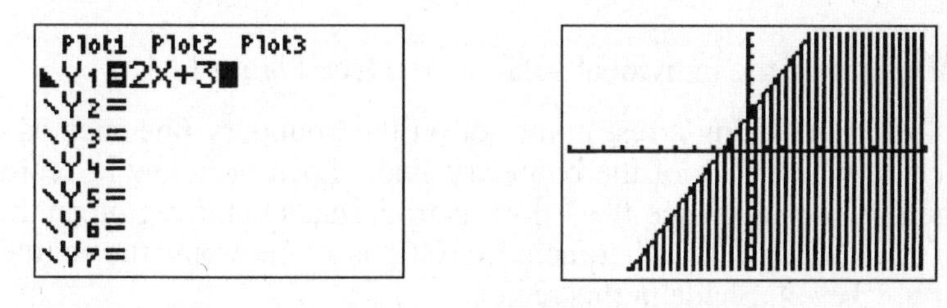

Figure 12.9 Entering $y < 2x + 3$

Figure 12.10 Graph of $y < 2x + 3$

Solving a System of Linear Inequalities

The solution set of a system of linear inequalities is the set of points in the region where their graphs overlap. To solve the system of inequalities

$$y > 3x - 4$$
$$x + 2y \leq 6$$

follow these steps:

STEP 1 Graph each boundary line (see Figure 12.11).

- Graph $y = 3x - 4$ using convenient points such as (0,–4) and (2,2). Since the solution set of $y > 3x - 4$ does *not* include the points on the boundary line, draw the graph of $y = 3x - 4$ as a *broken* line.
- Graph $x + 2y = 6$ using convenient points such as (0,3) and (6,0). As the solution set of $x + 2y \leq 6$ includes points on the boundary line, draw the graph of $x + 2y = 6$ as a *solid* line.

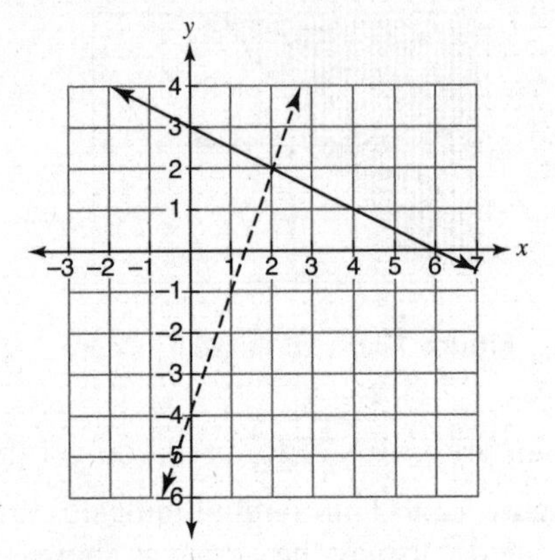

Figure 12.11 Graphing Boundary Lines for $y > 3x - 4$ and $x + 2y \leq 6$

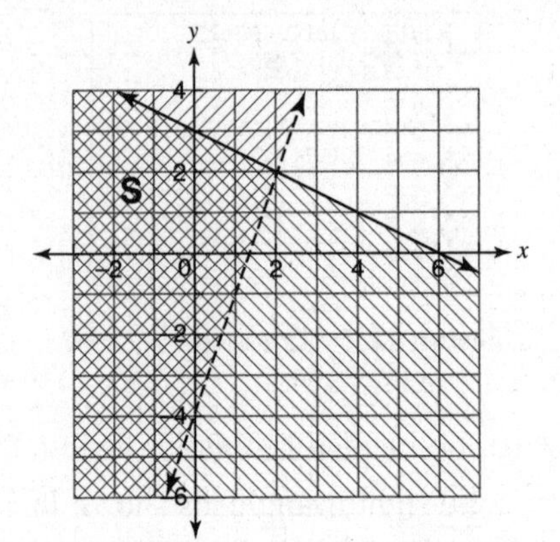

Figure 12.12 Finding the Solution Set for $y > 3x - 4$ and $x + 2y \leq 6$

STEP 2 Determine the individual solution sets (see Figure 12.12).

- $y > 3x - 4$: Using a test point not on the boundary line such as (0,0), decide which side of the boundary line, above or below it, represents the solution set. Since $0 > 3 \cdot 0 - 4$ is true, shade in the region on the side of the boundary line that includes (0,0) as it represents the solution set of $y > 3x - 4$. Shade in this region.
- $x + 2y \leq 6$: Use (0,0) as a test point. Since $0 + 2(0) \leq 6$ is true, shade in the region on the side of the boundary line that includes (0,0) as it represents the solution set of $x + 2y \leq 6$.

STEP 3 Indicate where the individual solution sets intersect.

The solution set **S** of the system of linear inequalities is the cross-hatched region where the solutions sets of the two inequalities overlap, as shown in Figure 12.12.

Check Your Understanding of Section 12.4

A. *Multiple Choice.*

1. Which ordered pair is in the solution set of $x < 6 - y$?
(1) (0,6) (2) (6,0) (3) (0,5) (4) (7,0)

2. Which is *not* a member of the solution set of $2x - 3y \geq 12$?
(1) (0,–4) (2) (–3,–6) (3) (10,3) (4) (6,0)

3. The accompanying diagram shows the graph of which inequality?

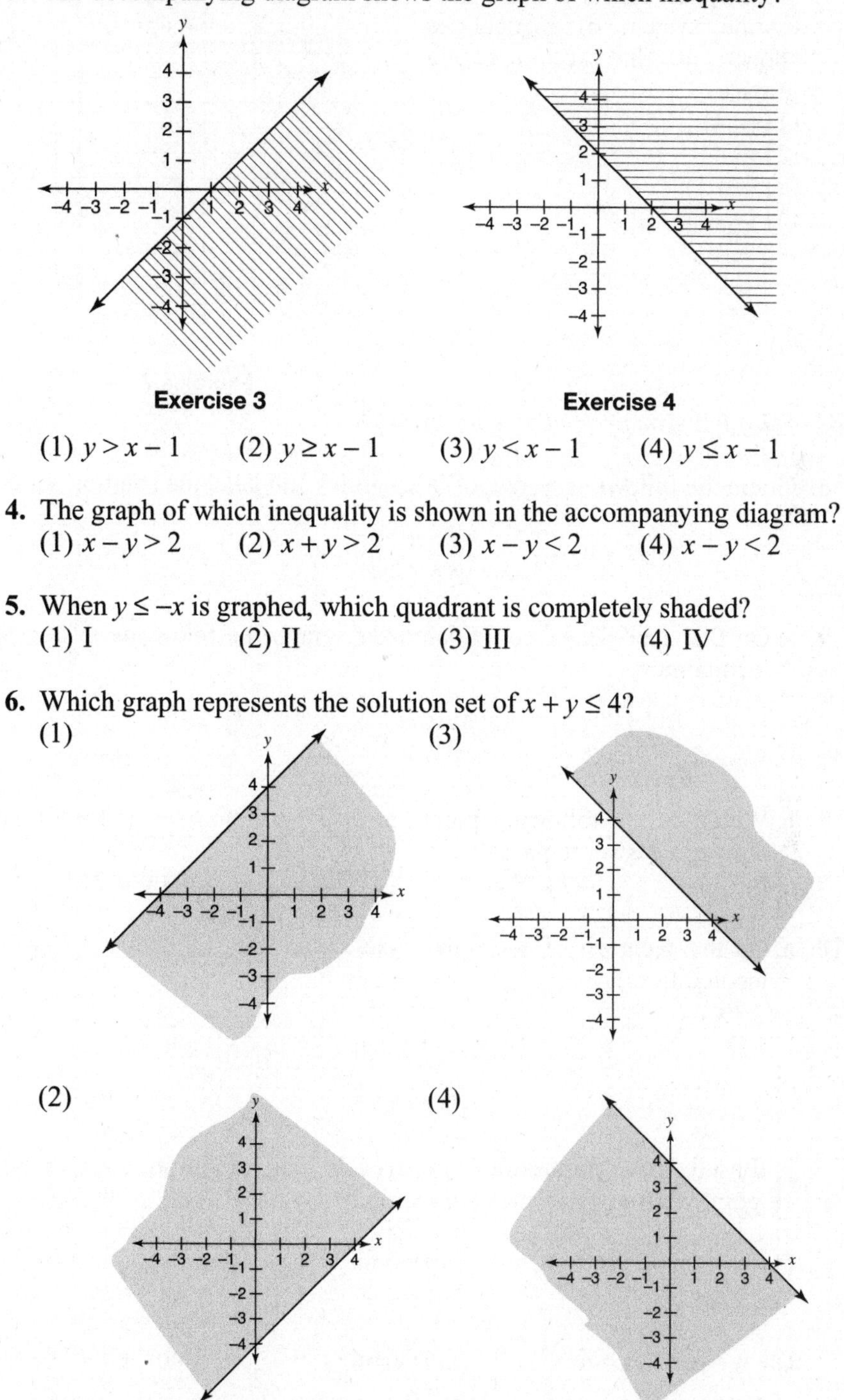

Exercise 3

Exercise 4

(1) $y > x - 1$ (2) $y \geq x - 1$ (3) $y < x - 1$ (4) $y \leq x - 1$

4. The graph of which inequality is shown in the accompanying diagram?
(1) $x - y > 2$ (2) $x + y > 2$ (3) $x - y < 2$ (4) $x - y < 2$

5. When $y \leq -x$ is graphed, which quadrant is completely shaded?
(1) I (2) II (3) III (4) IV

6. Which graph represents the solution set of $x + y \leq 4$?
(1) (3) (2) (4)

7. Which point is in the solution set of the system of inequalities shown in the accompanying graph?
(1) (0,4)
(2) (2,4)
(3) (−4,1)
(4) (4,−1)

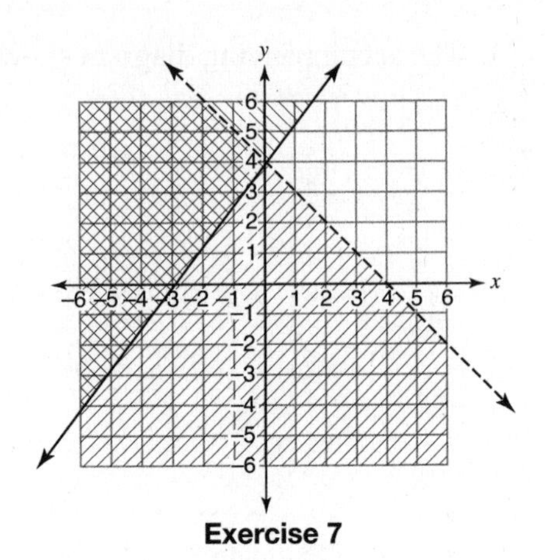

Exercise 7

B. *Show how you arrived at your answer.*

8. Graph the following system of inequalities and label the solution set ***S***:

$$y > x - 4$$
$$y + x \geq 2$$

9. a. On the same set of coordinate axes, graph the following system of inequalities:

$$3y \geq 2x - 6$$
$$x + y > 7$$

b. Which of the following points is in the solution set of the graphs drawn in answer to part a?
(1) (3,4) (2) (8,1) (3) (2,6) (4) (−2,7)

10. a. On the same set of coordinate axes, graph the following system of inequalities:

$$y \leq \frac{1}{2}x - 3$$
$$y > -2x + 4$$

b. Based on the graphs drawn in part a, in which solution set(s) does the point whose coordinates are (3,−1) lie?

(1) $y \leq \frac{1}{2}x - 3$ only
(2) $y > -2x + 4$ only
(3) both $y \leq \frac{1}{2}x - 3$ and $y > -2x + 4$
(4) neither $y \leq \frac{1}{2}x - 3$ nor $y > -2x + 4$

11. a. On the same set of axes, graph the inequalities $\frac{6-y}{4} \geq x$ and $5y + 25 > 2x$.

b. Write the coordinates of a point that is *not* in the solution set of the system of inequalities graphed in part a.

12. Marisa's music CD collection includes jazz CDs and classical CDs. She has, at most, 14 jazz CDs and, at most, 8 classical CDs. The combined number of jazz and classical CDs she has is not greater than 18. If x is the number of jazz CDs and y is the number of classical CDs, graph the region that contains the number of jazz and classical CDs in Marisa's music CD collection.

CHAPTER 13
QUADRATIC AND EXPONENTIAL FUNCTIONS

13.1 GRAPHING A QUADRATIC FUNCTION

KEY IDEAS

The graph of the quadratic function $y=ax^2+bx+c$ $(a\neq 0)$ is a **U**-shaped curve called a **parabola** with c as its y-intercept. A line can be drawn through every parabola that divides it into two mirror image parts. This line is called the **axis of symmetry**. It intersects the parabola at the **vertex**, which is either the lowest or the highest point on the curve depending on whether the coefficient of the x^2-term is positive or negative.

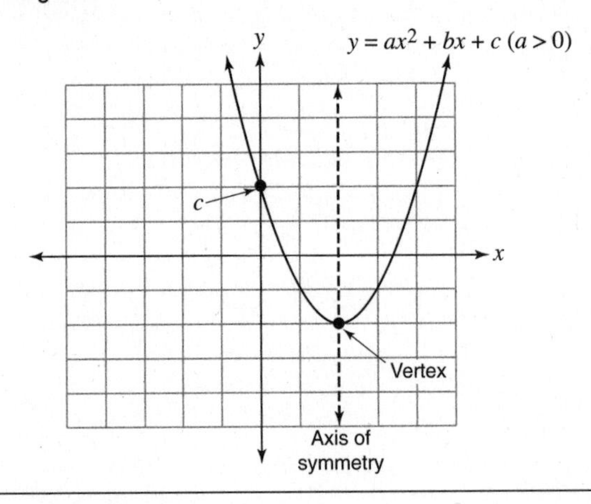

Vertex of a Parabola

A parabola may open up or open down depending on the sign of the coefficient of x^2 in the parabola equation. If in the parabola equation $y=ax^2+bx+c$,

- a is positive, as in Figure 13.1, the parabola "opens *up*," and the vertex is the *lowest* point on the curve where y takes on its *minimum* value.
- a is negative, as in Figure 13.2, the parabola "opens *down*," and the vertex is the *highest* point on the curve where y takes on its *maximum* value.

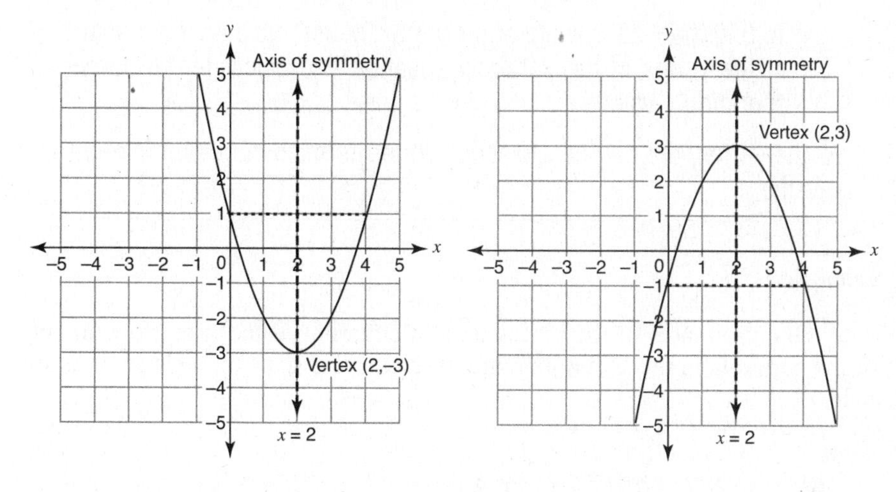

Figure 13.1 Graph of $y = x^2 - 4x + 1$ **Figure 13.2** Graph of $y = -x^2 + 4x - 1$

Some Facts About Parabolas

- When a parabola is folded along the axis of symmetry, its two parts exactly coincide. Thus, for equal x distances on either side of the axis of symmetry, the y-coordinates on the parabola are the same. This is illustrated in Figures 13.1 and 13.2 where the broken horizontal segments intersect points on the parabola that have matching y-coordinates.
- The axis of symmetry has an equation of the form $x = h$ where h is the x-coordinate of the vertex. In Figure 13.1, the vertex is at **(2**,–3) and an equation of the axis of symmetry is $x = 2$.
- An equation of the axis of symmetry can be determined without graphing the parabola $y = ax^2 + bx + c$ by using the formula, $x = -\dfrac{b}{2a}$.

  Because the axis of symmetry contains the vertex (turning point), the x-coordinate of the vertex can be obtained using the same formula as the axis of symmetry.

MATH FACTS

Given a parabola whose equation is $y = ax^2 + bx + c$.

- To find an equation of the axis of symmetry or the x-coordinate of the vertex, use the formula

$$x = -\frac{b}{2a}$$

- To find the y-coordinate of the vertex, substitute the x-coordinate of the vertex in the parabola equation and calculate the corresponding y-value.

At the vertex, the parabola reaches either its maximum or minimum y-value.

Example 1

State the coordinates of the vertex and an equation of the axis of symmetry for the parabola in the accompanying diagram. The scale on the axes is a unit scale.

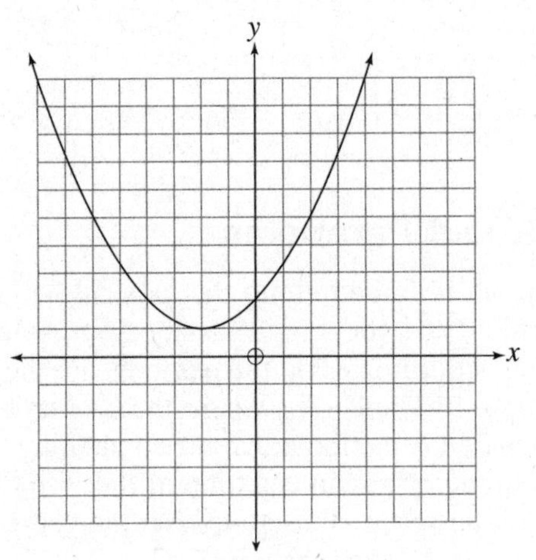

Solution: Draw a vertical line that divides the parabola into two equal parts, as shown in the accompanying diagram.

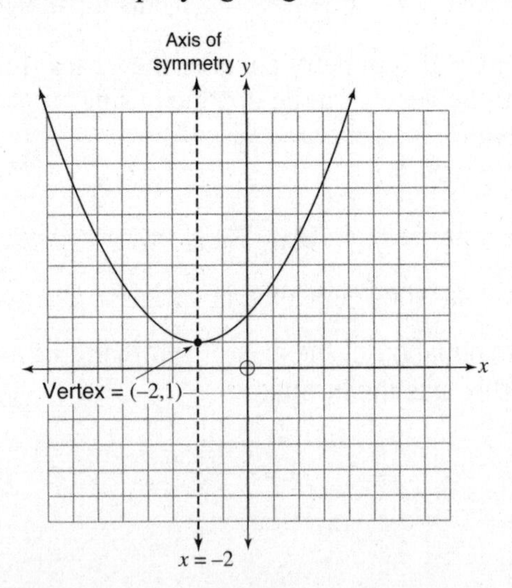

The axis of symmetry intersects the parabola at (–2,1). Since the x-coordinate of every point on the axis of symmetry is –2, its equation is $x = -2$.

The vertex is at **(–2,1)**, and an equation of the axis of symmetry is $\boldsymbol{x = -2}$.

Example 2

Find an equation of the axis of symmetry and the coordinates of the vertex of the parabola whose equation is $y = -2x^2 + 12x - 7$.

Solution: To find an equation of the axis of symmetry of $y = -2x^2 + 12x - 7$, use the formula $x = -\frac{b}{2a}$ where $a = -2$ and $b = 12$:

$$\begin{aligned} x &= -\frac{12}{2(-2)} \\ &= \frac{-12}{-4} \\ &= 3 \end{aligned}$$

Because the x-coordinate of the vertex is 3, find the y-coordinate of the vertex by substituting 3 for x in the parabola equation:

$$\begin{aligned} y &= -2(3)^2 + 12(3) - 7 \\ &= -18 \quad + 36 \quad - 7 \\ &= 11 \end{aligned}$$

An equation of the axis of symmetry is $\boldsymbol{x = 3}$ and the coordinates of the vertex are **(3,11)**.

Example 3

The profit in dollars, y, a coat manufacturer earns each day is given by the equation $y = -x^2 + 120x - 2000$, where x is the number of coats sold.

a. Find the number of coats the manufacturer must sell each day to earn a maximum profit for that day.
b. Find the maximum profit.

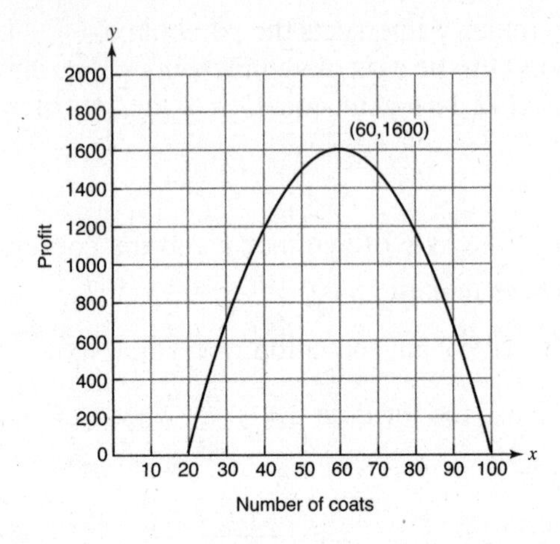

Solution: a. Because the leading coefficient of the parabola equation is negative, the vertex of the parabola is the highest point on the graph, as shown in the accompanying figure. The value of x that maximizes the profit y is the x-coordinate of the vertex:

$$\begin{aligned} x &= -\frac{b}{2a} \\ &= -\frac{120}{2(-1)} \\ &= 60 \end{aligned}$$

The coat manufacturer must sell **60** coats each day to earn a maximum profit for that day.

b. Because the maximum profit corresponds to the y-coordinate of the vertex, substitute 60 for x in the parabola equation to find the corresponding value of y:

$$\begin{aligned} y &= -60^2 + 120(60) - 2000 \\ &= -3600 + 7200 - 2000 \\ &= 1600 \end{aligned}$$

The maximum profit is **$1,600**.

Width of a Parabola

Changing the absolute value of the coefficient of the x^2 term in the parabola equation affects the width of the parabola. Of the three graphs in Figure 13.3, $y = 3x^2$ is the narrowest and $y = \frac{1}{3}x^2$ is the widest. As $|a|$ *increases* in the equation $y = ax^2 + bx + c$, the width of the parabola *decreases*.

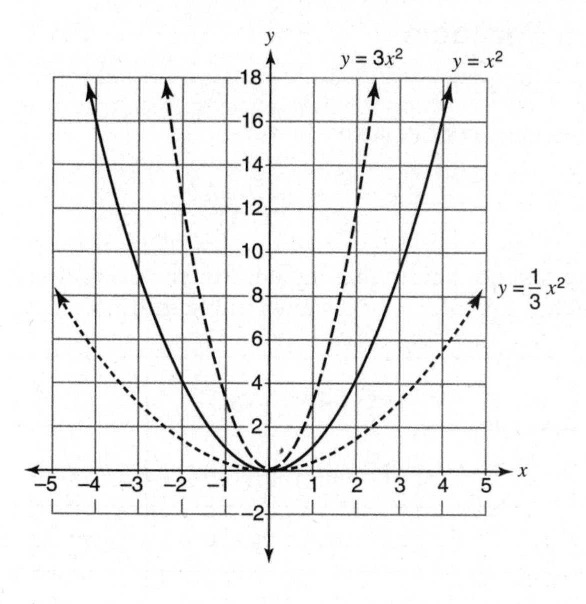

Figure 13.3 Comparing $y = x^2, y = 3x^2$, and $y = \frac{1}{3}x^2$

Vertical Shifts of $y = ax^2$

Based on the graphs of $y = x^2, y = x^2 + 3$, and $y = x^2 - 3$ in Figures 13.4 and 13.5, you can generalize that if k is a positive number,

- The graph of $y = ax^2 + k$ is the graph of $y = ax^2$ shifted *up* k units.
- The graph of $y = ax^2 - k$ is the graph of $y = ax^2$ shifted *down* k units.

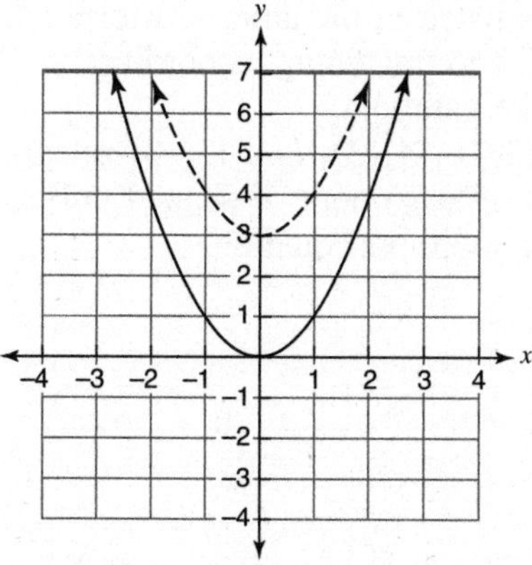

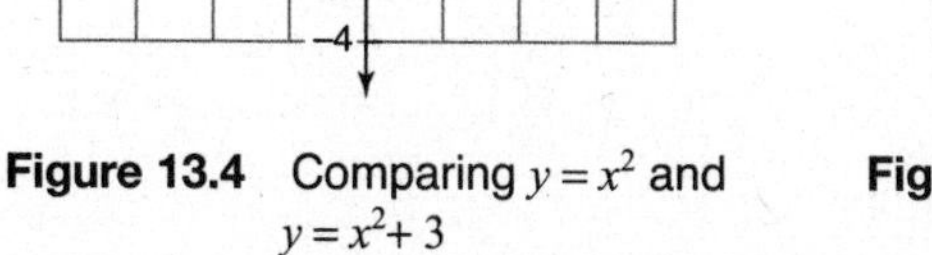
Figure 13.4 Comparing $y = x^2$ and $y = x^2 + 3$

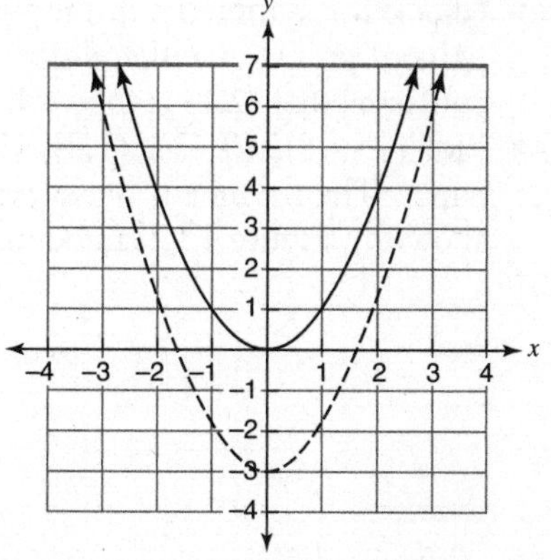

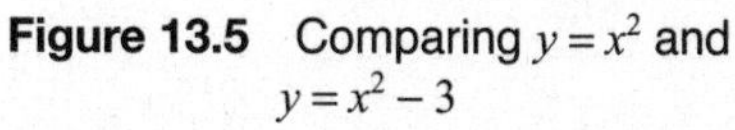
Figure 13.5 Comparing $y = x^2$ and $y = x^2 - 3$

Graphing a Parabola

To graph a parabola such as $y = -x^2 + 4x - 3$,

- Determine the x-coordinate of the vertex:

$$x = -\frac{b}{2a} = -\frac{4}{2(-1)} = 2.$$

- Prepare a table of values that includes three consecutive integer x-values on either side of $x = 2$, as shown in Figure 13.6.

x	$-x^2 + 4x - 3 = y$	(x, y)
-1	$-(-1)^2 + 4(-1) - 3 = -8$	$(-1,-8)$
0	$-0^2 + 4 \cdot 0 - 3 = -3$	$(0,-3)$
1	$-1^2 + 4 \cdot 1 - 3 = 0$	$(1,0)$
2	$-2^2 + 4 \cdot 2 - 3 = 1$	$(2,1)$
3	$-3^2 + 4 \cdot 3 - 3 = 0$	$(3,0)$
4	$-4^2 + 4 \cdot 4 - 3 = -3$	$(4,-3)$
5	$-5^2 + 4 \cdot 5 - 3 = -8$	$(5,-8)$

Figure 13.6 Table of Values

- Check for symmetry in the y-values listed in the table. Corresponding pairs of points on either side of $x = 2$ have matching y-coordinates. This confirms that (2,1) is the vertex of the parabola.
- Plot (–1,–8), (0,–3), (1,0), (2,1), (3,0), (4,–3), and (5,–8) on graph paper. Then connect these points with a smooth, **U**-shaped curve, as shown in Figure 13.7. Label the graph with its equation.

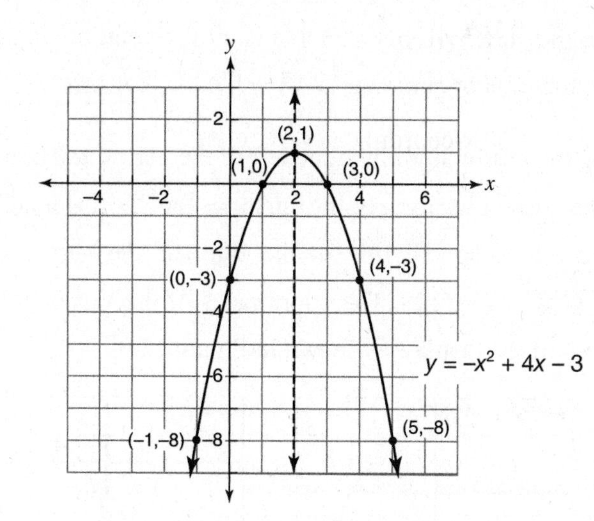

Figure 13.7 Graph of $y = -x^2 + 4x - 3$

Using Technology: Graphing a Parabola

To graph $y = -x^2 + 4x - 3$ using a graphing calculator, open the $Y =$ editor and set $Y_1 = -X \wedge 2 + 4X - 3$ by pressing

[(−)] [x, T, θ, n] [^] [2] [+] [4] [x, T, θ,n] [−] [3]

The variable x^2 can also be entered by pressing [x,T,θ,n] [x^2]. Display the parabola in a friendly viewing window such as [−4.7,4.7] × [−6.2,3.1], as shown in Figure 13.8.

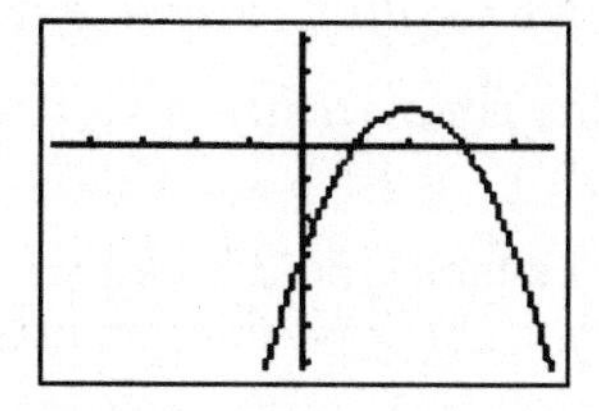

Figure 13.8 Graph of $y = -x^2 + 4x - 3$

The **maximum** (or **minimum**) function in the CALC menu can be used to determine the coordinates of the vertex of the parabola, if needed. To find the vertex of the parabola in Figure 13.8, move the cursor along the curve so that it is slightly to the left of the vertex.

- Since the vertex of this parabola is a maximum point, select the maximum function by pressing 2nd [CALC] 4 .
- Move the cursor slightly to the left of the vertex and press ENTER to set the current location of the cursor as the "Left Bound" of the vertex.
- Move the cursor slightly to the right of the vertex and then press ENTER two times. The coordinates of the vertex will appear at the bottom of the display, as shown in Figure 13.9.

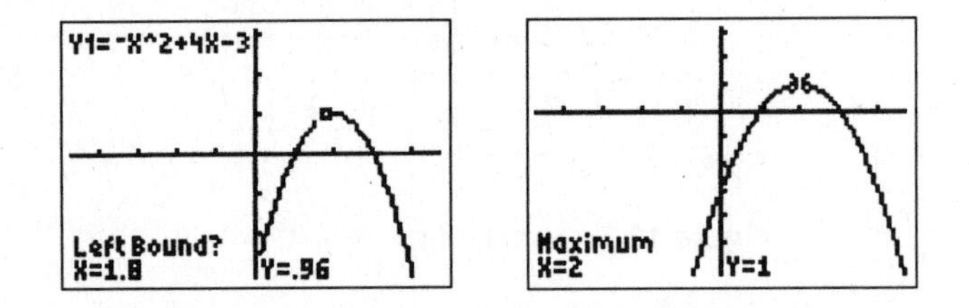

Figure 13.9 Finding the Vertex Using the Maximum Function

Using Technology: Preparing a Table of Values

A graphing calculator can also be used to build a table of values that corresponds to the table in Figure 13.6.

- Press 2nd [TBLSET] .
- Change the **TblStart** value to -1 since -1 is 3 units less than the x-coordinate of the vertex. If necessary, set $\Delta\text{Tbl} = 1$ so that x increases in steps of 1 unit.
- Press 2nd [TABLE] to see the table in Figure 13.9. If you need to look at table entries that are not currently in view, use a cursor key to scroll up or down.

By scrolling up or down the table you can easily spot the vertex by looking for the row that separates matching y-values on either side of it, as shown in Figure 13.10.

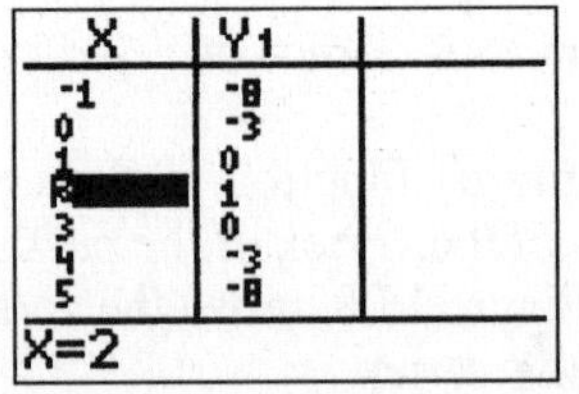

X	Y1
-1	-8
0	-3
1	0
2	1
3	0
4	-3
5	-8

X=2

Figure 13.10 Locating the Vertex

Check Your Understanding of Section 13.1

A. Multiple Choice.

1. What is the vertex, or turning point, of the parabola whose equation is $y=3x^2+6x-1$?

(1) (1,8) (2) (–1,–4) (3) (–3,8) (4) (3,44)

2. What is the minimum point of the graph of the equation $y = 2x^2 + 8x + 9$?

(1) (2,33) (2) (2,17) (3) (–2,–15) (4) (–2,1)

3. For which quadratic equation is the axis of symmetry $x = 3$?

(1) $y = -x^2 + 3x + 5$
(2) $y = -x^2 + 6x + 2$
(3) $y = x^2 + 6x + 3$
(4) $y = x^2 + x + 3$

4. What is an equation of the axis of symmetry of the parabola whose equation is $y=-\frac{1}{2}x^2+2x-7$?

(1) $x=-2$ (2) $x=2$ (3) $x=-4$ (4) $x=4$

5. What is the minimum point of the graph of the equation $y=2x^2+8x+9$?

(1) (2,33) (2) (2,17) (3) (–2,–15) (4) (–2,1)

6. Laura graphed the equation $y = -x^2$ and Alex graphed the equation $y = 3x^2$ on the same coordinate grid. What is the relationship between the two graphs?

(1) Alex's graph is wider and opens in the opposite direction from Laura's graph.
(2) Alex's graph is narrower and opens in the opposite direction from Laura's graph.
(3) Alex's graph is wider and is three units above Laura's graph.
(4) Alex's graph is narrower and is three units to the right of Laura's graph.

7. Which is an equation of the parabola that has the y-axis as its axis of symmetry and its minimum point at (0,3)?

(1) $y=x^2-3$ (2) $y=-x^2+3$ (3) $y=x^2+3$ (4) $y=x^2+3x$

8. Which is an equation of the parabola that has the y-axis as its axis of symmetry and its maximum point at (0,–5)?

(1) $y=x^2-5x$ (2) $y=-x^2+5$ (3) $y=x^2+5$ (4) $y=-x^2-5$

9. Which is an equation of a line that is parallel to the x-axis and intersects the parabola $y=-x^2+4x-3$ in one point?

(1) $y=1$ (2) $y=2$ (3) $x=1$ (4) $x=2$

10. Which is an equation of the parabola graphed in the accompanying diagram?

(1) $y=x^2-4$ (2) $y=x^2+4$ (3) $y=-x^2+4$ (4) $y=-x^2-4$

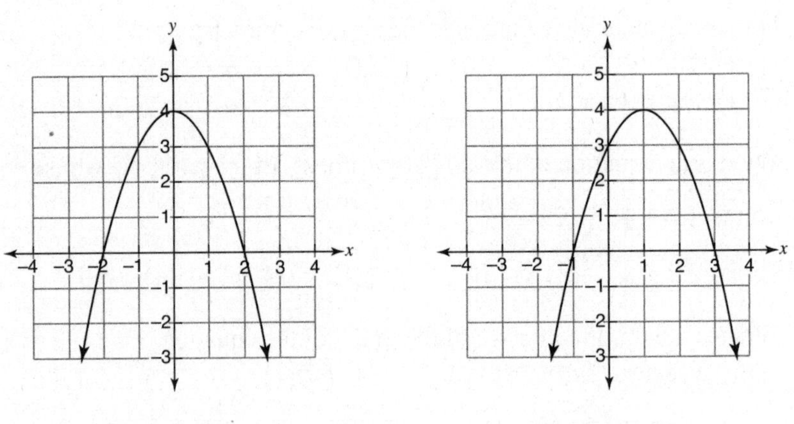

Exercise 10 **Exercise 11**

11. Which is an equation of the parabola shown in the accompanying diagram?

(1) $y=-x^2+2x+3$ (3) $y=x^2+2x+3$

(2) $y=-x^2-2x+3$ (4) $y=x^2-2x+3$

12. A parabola intersects the x-axis at $x=-3$ and $x=1$, and intersects the y-axis at (0,6). Which equation describes this parabola?

(1) $y=-2x^2+4x+6$ (3) $y=6x^2-2x+4$

(2) $y=2x^2+4x-6$ (4) $y=6x^2+2x-4$

13. What are the coordinates of the vertex and an equation of the axis of symmetry for the parabola in the accompanying figure? The scale on the axes is a unit scale.

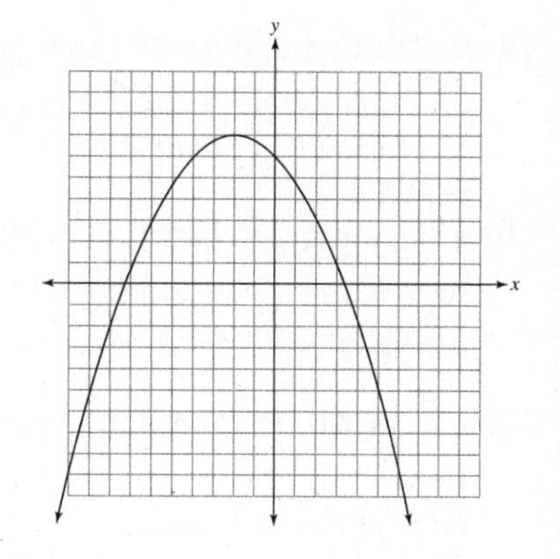

(1) vertex: (–2,7); axis of symmetry: $y = 7$
(2) vertex: (0,6); axis of symmetry: $x = 0$
(3) vertex: (7,–2); axis of symmetry: $y = -2$
(4) vertex: (–2,7); axis of symmetry: $x = -2$

B. *Show how you arrived at your answer.*

14. What is an equation of a line that contains the vertex of the parabola $y = 2x^2 + 8x + 3$ and is parallel to the line $y = 3x - 2$?

15. In the accompanying figure, parabola $y = -2x^2 + 12x$, $\overline{AB}$, and $\overline{BC}$ are drawn such that $\overline{AB}$ touches the parabola at its vertex with $\overline{AB} \perp \overline{BC}$. What is the perimeter of rectangle $OABC$?

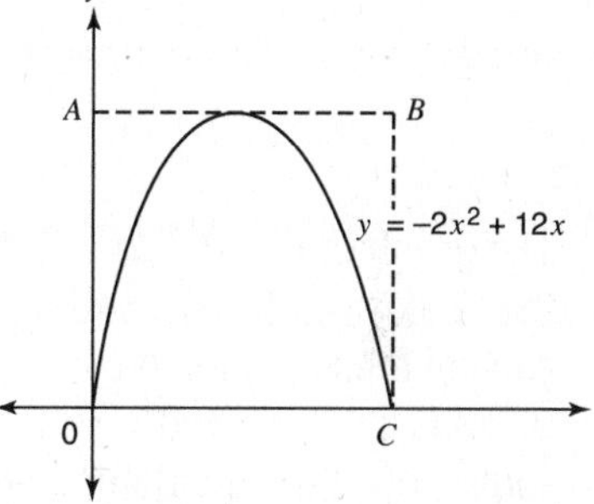

16. What is an equation of a line that contains the vertex of $y = -x^2 - 4x + 1$ and contains the point (–7,3)?

17. a. On graph paper, draw the graph of the equation $y = x^2 + 2x - 3$ for all values of x in the interval $-4 \le x \le 2$.
b. What is an equation of the line that is parallel to the x-axis and that intersects the parabola in one point?

18. a. On graph paper, draw the graph of $y = -\frac{1}{2}x^2 + 3x + 4$, including all values of x in the interval $0 \le x \le 6$.
b. What is the least integral value of k for which the line $y = k$ does *not* intersect the parabola?

19. a. Using graph paper, draw the graph of $y=2x^2-4x+6$.
b. What is the greatest integral value of k for which the line $y = k$ does *not* intersect the parabola?

20. a. Using graph paper, draw the graph of $y=-x^2+2x+5$.
b. What is the least integral value of k for which the line $y = k$ does *not* intersect the parabola?

21. The profit, P, for manufacturing a wireless device is given by the equation $P=-10x^2+750x-9000$, where x is the selling price, in dollars, for each wireless device.
a. Find the selling price that allows the manufacturer to maximize the profit on this wireless device.
b. Find the maximum profit.

13.2 SOLVING QUADRATIC EQUATIONS GRAPHICALLY

KEY IDEAS

The graph of $y = x^2 + 2x - 3$ crosses the x-axis at (–3,0) and (1,0). At each of these points, the value of y is 0. Thus, $x = -3$ and $x = 1$ are the roots of the quadratic equation $0 = x^2 + 2x - 3$.

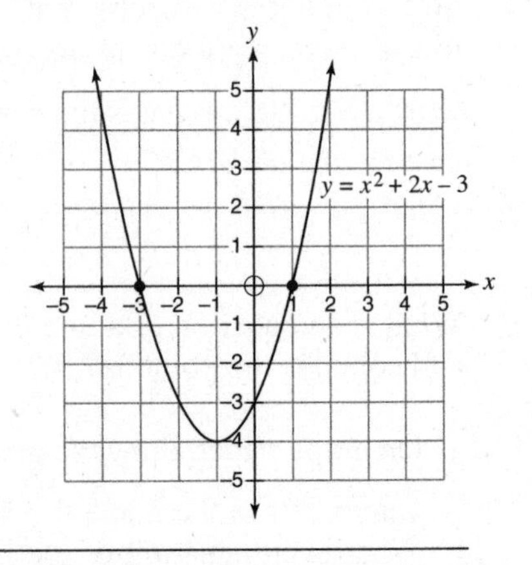

Finding the Roots of a Quadratic Equation from Its Graph

The real roots of a quadratic equation are the x-intercepts, if any, of its graph. If the graph of a quadratic equation has no x-intercepts, then the quadratic equation does not have any real roots.

MATH FACTS

To solve a quadratic equation of the form $ax^2 + bx + c = 0$ graphically, graph $y = ax^2 + bx + c$ and then read the x-intercepts from the graph.

Example 1

Graph $y = -x^2 - 4x + 5$. From the graph, determine the solution set of $-x^2 - 4x + 5 = 0$.

Solution: Find the x-coordinate of the vertex using the formula $x = -\frac{b}{2a}$, where $a = -1$ and $b = -4$:

$$\begin{aligned} x &= -\frac{b}{2a} \\ &= -\frac{-4}{2(-1)} \\ &= \frac{+4}{-2} \\ &= -2 \end{aligned}$$

Construct a table of values as shown below that includes three consecutive integer values of x on either side of $x = -2$.

x	$-x^2 - 4x + 5 = y$	(x, y)
–5	$-(-5)^2 - 4(-5) + 5 = 0$	(–5,0)
–4	$-0^2 - 4 \cdot 0 + 5 = 5$	(– 4,5)
–3	$-(-3)^2 - 4(-3) + 5 = 8$	(–3,8)
–2	$-(-2)^2 - 4(-2) + 5 = 9$	(–2,9)
–1	$-(-1)^2 - 4(-1) + 5 = 8$	(–1,8)
0	$-0^2 - 4 \cdot 0 + 5 = 5$	(0,5)
1	$-1^2 - 4 \cdot 1 + 5 = 0$	(1,0)

Graph the parabola using the points obtained in the last column of the table.

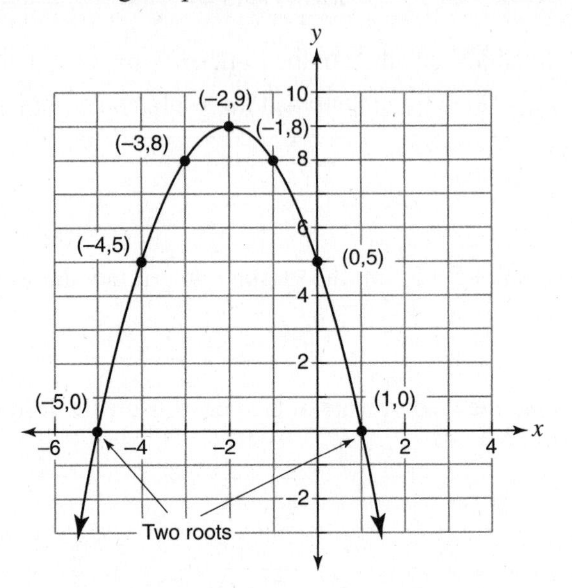

Because the x-intercepts are $x = -5$ and $x = 1$, the solution set is **{-5,1}** .

Using Technology: Solving a Quadratic Equation

To solve a quadratic equation such as $35 - x^2 = 2x$ using a graphing calculator, rewrite it in standard form as $x^2 + 2x - 35 = 0$. In the $Y =$ editor, set $Y_1 = x \wedge 2 + 2x - 35$.

- *Numerical solution*:
  Create a table by pressing 2nd [TABLE]. Scroll up and down the table until you locate the rows on which $Y_1 = 0$, as shown in Figure 13.11.

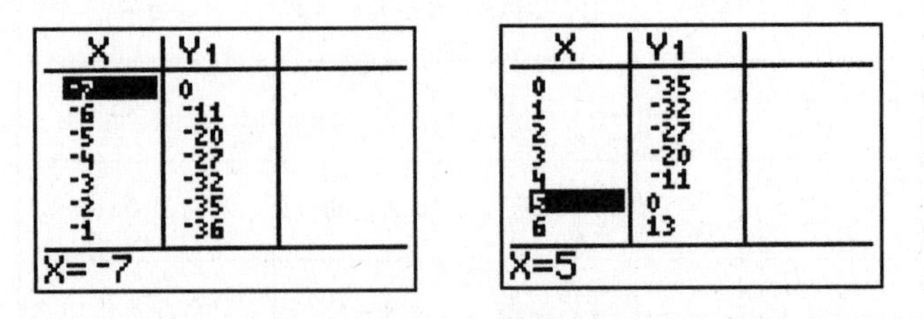

Figure 13.11 Numerical Solution: Finding Rows on Which $Y_1 = 0$

Thus, the solution set is $\{-7,5\}$. If the roots of the quadratic equation do not have integral values, you will need to adjust the value of ΔTbl in the table setup or use a different approach.

* *Graphical solution*:

  Graph Y_1 in a friendly window such as $[-9.4, 9.4] \times [-40, 40]$. Then find the zeros or x-intercepts of the function either by using the TRACE feature or by selecting **2:zero** from the CALCULATE menu and then working as follows:

  STEP 1 Move the cursor along the curve and stop when it is slightly to the left of the positive x-intercept, as shown in Figure 13.12. Press [ENTER] to set the value of the left bound for that intercept.

  STEP 2 Move the cursor slightly to the right of the same intercept and press [ENTER] to set the right bound. Press [ENTER] a second time thereby obtaining $x = 5$ as the positive x-intercept or root, as shown in Figure 13.13.

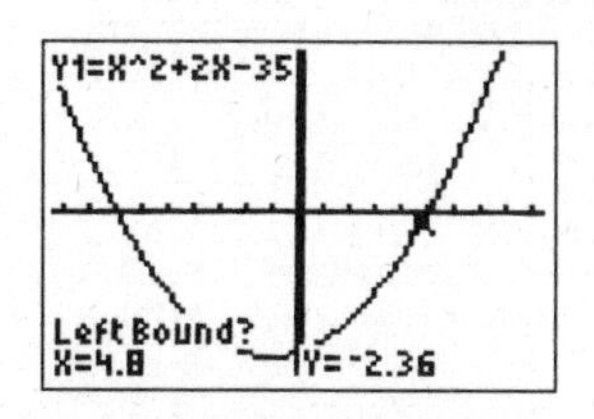

Figure 13.12 Finding the Left Bound of the Positive y-Intercept

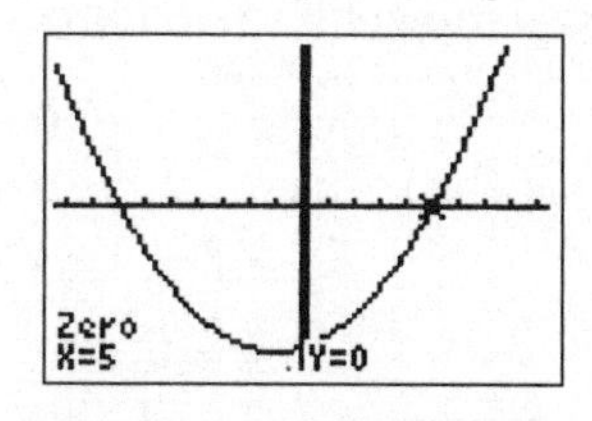

Figure 13.13 Finding the Positive x-Intercept

  STEP 3 Repeat steps 1 and 2 for the negative x-intercept obtaining $x = -7$ as the other root.

Modeling Motion of Objects

When an object like a ball is tossed in the air, its height as a function of time is a parabola in which the x-coordinate of each point on the curve represents how much time has elapsed after the object was launched, and the corresponding y-coordinate gives its height at that instant of time, as shown in Figure 13.14. The y-coordinate of the vertex of the parabola corresponds to the maximum height the object reaches before it begins to fall back to the ground.

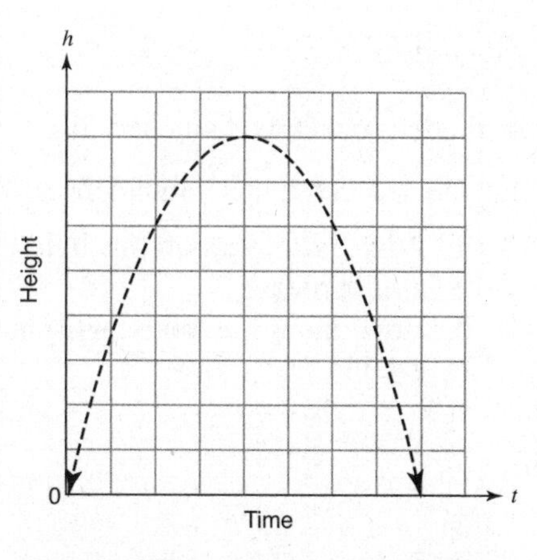

Figure 13.14 A Parabolic Function

Example 3

After t seconds, a ball tossed in the air from ground level reaches a height of h feet where $h = 144t - 16t^2$.

a. What is the height of the ball when t is 3 seconds?
b. After how many seconds will the ball hit the ground before rebounding?

Solution: a. Substitute 3 for t in the height equation:

$$\begin{aligned} h &= 144t - 16t^2 \\ &= 144\times 3 - (16\times 3^2) \\ &= 432 \quad - (16\times 9) \\ &= 432 - 144 \\ &= \mathbf{288\ feet} \end{aligned}$$

b. When the ball hits the ground, $h = 0$. Replace h with 0 in the height equation and solve for t:

$$\begin{aligned} 0 &= 144t - 16t^2 \\ 0 &= 4t(36 - 4t) \\ \textit{or}\ \ 4t(36-4t) &= 0 \end{aligned}$$

$$4t = 0 \quad or \quad 36 - 4t = 0$$

$$t = \frac{0}{4} \qquad\qquad -4t = -36$$

$$t = 0 \qquad\qquad t = \frac{-36}{-4} = 9 \text{ seconds}$$

The solution $t = 0$ seconds represents the instant of time at which the ball is tossed in the air. The solution $\boldsymbol{t = 9}$ seconds is the number of seconds it takes for the ball to reach the ground.

Example 4

A model rocket is launched from ground level. At t seconds after it is launched, it is h meters above the ground where $h = -4.9t^2 + 68.6t$.

a. What is the maximum height, to the *nearest meter*, attained by the model rocket?
b. How many seconds will have elapsed when the rocket hits the ground?

Solution 1:

a. The maximum height reached by the rocket is the y-coordinate of the vertex of the parabola.

- The x-coordinate of the vertex is

$$t=-\frac{b}{2a}=-\frac{68.6}{2(-4.9)}=-\frac{68.6}{-9.8}=7 \text{ seconds}$$

- Find the y-coordinate of the vertex by substituting $t = 7$ in the parabola equation:

$$\begin{aligned} h &= -4.9t^2 + 68.6t \\ &= -4.9(7)^2 + 68.6(7) \\ &= -240.1 + 480.2 \\ &= \mathbf{240.1\ meters} \end{aligned}$$

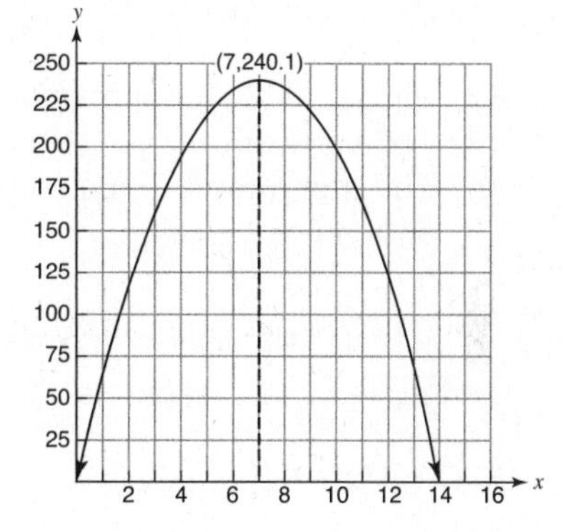

b. The x-intercepts of the parabola represent the times at which the rocket is at ground level. One x-intercept is at (0,0), which corresponds to the launch time. The other x-intercept is at $t = 2 \times 7 =$ **14** seconds since the axis of symmetry bisects the segment whose endpoints are the two x-intercepts.

Solution 2:

a. Graph $Y_1 = -4.9x^2 + 68.6x$ in a friendly viewing window such as $[0,18.8] \times [0,310]$. Using the **maximum** function from the CALC menu you can find that the x–coordinate of the turning point is **240.1** meters.

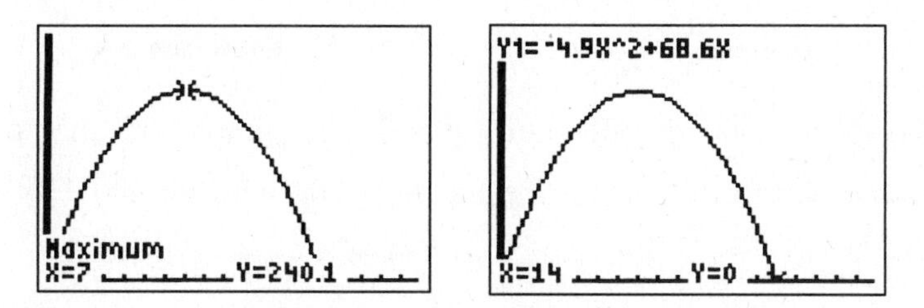

b. Use the TRACE feature to find that $x =$ **14** seconds when $y = 0$.

Check Your Understanding of Section 13.2

A. Multiple Choice.

1. The accompanying diagram shows a parabola. Which statement is *not* true?
(1) An equation of the axis of symmetry is $x = -2$.
(2) The parabola has a maximum point.
(3) The vertex of the parabola is $(-2,1)$.
(4) The graph represents a quadratic equation with no real roots.

2. The accompanying diagram shows the graph of the parabola $y = ax^2 + bx + c$. What is the *sum* of the two roots of the equation $ax^2 + bx + c = 0$?
(1) 1 (2) -1 (3) 0 (4) -12

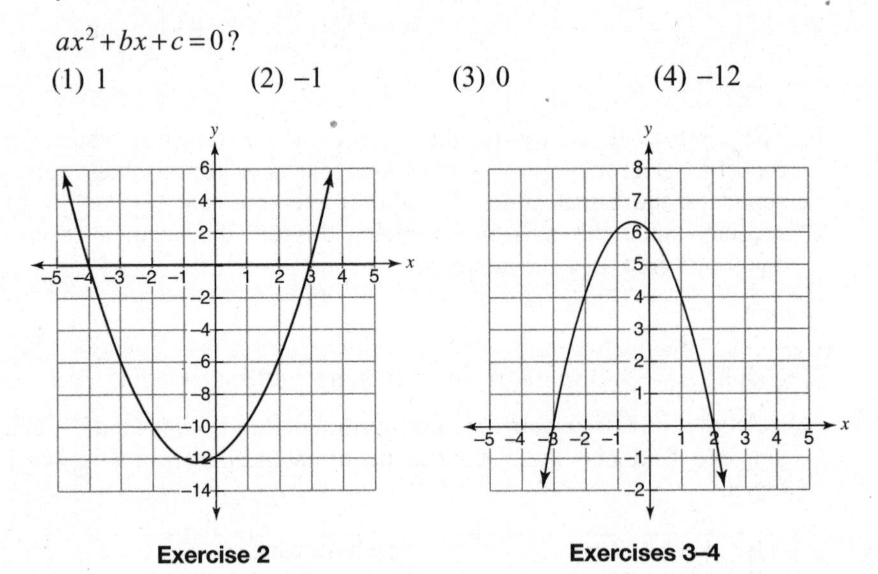

Exercise 2 **Exercises 3–4**

3. The accompanying diagram shows the graph of the parabola $y = ax^2 + bx + c$. What must be one root of the equation $ax^2 + bx + c = 0$?
(1) -3 (2) 6 (3) -6 (4) 0

4. For the parabola in Exercise 3, what is the least integral value of k for which the line $y = k$ does *not* intersect the parabola?
(1) 7 (2) 4 (3) 3 (4) 5

B. *Show how you arrived at your answer.*

5. The graph of a quadratic function is shown in the accompanying diagram. The scale on the coordinate axes is a unit scale. Write an equation of this graph.

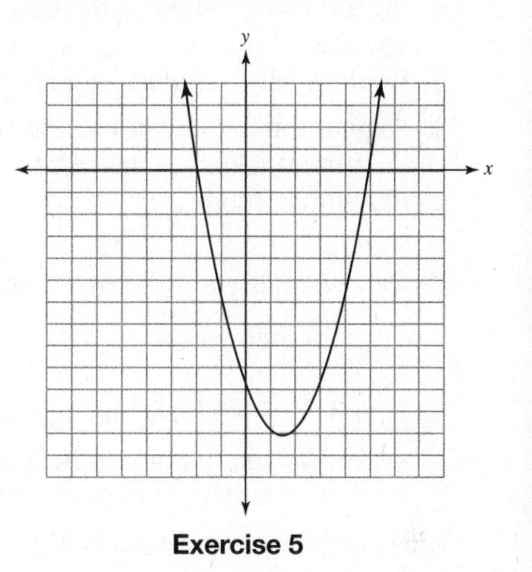

Exercise 5

6. A parabola that passes through the origin has its vertex at (−2,12). What is an equation of the parabola?

7. a. Draw the graph of the equation $y = x^2 - 8x + 15$ for all values of x in the interval $1 \leq x \leq 7$.

b. From the graph in part a, find the roots of the equation $x^2 - 8x + 15 = 0$.

c. Write an equation of the line that is parallel to the x-axis and intersects the graph drawn in part a in one point.

d. Determine the *greatest integral* value of k for which the line $y = k$ does *not* intersect the parabola.

8. a. Draw the graph of the equation $y = -x^2 - 5x + 6$ for all values of x in the interval $1 \leq x \leq 7$.

b. From the graph in part a, find the roots of the equation $-x^2 - 5x + 6 = 0$.

c. Determine the *least integral* value of k for which the line $y = k$ does *not* intersect the parabola.

9. Amy tossed a ball in the air in such a way that the path of the ball was modeled by the equation $y = -x^2 + 6x$. In the equation, y represents the height of the ball in feet and x is the time in seconds.

a. Using graph paper, graph $y = -x^2 + 6x$ for $0 \leq x \leq 6$.

b. From the graph, determine the maximum height of the ball and the number of seconds it takes to reach the maximum height.

10. a. Using the equation of motion given in Example 3 on page 364, find the number of seconds the ball is in the air when it reaches a height of 180 feet.
b. Explain why there are two answers in part a.
c. Explain how you can use the two solutions from part a to determine the time it takes for the ball to reach its maximum height. What is the maximum height?

11. After t seconds, a ball tossed in the air from ground level reaches a height of h feet, where $h = 144t - 16t^2$.
a. Graph $h = 144t - 16t^2$ using your graphing calculator. From the graph, determine the maximum height of the ball and the number of seconds it takes to reach the maximum height.
b. After how many seconds will the ball hit the ground before rebounding? Confirm your answer algebraically.

12. A swimmer dives from a diving board that is 24 feet above the water when $t = 0$ seconds. The distance d in feet that the diver falls is given by the equation $d = -16t^2 + 40t + 24$.
a. How many seconds does it take for the diver to hit the water?
b. What is the maximum height above the water the swimmer reaches during the dive?

13. An arch is built so that it has the shape of a parabola with the equation $y = -3x^2 + 21x$, where y represents the height of the arch.
a. What is the number of feet in the width of the arch at its base?
b. What is the number of feet in the maximum height of the arch?

13.3 SOLVING A LINEAR-QUADRATIC SYSTEM

KEY IDEAS

As with a linear system of equations, the solution of a linear-quadratic system of equations is the ordered number pair that satisfies both equations. A linear-quadratic system can be solved either graphically or algebraically.

Solving a Linear-Quadratic System Graphically

A system of two equations consisting of one linear equation and one quadratic equation can be solved graphically as is illustrated in Example 1.

Example 1

Solve the following system of equations graphically, and state the coordinates of all points in the solution set.

$$y = -x^2 - 4x + 3$$
$$x + y = -1$$

Solution: Draw the graphs of both equations on the same set of coordinate axes. Then note the coordinates of the points where the two graphs intersect.

STEP 1 Graph the parabola $y = -x^2 - 4x + 3$

The equation of the axis of symmetry of this parabola has the form $x = -\frac{b}{2a}$ where $a = -1$ and $b = -4$:

$$x = -\frac{-4}{2(-1)}$$
$$= -\left(\frac{-4}{-2}\right)$$
$$= -2$$

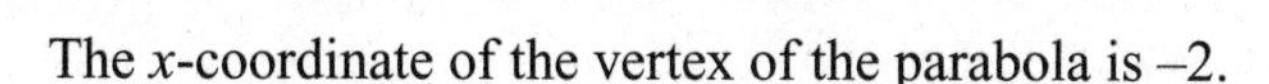

The x-coordinate of the vertex of the parabola is –2.

Create a table of values. Open the Y= editor of your graphing calculator, and set $Y_1 = -x^2 - 4x + 3$. Then use the table feature of your graphing calculator to display a table of values that includes three consecutive integer x-values on either side of $x = -2$:

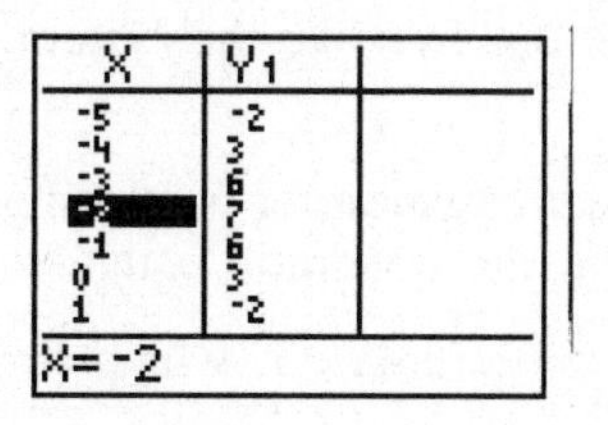

X	Y1
-5	-2
-4	3
-3	6
-2	7
-1	6
0	3
1	-2

X=-2

Plot the points in the table: (–5,–2), (–4,3), (–3,6), (–2,7), (–1,6), (0,3), and (1,–2). Then connect these points with a smooth U-shaped curve, as shown in the accompanying figure. Label the parabola with its equation.

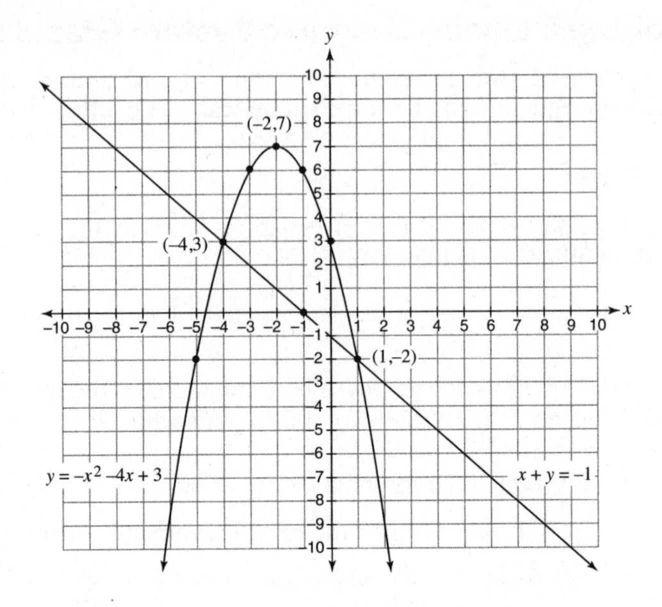

STEP 2 Graph the line $x + y = -1$. Plot any two convenient points that satisfy the equation, such as (0,–1) and (5,–6). Then connect these points with a straight line as shown in the accompanying figure. Label the line with its equation.

STEP 3 Determine the solution set by reading the coordinates of the points of intersection of the two graphs. The two graphs intersect at (–4,3) and (1,–2).

The points in the solution set are **(–4,3)** and **(1,–2)**.

Solving a Linear-Quadratic System Algebraically

A linear-quadratic system can be solved algebraically by solving the first-degree equation for either variable and then substituting for that variable in the quadratic equation. To solve the system $y = -x^2 + 4x - 3$ and $x + y = 1$ algebraically,

- Solve the linear equation for y, which gives $y = 1 - x$.
- Eliminate y in the quadratic equation by replacing it with $1-x$. The result is $1 - x = -x^2 + 4x - 3$, which simplifies to $x^2 - 5x + 4 = 0$. Solve $x^2 - 5x + 4 = 0$ by factoring:

$$\begin{aligned} (x-1)(x-4) &= 0 \\ x-1=0 \;\; or \;\; x-4 &= 0 \\ x=1 \;\; or \;\; x &= 4 \end{aligned}$$

- Find the corresponding values of y by substituting each solution for x in the linear equation:
  - If $x = 1$, then $1 + y = 1$ and $y = 0$ so (1,0) is a solution.
  - If $x = 4$, then $4 + y = 1$ and $y = -3$ so (4,–3) is a solution.

The solution set is **{(1,0), (4,–3)}**.

Using Technology: Solving a Linear-Quadratic System

To solve the system $y = -x^2 - 4x + 3$ and $x + y = -1$ using a graphing calculator, open the Y= editor and set $Y_1 = -x^2 - 4x + 3$ and $Y_2 = -x - 1$.

- To solve the system graphically, select a friendly window in which the graph fits. Then use the intersect feature from the **CALC** menu to find that the graphs intersect at (–4,3) and (1,–2). Figure 13.15 displays the graphs in a [–9.4, 9.4] × [–9.3, 9.3] viewing window.

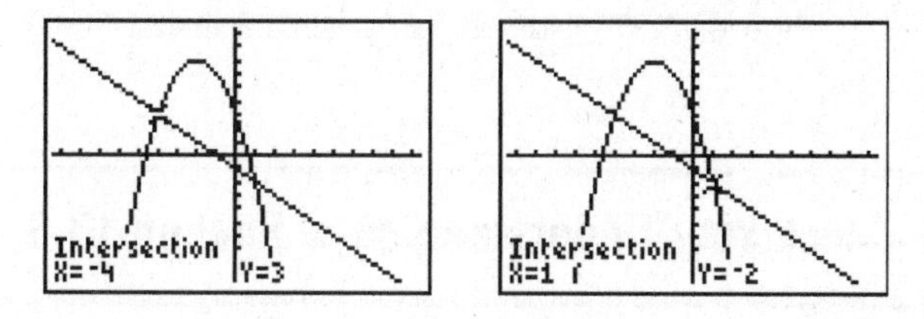

Figure 13.15 Solving $y = -x^2 - 4x + 3$ and $x + y = -1$ Graphically

- To solve the same system numerically, press [2nd] [TABLE] to display a table of values for the two stored functions. Scroll up or down the table until you locate the rows on which $Y_1 = Y_2$ as shown in Figure 13.16.

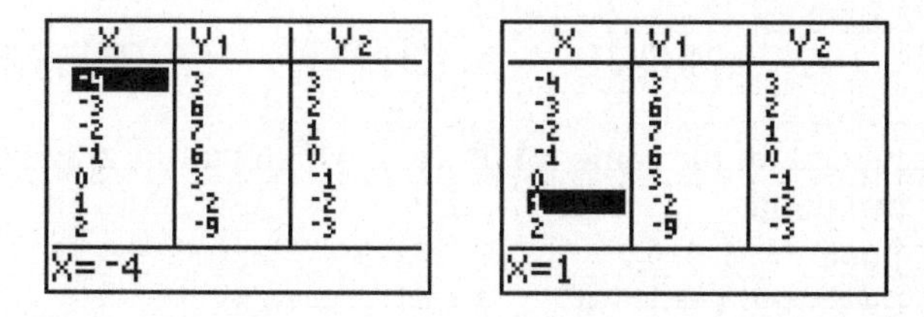

X	Y1	Y2
-4	3	3
-3	6	2
-2	7	1
-1	6	0
0	3	-1
1	-2	-2
2	-9	-3

X= -4

X	Y1	Y2
-4	3	3
-3	6	2
-2	7	1
-1	6	0
0	3	-1
1	-2	-2
2	-9	-3

X=1

Figure 13.16 Solving $y = -x^2 - 4x + 3$ and $x + y = -1$ Numerically

Be aware that if the solutions do not have integral values, you will need to adjust the value of ΔTbl in the table setup or use a different approach.

Possible Number of Solutions

As illustrated in Figure 13.17, the solution set of a linear-quadratic system of equations may consist of two solutions (see line A); one solution (see line B); or no solution (see line C).

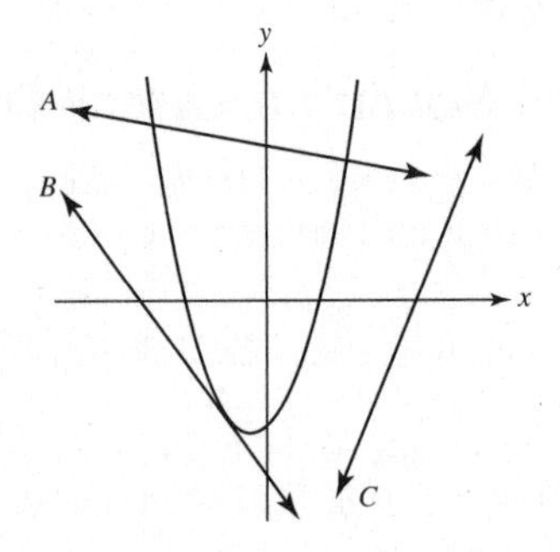

Figure 13.17 Number of Possible Solutions

Check Your Understanding of Section 13.3

A. *Multiple Choice.*

1. The graphs of the equations $y = x^2$ and $x = 2$ intersect in
(1) 1 point (2) 2 points (3) 3 points (4) 0 point

2. Which is a point of intersection of the graphs of the line $y = x$ and the parabola $y = x^2 - 2$?
(1) (1,1) (2) (2,2) (3) (0,0) (4) (2,1)

3. When graphed on the same set of axes, which pair of graphs intersect at exactly two points?
(1) $y = 4$ and $y = x^2 + 5$
(2) $y = 1 + x^2$ and $y = 1 - x^2$
(3) $y = 5$ and $y = -x^2 + 5$
(4) $y = x$ and $y = -x^2$

4. What is a solution for the following system of equations?

$$y = x^2$$
$$y = -4x + 12$$

(1) (–2,4) (2) (6,36) (3) (2,4) (4) (6,24)

5. Which is a solution for the system of equations $y = 2x - 15$ and $y = x^2 - 6x$?
(1) (3,–9) (2) (0,0) (3) (5,5) (4) (6,0)

6. When the graphs of the equations $y = x^2 - 5x + 6$ and $x + y = 6$ are drawn on the same set of axes, at which point do the graphs intersect?
(1) (4,2) (2) (5,1) (3) (3,3) (4) (2,4)

B. *Show how you arrived at your answer.*

7–12. *Determine graphically the solution set for each system of equations. Then confirm your answer by solving the same system of equations algebraically.*

7. $y = x^2 + 4x - 1$
$y + 3 = x$

9. $y = -x^2 - 2x + 8$
$y = x + 4$

11. $y = x^2 - 6x + 5$
$y + 7 = 2x$

8. $y = x^2 - 4x + 9$
$y - x = 5$

10. $y = x^2 - 6x + 6$
$y - x = -4$

12. $y = x^2 - 5x + 11$
$y + 5 = 3x$

13. a. Solve the system $y = -x^2 + 6x - 3$ and $x + y = 3$ graphically using graph paper.

b. Check your answer by solving the system graphically and numerically using your graphing calculator.

14. a. Solve the system $y = x^2 + 4x + 1$ and $y = 5x + 3$ graphically using graph paper.

b. Check your answer by solving the system graphically and numerically using your graphing calculator.

13.4 EXPONENTIAL GROWTH AND DECAY

KEY IDEAS

The function $y = 2^x$ is an exponential function with base 2. In an exponential function, the base is always a positive number other than 1 and the exponent contains the variable. An exponential function always has a positive value as a positive base raised to any nonzero power is positive. As x increases, the graph of $y = 2^x$ rises more and more rapidly. As x decreases, the graph gets closer and closer to the x -axis without touching it.

Exponential functions are used to describe growth (or decay) processes such as population or investment growth in which a quantity increases (or decreases) by a fixed percent of its current value over equal periods of time.

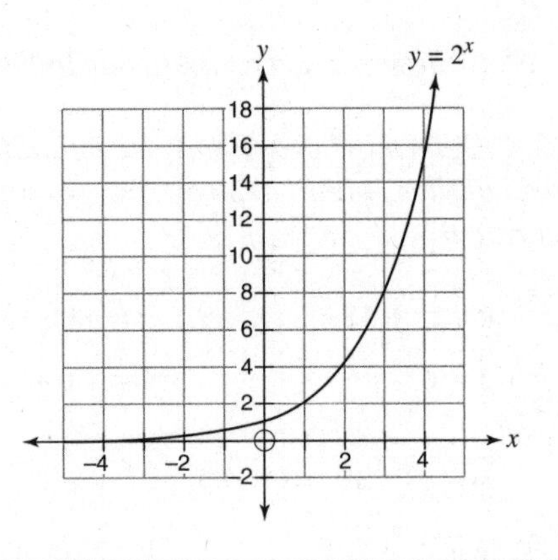

The Exponential Function

An **exponential function** is an equation of the form $y = b^x$ provided b is a positive number other than 1. The value of the base b determines whether y increases or decreases as x increases.

As shown in Figure 13.18:

- The graph of $y = b^x$ is confined to Quadrants I and II where y is always positive.
- As x increases when $b > 1$, y increases and the graph rises at an increasingly faster rate.
- As x increases when $0 < b < 1$, y decreases and the graph falls but at a slower and slower rate.

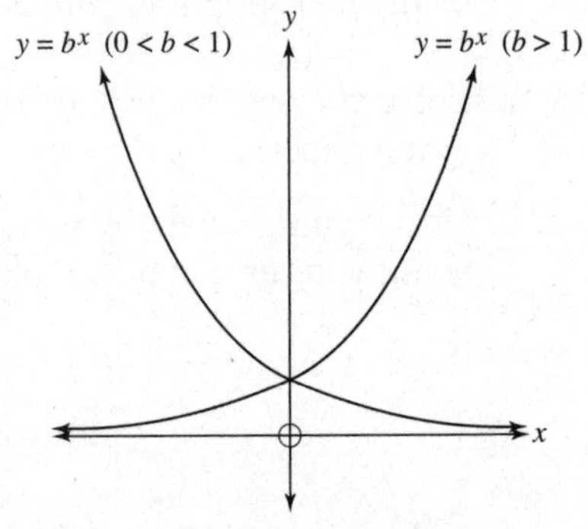

Figure 13.18
The Exponential Function $y = b^x$.

Using $y = Ab^x$ to Model Growth

Exponential functions are often used to represent real-life processes such as population growth or decline. In an exponential function of the form $y = Ab^x$, x typically represents the number of time periods over which a growth process is being studied. The constant A represents the initial or starting value of y before the growth (or decay) process begins. As time passes, y changes but not at a constant rate. In exponential growth, as the time periods increase, y increases at a faster and faster rate. In exponential decay, as the time periods increase, y decreases but at a slower and slower rate.

Linear Versus Exponential Functions

A linear function changes at a constant rate. Each time x changes by 1 unit, y changes by the same *fixed* amount. If $y = 3x + 1$, then y increases 3 units each time x increases by 1 unit. The rate at which y changes is the slope of the line, which remains constant along the line.

In an exponential function, however, each time the exponent x increases by 1 unit, y changes by a *different* amount that is a set percent or fixed multiplying factor of its previous value. As a result, the slope or the rate at which y changes varies along an exponential curve. To illustrate this process, consider the accompanying table of values that represent points from the graph of $y = 5 \cdot 2^x$ shown in Figure 13.19:

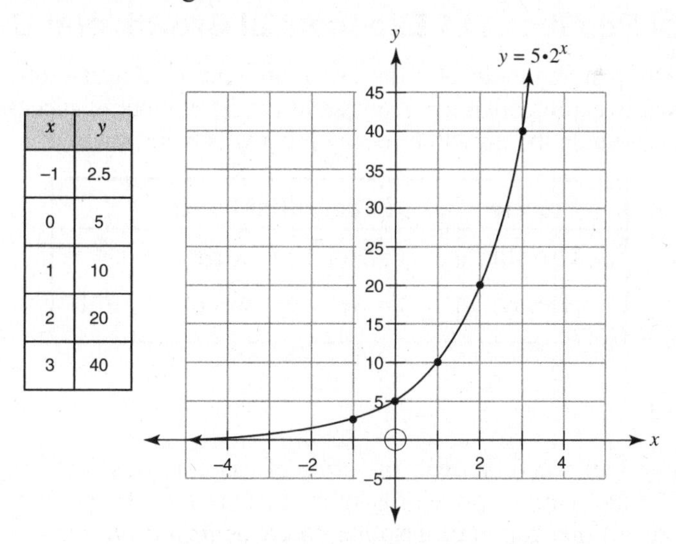

x	y
–1	2.5
0	5
1	10
2	20
3	40

Figure 13.19 Graph of $y = 5 \cdot 2^x$

You should verify that each time x increases by 1 in Figure 13.19, y increases to twice its previous value. In terms of percent, y increases by

100% of its previous value each time x increases by 1. This type of behavior represents exponential growth.

An Example of Exponential Growth

Suppose \$320 is deposited in a bank account that earns 7% interest compounded annually.

- After one year, the balance in the account is:

$$\underbrace{320}_{\text{initial amount}} + \underbrace{320\times 0.07}_{\text{interest}} = 320(1+0.07) = 320(1.07)^1$$

- After the second year, the balance in the account is

$$\underbrace{320(1.07)}_{\text{old amount}} + \underbrace{(320(1.07))\times 0.07}_{\text{interest}} = 320(1.07)(1+0.07) = 320(1.07)^2$$

- After x years, the initial balance has been multiplied x times by 1.07:

$$y = 320\underbrace{(1.07)(1.07)\ldots(1.07)}_{x\text{ factors}} = 320(1.07)^x$$

The equation $y = 320(1.07)^x$ represents exponential growth in which 320 is the initial amount and 1.07 is 1 plus the rate of growth of 7%.

General Equations of Exponential Growth and Decay

In the accompanying table, A = the initial amount, r = the rate of growth or decay (usually expressed as a percent), and x = the number of successive time periods over which the growth or decay process is being studied.

Process	Exponential Function
Growth	$y = A(1 + r)^x$, where $r > 0$
Decay	$y = A(1 - r)^x$, where $0 < r < 1$

Example 1

On January 1 of a certain year, the price of gasoline was \$3.50 per gallon. Throughout the year, the price of gasoline increased by 3% per month. What was the cost of one gallon of gasoline, to the *nearest cent*, one year later on January 1?

Solution: Use the equation for exponential growth $y = A(1 + r)^x$, where $A = 2.75$, $r = 3\% = 0.03$, and $x = 12$:

$$\begin{aligned} y &= A(1 + r)^x \\ &= 2.75(1 + 0.03)^{12} \\ &= 2.75(1.03)^{12} \\ &\approx 3.920842439 \end{aligned}$$

The cost of one gallon of gasoline, to the *nearest cent*, was **\$3.92**.

Example 2

Raymond buys a new car for \$24,500. The car depreciates (loses its value) by about 11% per year. What is the value of the car, to the *nearest dollar*, after five years?

Solution: Use the equation for exponential decay $y = A(1 - r)^x$, where $A = 24{,}500$, $r = 11\% = 0.11$, and $x = 5$:

$$\begin{aligned} y &= A(1 - r)^x \\ &= 24{,}500(1 - 0.11)^5 \\ &= 24{,}500(0.89)^5 \\ &\approx 13{,}680.94565 \end{aligned}$$

The value of the car, to the *nearest dollar*, is **\$13,681**.

Check Your Understanding of Section 13.4

A. Multiple Choice.

1. Kathy deposits \$250 into an investment account with an annual rate of 5%, compounded annually. The amount in her account can be determined by the formula $A = P(1+R)^t$, where P is the amount deposited, R is the annual interest rate, and t is the number of years the money is invested. If she makes no other deposits or withdrawals, how much money will be in her account at the end of 15 years?
(1) \$257.50 (2) \$437.53 (3) \$519.73 (4) \$3,937.50

2. Daniel's Print Shop purchased a new printer for \$35,000. Each year it depreciates at an annual rate of 5%. What will its approximate value be at the end of the fourth year?
(1) \$33,250.00 (2) \$30,008.13 (3) \$28,507.72 (4) \$27,082.33

3. The population of Clarkstown was 3,381,000 in 2010 and is growing at an annual rate of 1.8%. Assuming this growth rate continues, what is the projected population of Clarkstown in the same month in the year 2019?
(1) 3,899,666 (2) 3,969,860 (3) 3,990,589 (4) 4,041,317

4. A sum of $14,000 is being held in a non-interest-bearing account. If 15% of the current balance is being withdrawn from the account every year, how much money will be in the account 5 years from now?
(1) $3,500 (2) $6,211.87 (3) $7,788.13 (4) $10,500.00

5. The accompanying diagram represents the biological process of cell division.

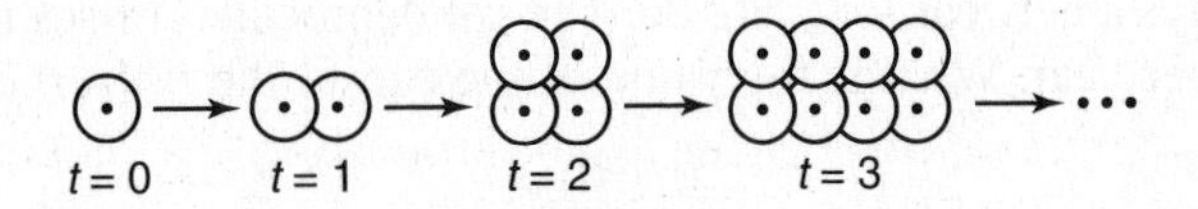

Exercises 5

If this process continues, which equation best represents the number of cells, y, at any time, t?
(1) $y = t + 2$ (2) $y = 2t$ (3) $y = t^2$ (4) $y = 2^t$

6. The value, y, of a $24,000 investment over x years is represented by the equation $y = 24{,}000(1.2)^{\frac{x}{2}}$. What is the profit on a 6-year investment?
(1) $17,472 (2) $21,400 (3) $41,472 (4) $45,400

7. A culture of 5,000 bacteria triples every 20 minutes. If y represents the size of the bacteria population after m minutes have elapsed, which equation can be used to represent the growth of the bacteria?
(1) $y = 5{,}000(20)^{3m}$
(2) $y = 5{,}000(3)^{20m}$
(3) $y = 5{,}000(3)^{\frac{m}{20}}$
(4) $y = 5{,}000(20)^{\frac{m}{3}}$

8. Which equation best represents the accompanying graph?
(1) $y = 2^x$
(2) $y = x^2 + 2$
(3) $y = \left(\frac{1}{2}\right)^x$
(4) $y = -2^x$

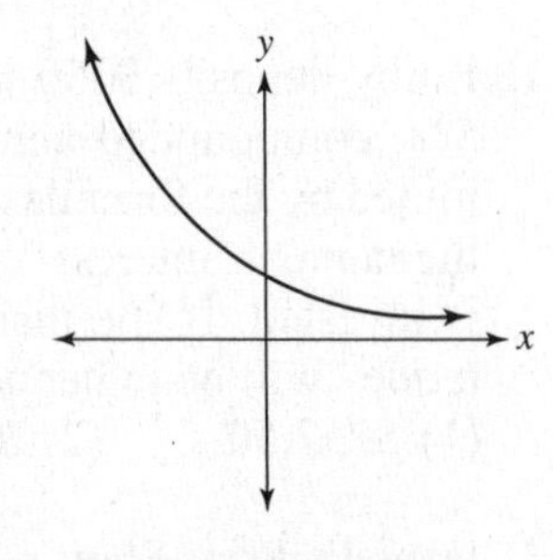

Exercise 8

9. In the equation $y = 4^x$, by what percent does y increase when x increases by 1 unit?
(1) 100 (2) 200 (3) 300 (4) 400

10. Which equation represents the data in the accompanying table?

Time in hours, x	0	1	2	3	4	5	6
Population, y	5	10	20	40	80	160	320

(1) $y = 2x + 5$ (2) $y = 2^x$ (3) $y = 5x^2$ (4) $y = 5(2^x)$

11. The New York Volleyball Association invited 64 teams to compete in a tournament. After each round, half of the teams were eliminated. Which equation represents the number of teams, t, that remained in the tournament after r rounds?
(1) $t = 64(r)^{0.5}$
(2) $t = 64(-0.5)^r$
(3) $t = 64(1.5)^r$
(4) $t = 64(0.5)^r$

12. The height, h, of a bouncing ball after x bounces is represented by the equation $h = 80(0.5)^x$. How many times higher is the first bounce than the fourth bounce?
(1) 8 (2) 2 (3) 16 (4) 4

13. A radioactive substance has an initial mass of 100 grams, and its mass halves every 4 years. Which expression shows the number of grams remaining after x years?

(1) $100(4)^{\frac{x}{4}}$ (2) $100\left(\frac{1}{4}\right)^{2x}$ (3) $100\left(\frac{1}{2}\right)^{\frac{x}{4}}$ (4) $100\left(\frac{1}{2}\right)^{4x}$

B. *Show how you arrived at your answer.*

14. The Booster Club raised \$30,000 for a sports fund. No more money will be placed into the fund. Each year, the fund will decrease by 8%. Determine the amount of money, to the *nearest cent*, that will be left in the sports fund after 5 years.

15. Louis inherits \$20,000. He invests half of it with a guaranteed rate of return of 6% per year. If he spends the remaining amount of his inheritance at a rate of 10% per year, how much will Louis's original inheritance be worth at the end of 5 years to the *nearest cent*?

16. The current population of Little Pond, New York, is 20,000. The population is decreasing, as represented by the equation $P = A(1.3)^{-0.25t}$, where P = final population, t = time in years, and A initial population. Estimate *to the nearest hundred people* the population of Little Pond 10 years from now.

17. Depreciation (the decline in cash value) on a car can be determined by the formula $V = C(1 - r)^n$, where V is the value of the car after n years, C is the original cost, and r is the rate of depreciation. A used car was purchased in February 2012 for $15,900. If the car depreciates 13% of its value each year, what will be the estimated value of the car, to the *nearest hundred dollars*, 5 years later?

18. When a principal of P dollars is invested at an annual rate of interest r, compounded n times per year, the principal grows at an exponential rate. After x years, the dollar amount of the investment, A, is given by the compound interest formula, $A = P\left(1+\frac{r}{n}\right)^{nx}$. The Franklins inherited $20,000 that they want to invest for their child's future college expenses. If they invest it at an annual rate of 5% with interest compounded quarterly, determine the value of the account, *in dollars*, after 10 years.

19. A rubber ball is dropped from 12 feet above the gymnasium floor and is allowed to bounce up and down. After each bounce, it rises 75% of the height from which it fell.

a. Write an exponential function that expresses the relationship between height, h, and the number of bounces, n.

b. Using your answer from part a and the table feature of your graphing calculator, determine the least number of bounces for the ball to achieve a height of less than 1 foot from the floor.

20. A certain drug raises a patient's heart rate, h, in beats per minute, according to the equation $h = 70 + 0.2x$, where x is the number of milligrams of the drug in the patient's bloodstream. After t hours, the level of the drug in the patient's blood stream is given by the equation $x = 300(0.8)^t$. After 5 hours, what is the number of beats per minute in the patient's heart rate, correct to the *nearest whole number*?

CHAPTER 14

STATISTICS AND VISUAL REPRESENTATIONS OF DATA

14.1 MEASURES OF CENTRAL TENDENCY

KEY IDEAS

A **measure of central tendency** for a set of data values is a single number that represents a central or middle value about which the data values are clustered. The *mean* (average), *mode*, and *median* are measures of central tendency that collectively help to describe how the individual data values are distributed.

The Mean or Average

For a set of scores, the **mean** or average is the sum of the scores divided by the number of scores. The mean for 76, 84, and 92 is

$$\text{mean} = \frac{76+84+92}{3} = \frac{252}{3} = 84$$

When finding an average, individual scores may be weighted differently. Suppose Sally received test grades of 76, 84, and 92, where 92 represents her grade on the final exam. If the final exam has a weighting factor of 2, then the final exam score is counted two times when using the average formula so her *weighted* test average is

$$\frac{76+84+2(92)}{4} = \frac{344}{4} = 86$$

Example 1

If 9 and y have the same average as 3, 6, and 27, what is the value of y?

Solution:

- The average of 9 and y is $\frac{y+9}{2}$.
- The average of 3, 6, and 27 is $\frac{3+6+27}{3} = \frac{36}{3} = 12$.

- Set the two averages equal to each other:

$$\frac{y+9}{2}=12$$

$$\overset{1}{\cancel{2}}\left(\frac{y+9}{\cancel{2}}\right)=2(12)$$

$$y=24-9$$

$$\mathbf{y=15}$$

Example 2

Raymond's first four test grades are 85, 89, 87, and 96. What is the lowest grade Raymond can get on his next test so that the average of the five test grades will be at least 90?

Solution: For the average to be *at least* 90, the average must be equal to or greater than 90. If x represents Raymond's grade on the next test, then

$$\frac{x+85+89+87+96}{5}\geq 90$$

$$5\left(\frac{x+357}{5}\right)\geq 5(90)$$

$$x+357\geq 450$$

$$x\geq 450-357$$

$$x\geq 93$$

Since the least possible value of x is 93, the lowest grade Raymond can receive on the next test is **93**.

Example 3

In an algebra class of 30 students, the average midterm grade for the entire class was 82.4. The twelve students who attended after-school tutoring had an average midterm grade of 92. What was the average midterm grade of the remaining 18 students if every student in the class had a midterm grade?

Solution: If x represents the average midterm grade of the remaining 18 students, then

$$82.4 = \frac{(12 \times 92) + (18x)}{12 + 18}$$
$$82.4 = \frac{1104 + 18x}{30}$$
$$1104 + 18x = (30)(82.4)$$
$$18x = 2472 - 1104$$
$$\frac{18x}{18} = \frac{1368}{18}$$
$$x = 76$$

The average midterm grade of the remaining 18 students was **76**.

Median, Mode, and Range

The **mode** is the score that appears the most number of times. A set of scores may have more than one mode. The **median** is the center value when the scores are arranged in size order. The **range** is the difference between the highest and lowest scores. Consider the set of scores that includes 93, 69, 75, 81, 93, 84, and 75.

- The set has two modes, 93 and 75.
- To find the median, arrange the 7 scores in size order: 69,75,75, **81**, 84, 93, 93. The median is 81 since the same number of scores are below it as are above it. When an ordered list of numbers contains an *even* number of values, the median is the average of the two middle values, as illustrated in Example 4.
- The range is 93 – 69 = 24.

Example 4

The ages of the first eight people entering a zoo were 37, 46, 7, 23, 29, 18, 50, and 15. What was the median age?

Solution: First arrange the numbers in size order. Because there are an even number of data values, the median is the average of the two middle values:

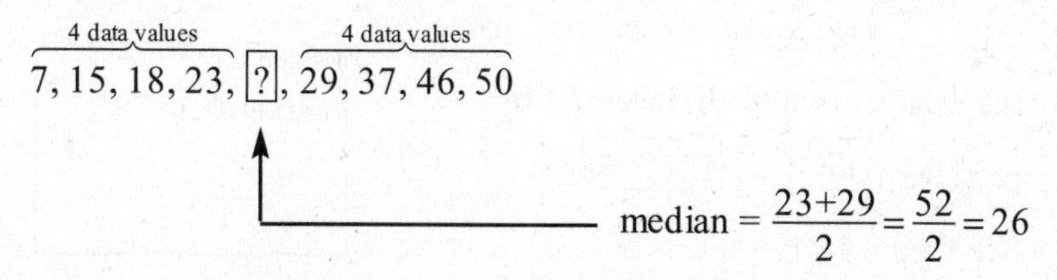

The median age was **26**.

Example 5

A number is selected at random from the set that includes 8, 5, 13, 5, 7, and 10. Find the probability that the number will be

a. Greater than the mean b. Equal to the median

Solution: First, find the mean:

$$\text{mean} = \frac{8+5+13+5+7+10}{6} = \frac{48}{6} = 8$$

a. The set of data values consists of six numbers, two of which (10 and 13) are greater than 8. Hence, the probability that the number selected will be greater than the mean is $\mathbf{\frac{2}{6}}$.

b. To find the median, first arrange the given set of numbers in size order: 5, 5, 7, 8, 10, 13. Since there is an even number of scores, the median is the average of the two middle scores, 7 and 8. Thus,

$$\text{median} = \frac{7+8}{2} = 7.5$$

Selecting a number from the set that is equal to 7.5 is an impossible event. Hence, the probability that the number selected will be equal to the median is **0**.

Using Technology: Finding the Mean and Median

To use a graphing calculator to find the mean of a list of numbers such as 8, 5, 13, 5, 7, and 10,

- Store the data values in a list, say L1, as described in Section 11.5.
- Open the LIST MATH menu by pressing [2nd] [LIST] [◁] where the label LIST is above the [STAT] key and [◁] represents the left cursor arrow key.
- Press [3] to select the function that returns the mean when the name of the data list is entered. Insert L1 by pressing [2nd] [1] [ENTER]

  See Figure 14.1.

```
mean(L1
                 8
median(L1
               7.5
```

Figure 14.1 Finding the Mean and Median of Data Stored in L1

You can also find the mean, $\bar{x}$, by pressing [STAT] [▷] [1]. To find the median of the same list of stored data values, as shown in Figure 14.1, press

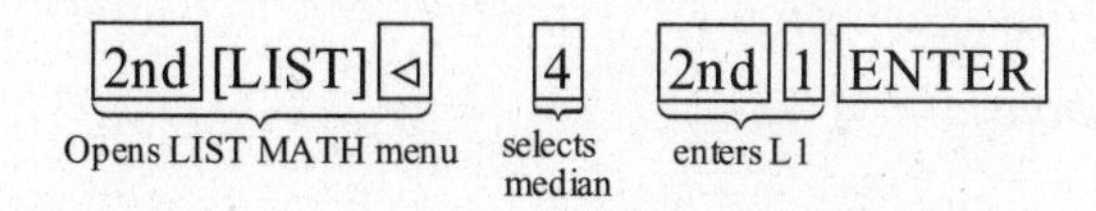

Comparing Measures of Central Tendency

The mean, median, and mode may or may not be good measures of central tendency.

- The *mean* can be helpful as a measure of central tendency when the data values are reasonably close in value. The mean of 76, 82, and 85 is 81, which is fairly representative of the members in the set so it is a good measure of central tendency. If a set of scores includes an **outlier**—a value that differs greatly from the other numbers in the set, the mean may not be a good measure of central tendency. For example, the mean of 1, 2, 5, and 200 is 52 which is not representative of the four data values. Because 200 is an outlier in a set with only a few data values, the mean is not a good measure of central tendency.
- The *mode* can be useful as a measure of central tendency when a set of data includes many of the same numbers.
- The *median* may be helpful as a measure of central tendency when a set of data includes numbers that are much smaller or much greater than most of the other numbers in the set of data.

Transforming an Entire Set of Data

If each score in a set of data values is changed in the same way, you can predict how the different statistical measures are affected without recalculating these statistics.

- When each value in a set of data is increased (or decreased) by the same nonzero number, then the mean, median, and mode are each increased (or decreased) by that number, but the range is *not* affected. If the mean of a set of test scores is 78 and *each* test score is increased by 6, then the mean of the revised set of test scores is $78 + 6 = 84$.
- When each value in a set of data is multiplied (or divided) by the same nonzero number, then the mean, median, mode, and range are each multiplied (or divided) by that number. Suppose the mean number of music CDs that five friends own is 13. Mary predicts that six months from now each of the five friends will own twice as many music CDs as they do now. Based on Mary's prediction, the mean number of music CDs they will own is $13 \times 2 = 26$.

Check Your Understanding of Section 14.1

A. *Multiple Choice.*

1. The mean (average) of a set of seven numbers is exactly 81. If one of the numbers is discarded, the average of the remaining numbers is exactly 78. Which number was discarded?
(1) 89 (2) 92 (3) 97 (4) 99

2. On a trip, a student drove 40 miles per hour for 2 hours and then drove 30 miles per hour for 3 hours. What is the student's average rate of speed, in miles per hour, for the whole trip?
(1) 32.5 (2) 34 (3) 35 (4) 37.2

3. A man drove a car at an average rate of speed of 45 miles per hour for the first 3 hours of a 7-hour trip. If the average rate of speed for the entire trip was 53 miles per hour, what was the average rate of speed in miles per hour for the remaining part of the trip?
(1) 55 (2) 57 (3) 59 (4) 62

4. Rick's recorded times in four 1-mile runs are 4.8 minutes, 5.3 minutes, 4.7 minutes, and 5.4 minutes. For Rick's next run, which time will give him a mean of 5.0 minutes?
(1) 4.8 min (2) 5.3 min (3) 5.7 min (4) 6.0 min

5. The average score on the first two tests that Marisol took was 88. On the third test she received a score of 94. What was her average for the three tests?
(1) 90 (2) 91 (3) 88 (4) 92

6. During each grade marking period, there are five tests. If Vanita needs a 65 average to pass this marking period and her first four grades are 60, 72, 55, and 80, what is the lowest score she can earn on the last test to have a passing average?
(1) 58 (2) 65 (3) 80 (4) 100

7. The accompanying chart shows how the cost of a specific notebook varied over a 5-week period. Based on the chart, which statement is true about the cost of this notebook over this period?

(1) The mode was \$3.00.
(2) The mean was \$4.30.
(3) The median was \$4.50.
(4) The median was \$3.00.

Week	Cost
1	\$5.00
2	\$5.25
3	\$3.00
4	\$3.50
5	\$4.75

8. The average of a set of 20 test scores is represented by x. If each score is increased by y points, which expression represents the average of the revised set of test scores?

(1) $x+y$ (2) $x+20y$ (3) $x+\frac{y}{20}$ (4) $\frac{x+y}{20}$

9. The accompanying table shows the number of books read during the summer vacation by six students.

Number of Books Read					
John	Randy	Vincent	Sara	Judy	Syed
8	1	5	9	5	7

If x represents the mean number of books read, m represents the median number of books read, and f represents the most frequently occurring number of books read, which of the following statements is correct for this set of data?

(1) $m<x<f$ (2) $f<x<m$ (3) $f<m<x$ (4) $m<f<x$

10. In a set of n data values, m represents the median. If each number in the set is decreased by 3, which expression represents the median of the transformed set of data values?

(1) m (2) $m-3$ (3) $m-\frac{3}{2}$ (4) $\frac{m-3}{n}$

B. Show how you arrived at your answer.

11. Bonnie receives grades of 79, 83, and 86 on her first three math tests. What grade must she receive on her next test in order to have an average grade of exactly 85 for the four exams?

12. If 11 and n have the same average as 5, 8, 17, and 32, what is the value of n?

13. Vanessa's mathematics teacher computes each student's test average by weighting the final exam three times as much as a class test. If Vanessa received grades of 87, 86, 88, 85, and 89 on her five class tests, what grade must Vanessa receive on the final exam so that her test average is 90?

14. Carol's average driving speed for a 4-hour trip was 48 mph. During the first 3 hours, her average rate was 50 mph. What was her average rate for the last hour of the trip?

15. The mean (average) weight of three dogs is 38 pounds. One of the dogs, Sparky, weighs 46 pounds. The other two dogs, Spot and Eddie, have the same weight. Find Eddie's weight.

16. Susan received 78, 89, and 82 on her first three exams. What is the lowest score she can receive on her next exam and have an exam average for the four exams of at least 85?

17. What is the area of the circle whose radius is the mean of the radii of two circles with areas of 16π and 100π square units?

18. The accompanying table shows the number of e-mails received by each of six friends in the first week of June.

a. Find the mean, median, and range of this set of data.

b. If the number of e-mails received by each person increases by 8 in the week that follows, what will be the mean, median, and range of the new set of data?

c. If in the same week as in part b, each person sent 15 more than one-half the number of e-mails he or she received, what is the mean number of e-mails sent by these six friends during this week?

Name	E-mails Received in First Week of June
Erin	64
Jose	52
Ben	76
Frank	36
Alyssa	72
Judy	84

Exercise 18

19. On the first six tests in her social studies course, Jerelyn's scores were 92, 78, 86, 92, 95, and 91. If Jerelyn took a seventh test and raised the mean of her scores exactly 1 point, what was her score on the seventh test?

20. The average of four numbers is exactly 32. The largest of the four numbers exceeds twice the smaller number by 10 and is 5 more than one of the other numbers. The remaining number is one-half as great as the largest of the four numbers. What is the largest of the four numbers?

21. Five data values are represented by $3x-1, 2(x+1), x^2-2x, \frac{1}{2}x+7$, and $\sqrt{x^3}$. If $x = 4$, what is the mode?

22. A number is selected at random from the set that includes 2, 2, 2, 2, 3, 3, 3, and 7. Find the probability that the number selected will be

a. Greater than the mean b. Equal to the median

23. In order to compensate for a difficult midterm exam, Danielle's mathematics teacher adjusted each student's midterm exam score by replacing it by one-half of the original scored increased by 50.

a. If the mean of the original set of midterm scores was 82, what is the mean of the set of revised scores?
b. If Danielle's adjusted midterm score is 87, what was her original midterm score?

24. Tamika could not remember her scores from five mathematics tests. She did remember that the mean was *exactly* 80, the median was 81, and the mode was 88. If all her scores were integers with 100 the highest possible score and 0 the lowest score possible, what was the *lowest* score she could have received on any one test?

25. The median of seven test scores is 82, the mode is 87, the lowest score is 70, and the mean (average) is 80. If each of the scores is a whole number, what is the greatest possible test score? Explain your reasoning.

26. Find all possible values of x such that when the five numbers x, 14, 11, 6, and 17, are arranged in order, the mean is equal to the median.

27. In a set of data values, m represents the least value, and M represents the greatest value. Prove or disprove that the range does not change if each value in the set is

a. Increased by the same positive number k
b. Multiplied by the same positive number k

14.2 BOX-AND-WHISKER PLOTS

KEY IDEAS

A **median** separates an ordered list of data into two equal groups. **Quartiles** are values that separate an ordered list of data into four equal groups. A **box-and-whisker plot** graphically summarizes how spread out the data are about the median using 5 key numbers: (1) minimum data value; (2) lower quartile; (3) median; (4) upper quartile; and (5) maximum data value. A box-and-whisker plot does not show all of the individual data values or tell you how many data values are included in the set.

Quartiles and Range

Consider the ordered data:

5, 8, 10, 13, 14, 17, 19

- Since the list contains 7 data values, the median is the fourth or middle value, which is 13.
- The lower half of the data includes 5, 8, and 10. The lower quartile is the median of these three values, which is 8. See Figure 14.2.

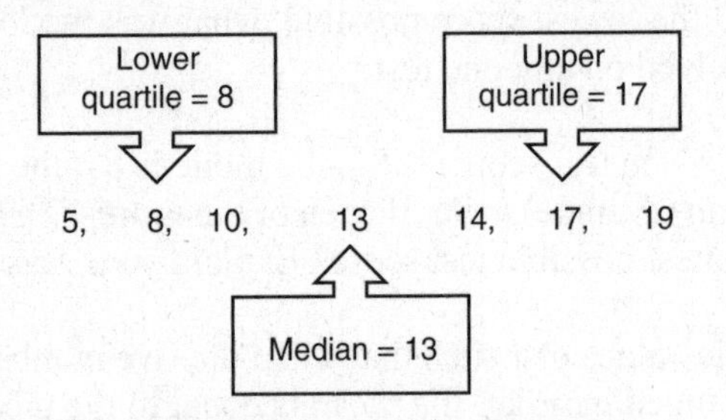

Figure 14.2 Finding Quartiles

- The upper half of the data contains 14, 17, and 19. The upper quartile is the median of these three values, which is 17.
- The range of the data is 19 – 5 = 14.
- The interquartile range is the difference between the lower and upper quartiles: 17 – 8 = 9.

MATH FACTS

The **median** of a data set separates the data into two equal groups: a lower half and an upper half. The median is not included in either of these two groups.

- The **lower quartile** (Q_1) is the median of the lower half of data values. The lower quartile is the number below which 25% of the data values fall.
- The **second** or **middle quartile** (Q_2) corresponds to the median. The median is the number below which 50% of the data values fall.
- The **upper quartile** (Q_3) is the median of the upper half of data values. The upper quartile is the number below which 75% of the data values fall.
- The **range** of the data is the difference between the maximum and minimum data values.
- The **interquartile range** is the difference between the third and first quartiles ($Q_3 - Q_1$).

Example 1

Find the lower and upper quartiles for the data:

48, 10, 78, 27, 45, 21, 55, 33, 57, 65, 75, 80, 65

Solution: First arrange the data in size order in order to find the overall median of the data:

10, 21, 27, 33, 45, 48, 55, 57, 65, 65, 75, 78, 80

- Since the number of data values is an odd number, the median is the middle or seventh value. The median is 55, as shown in the accompanying diagram.

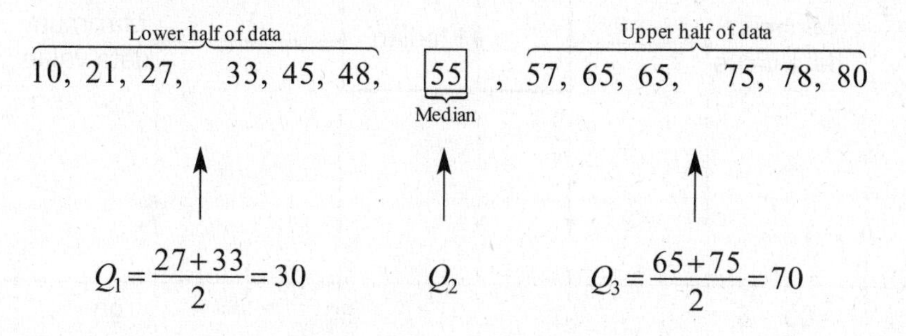

- Find the median of the lower half of the data, which consists of the six values below 55: 10, 21, 27, 33, 45, and 48. Since the number of data values is an even number, the lower quartile is the average of the two middle values:

$$Q_1 = \frac{27+33}{2} = \mathbf{30}$$

- Find the median of the upper half of the data, which consists of the six values above 55: 57, 65, 65, 75, 78, 80. The upper quartile is the average of the two middle values:

$$Q_3 = \frac{65+75}{2} = \mathbf{70}$$

Notice that when figuring out the quartiles, the overall median, 55 in this case, is not included in either the lower half or the upper half.

Box-and-Whisker Plots

Suppose that for a certain set of data you know that:

- minimum data value = 64
- lower quartile = 70
- median = 74
- upper quartile = 80
- maximum data value = 88

The box-and-whisker plot in Figure 14.3 uses these five key values to show at a glance how spread out the data are around the median.

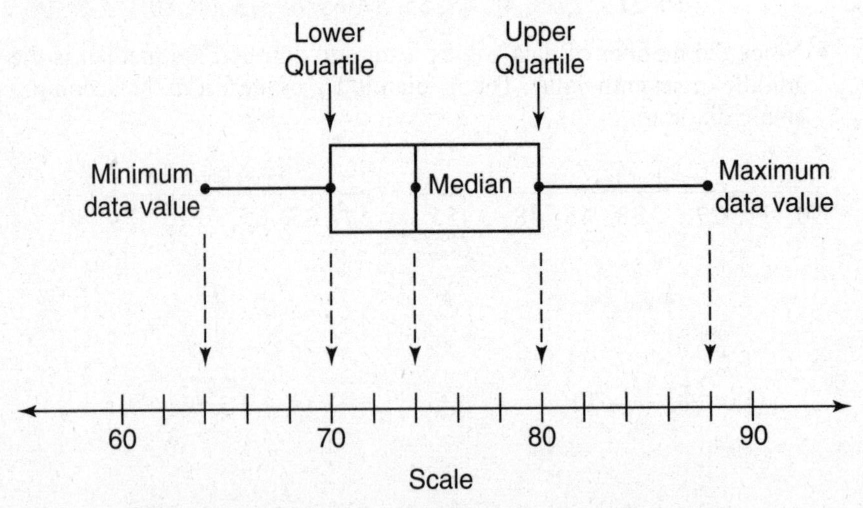

Figure 14.3 Box-and-whisker plot using 5 key data values.

In a box-and-whisker plot, the lower and upper quartiles are at the vertical sides of the rectangular box. The median is represented by a vertical line in the interior of the box.

Horizontal segments, called whiskers, are drawn from the two vertical ends of the box to points that represent the minimum and maximum data values.

The rectangular box and whiskers are aligned above a horizontal axis labeled with an appropriate scale so that the five key statistical values can be read from the scale, as shown in Figure 14.3.

Example 2

The accompanying box-and-whisker plot represents the scores earned on a science test taken by 28 students.

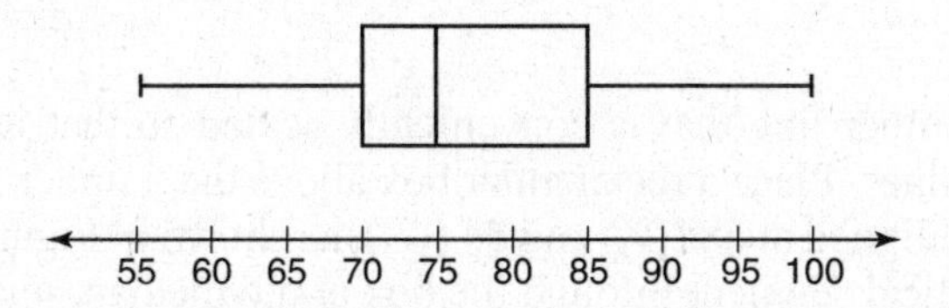

a. What is the median score?
b. What is the range of the scores?
c. How many students earned scores less than 70?
d. What percent of the students earned scores of at least 85?

Solution:

a. The vertical segment inside the box represents the median. Since it is aligned over the 75 mark, the median score is **75**.
b. The range of a set of scores is the difference between the lowest and highest scores. Reading from the diagram, the lowest score is 55 and the highest score is 100. So the range of scores is 100 – 55 = **45**.
c. A score of 70 is aligned with the left edge of the box, so it corresponds to the lower quartile mark. Since 25% of the 28 scores fall within the lower quartile and $0.25 \times 28 = 7$, **7** students scored less than 70.
d. Since a score of 85 represents the upper quartile, 75% of the students had scores below 85. This means **25%** of the students had scores of at least 85.

Constructing a Box-and-Whisker Plot from Data

To construct a box-and-whisker plot using the 14 scores

79, 84, 64, 69, 65, 66, 94, 68, 73, 75, 85, 77, 78, 90

work as follows.

- Arrange the numbers in ascending order, and determine the five key numbers.

64, 65, 66	68,	69, 73, 75,	?	77, 78, 79,	84,	85, 90, 94
↑	↑		↑		↑	↑
Min	$Q_1 = 68$		Median (Q_2)		$Q_3 = 84$	Max
			$= \frac{75+77}{2} = 76$			

- Draw a number line that is conveniently scaled so that it includes the five key values. Place a rectangular box above the number line with vertical sides aligned at 68 (Q_1) and 84 (Q_3), as shown in Figure 14.4. Then draw a vertical segment through the box at the median, making sure the segment is aligned with 76.

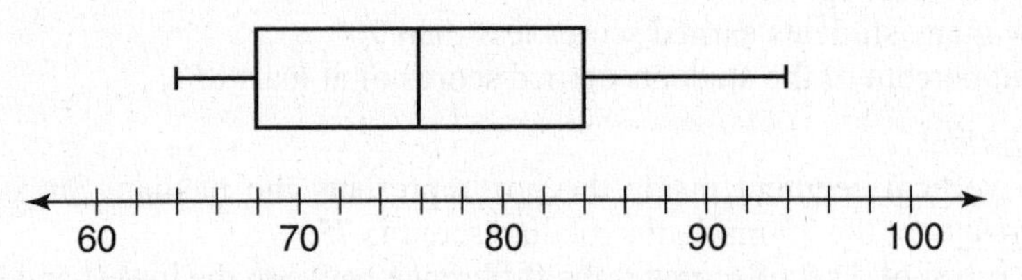

Figure 14.4 Drawing a box-and-whisker plot from data.

- Draw the two whiskers extending from the midpoints of the vertical sides of the box and stopping on the left side at 64 (minimum value) and ending on the right side at 96 (maximum value) as shown in Figure 14.4.

Using Technology: Finding the 5 Key Values

You can use your graphing calculator to find the five key statistical values needed to construct a box-and-whisker plot.

Press [STAT] and [1] to open the Stat/List editor. Enter the individual data values from the previous example as list L1:

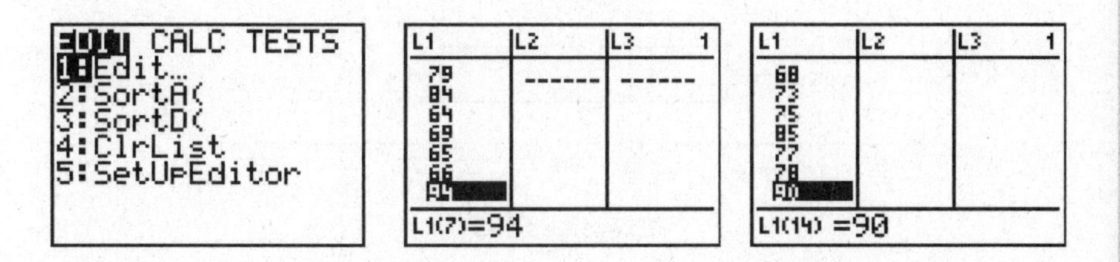

Display the quartile values and other one-variable statistics by pressing

[STAT] [▷] [1] [2ND] [L1] [ENTER].

Press the down arrow key until you see all five of the key statistics values needed to construct a box-and-whisker plot: minimum data value = 64, $Q_1 = 68$, median = 76, $Q_3 = 84$, and maximum data value = 94:

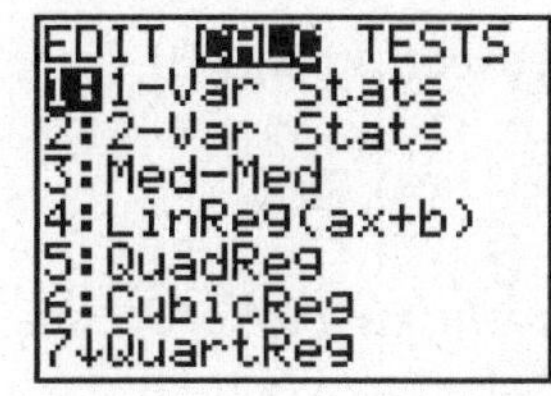

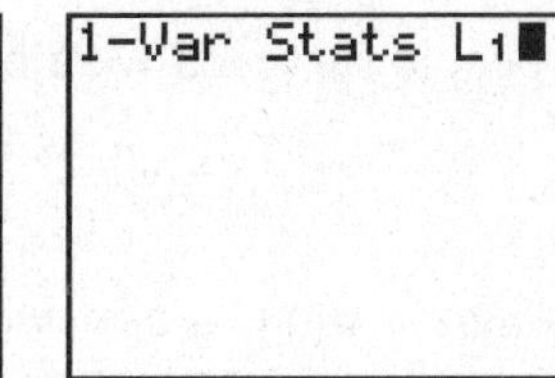

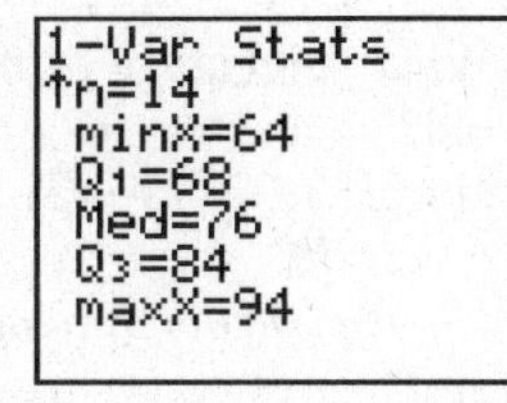

If you want to see the original list of data values arranged in size order, press

[STAT] [2] [2ND] [L1] [ENTER].

that selects the SortA(function and sorts the numbers stored in L1 in ascending order. Open the Stat List editor, and scroll down to see the sorted list of data values:

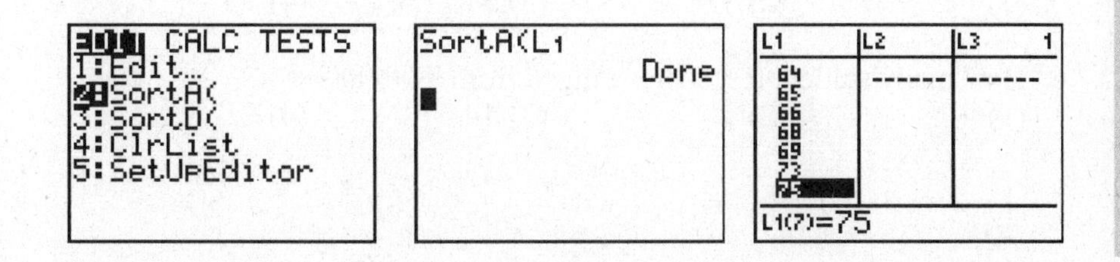

Check Your Understanding of Section 14.2

A. Multiple Choice.

1–3. The box-and-whisker plot shown in the accompanying diagram represents the ages of a survey taken of 80 people.

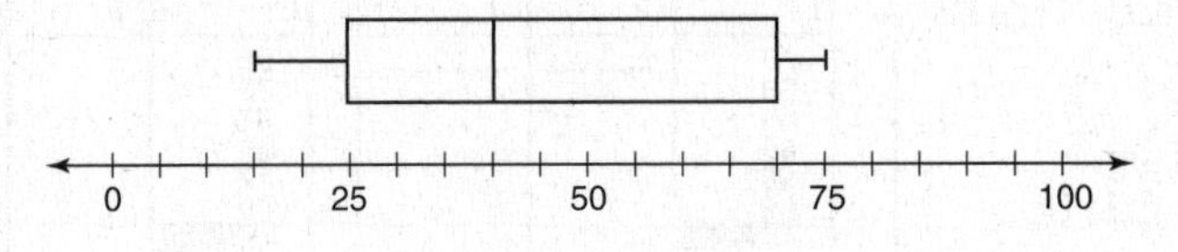

Exercises 1–3

1. What is the range of the data represented in the box-and-whisker plot?
(1) 40 (2) 45 (3) 60 (4) 100

2. What percentage of the people surveyed were less than 40 years old?
(1) 25% (2) 50% (3) 75% (4) 100%

3. How many of the people who were surveyed were less than 25 years old?
(1) 20 (2) 25 (3) 40 (4) 60

4. What percentage of the people surveyed were between 40 and 70 years old?
(1) 25% (2) 30% (3) 50% (4) 75%

5–7. The accompanying diagram shows a box-and-whisker plot of 28 student test scores on an algebra midterm examination.

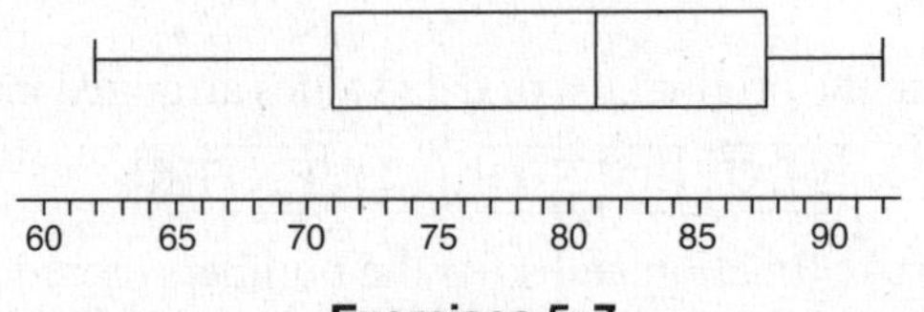

Exercises 5–7

5. What is the median score?
(1) 62 (2) 71 (3) 81 (4) 92

6. How many student test scores ranged from 81 to 88?
(1) 4 (2) 7 (3) 14 (4) 21

7. A test score of 72 could correspond to which of the following percentiles?
(1) 24 (2) 30 (3) 52 (4) 77

8. A movie theater recorded the number of tickets sold daily for a popular movie during the month of June. The box-and-whisker plot shown below represents the data for the number of tickets sold, in hundreds.

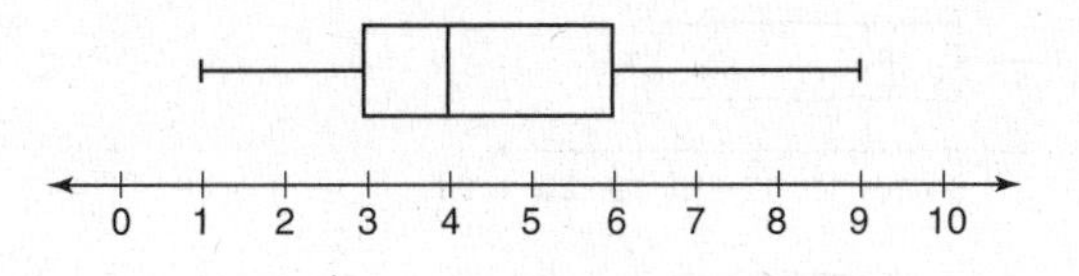

What conclusion can be made using this plot?

(1) The second quartile is 600.
(2) The mean of the attendance is 400.
(3) The range of the attendance is 300 to 600.
(4) Twenty-five percent of the attendance is between 300 and 400.

9. The accompanying box-and-whisker plots can be used to compare the annual incomes of three professions.

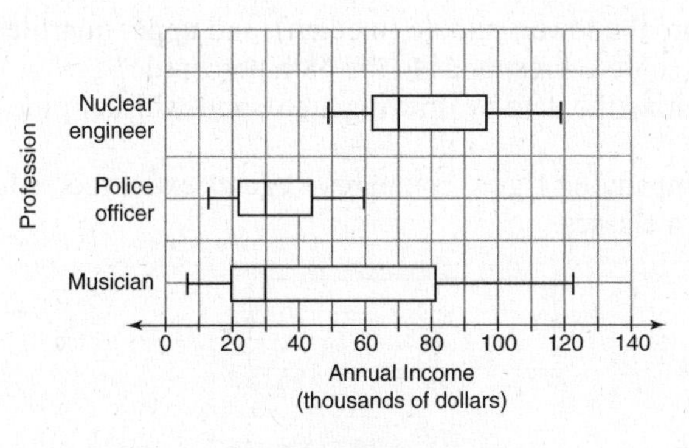

Exercise 9

Based on the box-and-whisker plots, which statement is true?

(1) The median income for nuclear engineers is greater than the income of all musicians.
(2) The median income for police officers and musicians is the same.
(3) All nuclear engineers earn more than all police officers.
(4) A musician will eventually earn more than a police officer.

10. The data set 5, 6, 7, 8, 9, 9, 9, 10, 12, 14, 17, 17, 18, 19, 19 represents the number of hours spent on the Internet in a week by students in a mathematics class. Which box-and-whisker plot represents the data?

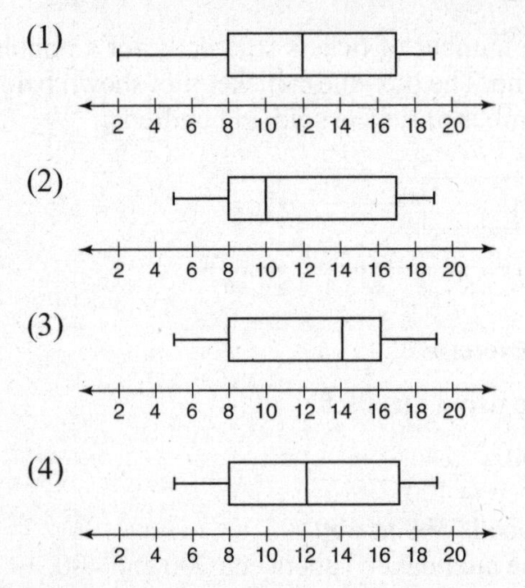

B. *Show how you arrived at your answer.*

11. The midterm exam scores for a class were 60, 65, 65, 67, 71, 70, 73, 75, 76, 76, 79, 81, 83, 84, 85, 85, 88, 89, 90, 92, 95, 96, 99, 100, and 100.

a. What are the lower, middle (median), and upper quartiles?
b. Which score corresponds to the 68th percentile?
c. Summarize the data by drawing a box-and-whisker plot.

12. The accompanying figure compares Regents exam scores for two different algebra classes.

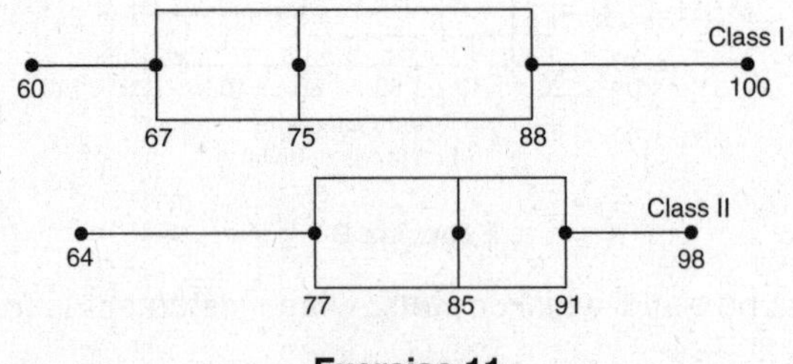

Exercise 11

a. By how many points does the median Regents test score for Class II exceed the median Regents test score for Class I?
b. Which class had the greatest range in Regents test scores? Justify your answer.
c. If a student is picked at random from Class II, what is the probability the student's Regents exam score will be at least 92?

14.3 HISTOGRAMS

KEY IDEAS

The **frequency** of a score is the number of times that score appears in a data list. A **histogram** is a bar graph in which the height of each rectangular bar represents the frequency or number of data items contained in the interval indicated at the bottom of the bar.

Making a Frequency Table Using Intervals

A class of 25 students received the following scores on an Algebra test: 58, 70, 60, 65, 68, 70, 90,70, 72, 74, 70, 70, 75,78, 80, 96, 75, 80, 83, 80, 88, 83, 90, 65, and 75

Figure 14.5 shows how the data can be tabulated and organized by using a three-column table.

Interval	Tally	Frequency
50–59	/	1
60–69	////	4
70–79	~~////~~ ~~////~~ /	11
80–89	~~////~~ /	6
90–99	///	3
		Sum = 25

Figure 14.5 Frequency Table

The intervals were selected so that they have a convenient uniform width of 10 while accommodating both the lowest and highest scores. Since the lowest score is 58, the first interval is 50–59. The highest score is 96, so the last interval is 90–99. A quick glance at the frequency table in Figure 14.5 gives a good idea of how the test scores are distributed, with most of the scores falling between 70 and 79.

Drawing a Frequency Histogram

To draw a *frequency histogram* based on the frequency table in Figure 14.5,

- Draw and label axes using graph paper. Label the vertical axis "Frequency" and the horizontal axis "Test scores."

- Choose a scale. Since the frequency ranges from 1 to 11, mark off the vertical axis in units of 1. Mark off the horizontal axis so that each interval has the same width. The width may be any convenient number of unit boxes.
- For each interval, draw vertical bars next to one another. The frequency of each interval determines the bar height, as shown in Figure 14.6.

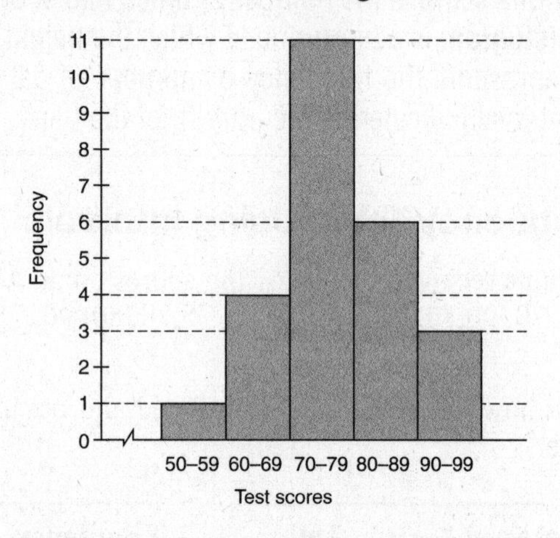

Figure 14.6 Drawing a Frequency Histogram

Example 1

Based on the accompanying frequency table, determine the intervals that contain the median, the lower quartile, and the upper quartile.

Interval	Frequency
1–15	1
16–30	2
31–45	6
46–60	4
61–75	2
76–90	5

Solution: Adding the entries in the frequency column gives a total of 20 scores.

- The median is the middle value, so it lies between the 10th and 11th scores. By accumulating frequencies, you know that the fourth interval, **46–60,** contains the 10th, 11th, 12th, and 13th scores so this interval contains the median.
- The lower quartile is the number at or below which 25% of the scores fall. Since there is a total of 20 scores and

$$25\% \text{ of } 20 = \frac{1}{4} \times 20 = 5$$

the lower quartile is the 5th score counting up from the bottom interval which is located in the interval, **31–45.**

- The upper quartile is the number at or below which 75% of the scores fall. Since there is a total of 20 scores and

$$75\% \text{ of } 20 = \frac{3}{4} \times 20 = 15$$

the upper quartile is the 15th score counting up from the bottom interval which is located in the interval, **61–75**.

Example 2

The accompanying table shows the distribution of ages of principals in a school district.

a. Using this table, draw a frequency histogram.
b. Which interval contains the lower quartile?
c. What percent of the principals are older than 43?

Interval	Frequency
68–75	2
60–67	6
52–59	5
44–51	8
36–43	5
28–35	2

Solution: a. See the accompanying figure.

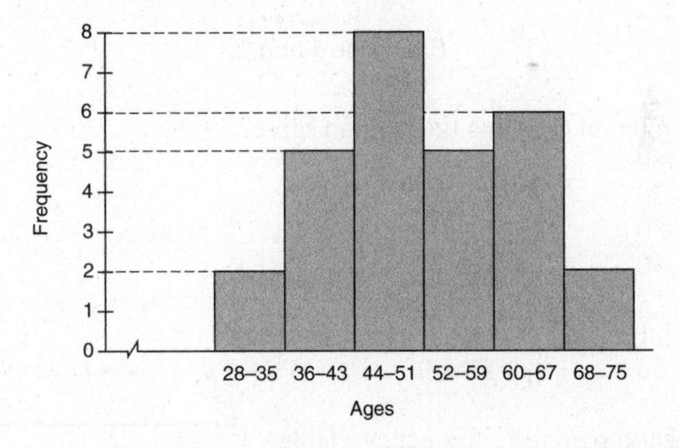

b. The frequency table contains 28 data values. In an ordered list of data, the lower quartile is the value at or below which $\frac{1}{4}$ of the data values fall. Since $\frac{1}{4} \times 28 = 7$, the lower quartile is the 7th lowest data value in the frequency table. Adding the frequencies in the lowest two intervals of the table gives 7. Thus, the lower quartile is contained in the second interval, **36–43.**

c. Summing the frequencies in the intervals above 36–43 gives $8 + 5 + 6 + 2 = 21$. Thus, the ages of 21 of the 28 principals fall in the intervals above 36–43. Since $\frac{21}{28} = \frac{3}{4} = 75\%$, **75%** of the principals are older than 43.

Check Your Understanding of Section 14.3

A. Multiple Choice.

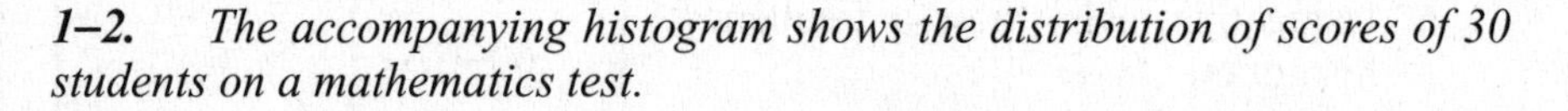

1–2. *The accompanying histogram shows the distribution of scores of 30 students on a mathematics test.*

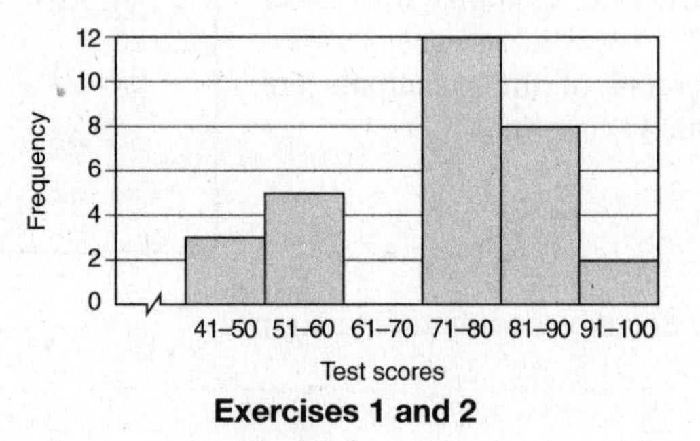

Exercises 1 and 2

1. Which interval contains the median score?
(1) 51–60
(2) 61–70
(3) 71–80
(4) 81–90

2. Which interval contains the lower quartile?
(1) 41–50 (2) 51–60 (3) 71–80 (4) 81–90

3. The accompanying frequency table shows data collected by the police in an automobile speed check. Which interval contains the upper quartile?
(1) 36–45
(2) 46–55
(3) 56–65
(4) 66–75

Speed Interval	Frequency
66–75	45
56–65	110
46–55	120
36–45	25

Exercise 3

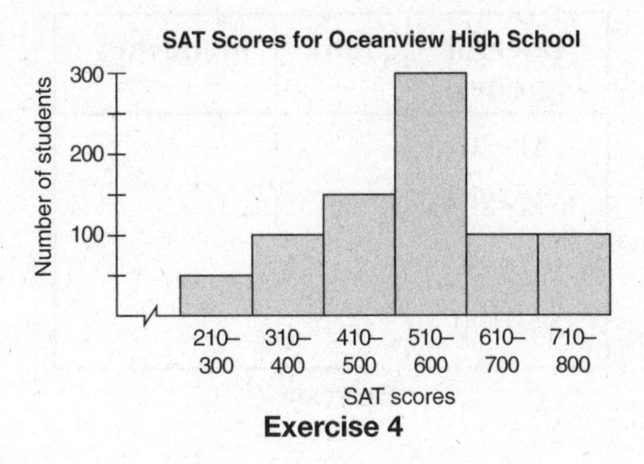

Exercise 4

4. The accompanying histogram shows the distribution of SAT scores at Oceanview High School. What percent of students in this school scored at least 610?
(1) 25 (2) 37.5 (3) 62.5 (4) 75

5. The accompanying frequency table shows data collected by the weather bureau for the daily high temperatures in January in Buffalo. Which interval contains the median temperature?
(1) 0–9 (3) 20–29
(2) 10–19 (4) 30–29

Interval (temperature)	Frequency
30–39	13
20–29	6
10–19	5
0–9	7

Exercise 5

B. *Show how you arrived at your answer.*

6. The accompanying table summarizes the results of a test.

a. Which interval contains the upper quartile?
b. If a score is picked at random, what is the probability that the score will be at most 80?

Interval	Frequency
91–100	3
81–90	5
71–80	4
61–70	5
51–60	3

Exercise 6

Interval (grades)	Tally	Frequency
61–70		
71–80		
81–90		
91–100		

Exercise 7

7. Sarah's mathematics grades for one marking period were 85, 72, 97, 81, 77, 93, 100, 75, 86, 70, 96, and 80.

a. Copy and then complete the accompanying frequency table.
b. On graph paper, construct a frequency histogram for Sarah's grades.
c. Which interval contains the 75th percentile (upper quartile)?

8. For the data in the accompanying table,

a. What is the median score?
b. What percentile corresponds to a score of 88?

Score	Frequency
68	3
75	9
82	4
88	2
93	6

Exercise 8

9. The accompanying histogram shows the distribution of test scores for a class.

a. What percent of students had test scores lower than 76?
b. Which interval contains the 80th percentile?
c. What is the probability that a score picked at random will be between 65 and 70?

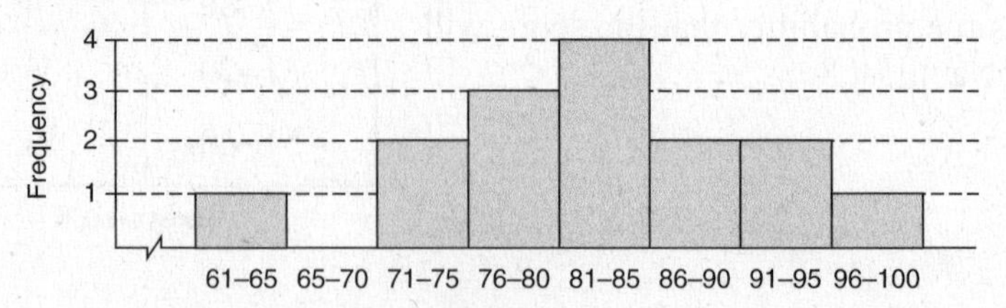

Exercise 9

Misspelled Words	Frequency (Number of Essays)
0	1
1	0
2	3
3	5
4	4
5	9
6	3

Exercise 10

10. The accompanying table shows the distribution of misspellings in 25 student essays.

a. On graph paper, construct a frequency histogram based on the data.
b. Find the *mode* number of misspelled words.
c. Find the *mean* number of misspelled words.
d. Find the *median* number of misspelled words.

Interval	Frequency
91–100	
81–90	
71–80	
61–70	
51–60	

Exercise 11

11. The following data are test scores for a class of 16 students:

96, 83, 91, 77, 58, 88, 80, 62, 89, 100, 87, 93, 64, 98, 88, 86

a. Copy and complete the accompanying table.
b. On graph paper, construct a frequency histogram based on the data.
c. Which interval contains the median?
d. Which interval contains the lower quartile?

Test Score	Tally	Frequency
53–58		
59–64		
65–70		
71–76		
77–82		
83–88		
89–94		
95–100		

Exercise 12

12. The distribution of math test scores for 25 students was 92, 87, 60, 76, 90, 83, 99, 55, 82, 76, 90, 71, 88, 94, 87, 75, 94, 87, 98, 62, 80, 97, 86, 78, and 65.

a. Copy and complete the accompanying frequency table for these scores.
b. If a score of at least a 65 represents a passing grade, what is the probability that a student chosen at random passed the test?
c. Which interval contains the upper quartile?
d. What percent of the students scored in the interval containing the median?

13. The accompanying table represents the distribution of 29 course grades in a college mathematics class. Grades of A, B, C, and D are passing, and x represents the number of students who received a grade of A.

Grade	Frequency
A	x
B	$2x - 3$
C	$x + 1$
D	5
F	$x - 4$

Exercise 13

a. Find the number of students who received a grade of either A or B.
b. On graph paper, construct a frequency histogram assuming a grade of A corresponds to 90–100, B to 80–89, C to 70–79, D to 60–69, and F to 50–59.
c. If a student is selected at random from the class, find the probability that the student's grade is

(1) *Not* a passing grade
(2) Equal to the mode
(3) *At least* a C

14.4 CUMULATIVE FREQUENCY HISTOGRAMS

KEY IDEAS

For each interval of scores in a frequency table, the corresponding *cumulative* frequency table and histogram display the *sum* of the scores in all preceding intervals up to and including the scores contained in that interval.

Creating a Cumulative Frequency Table

The first two columns of the table in Figure 14.7 correspond to a frequency table that shows the distribution of 25 student test scores. The last two columns represent the cumulative frequency table for this set of scores.

Interval	Frequency	Cumulative Frequency		Interval
50–59	1	1		50–59
60–69	4	5	$(4 + 1 = 5)$	50–69
70–79	11	16	$(11 + 5 = 16)$	50–79
80–89	6	22	$(6 + 16 = 22)$	50–89
90–99	3	25	$(3 + 22 = 25)$	50–99

Figure 14.7 Completing a Cumulative Frequency Table

For each interval after the first, the entries in the cumulative frequency column are obtained by adding the frequency entry that appears on the same line to the cumulative frequency of the preceding line. Since the first frequency has no entry before it, nothing is added to 1 to obtain the cumulative frequency for the first interval.

Drawing a Cumulative Frequency Histogram

Figure 14.8 shows the cumulative frequency histogram for the data contained in the cumulative frequency table in Figure 14.9. The difference in the heights of consecutive rectangles indicates the number of scores in the frequency interval with the same right endpoint. For example, since the difference in the heights of the rectangles for the cumulative frequency intervals 50–79 and 50–89 is $22 - 16 = 6$, there are six scores in the frequency interval 80–89.

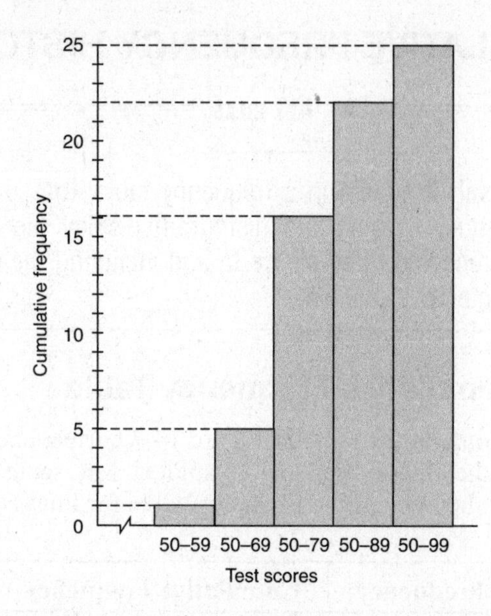

Figure 14.8 Cumulative Frequency Histogram

Example 1

Table 1 represents the distribution of SAT math scores for 60 students at State High School. Table 2 is the cumulative frequency table for the same set of scores.

Table 1

Scores	Frequency
710–800	4
610–700	10
510–600	15
410–500	18
310–400	11
210–300	2

Table 2

Scores	Cumulative Frequency
210–800	
210–700	
210–600	
210–500	
210–400	
210–300	2

a. Copy and then complete the cumulative frequency table (Table 2).
b. Using the table completed in part a, draw a cumulative frequency histogram on graph paper.
c. What percent of students scored above 700 or below 310?

Table 2

Scores	Cumulative Frequency
210–800	$56 + 4 = 60$
210–700	$46 + 10 = 56$
210–600	$31 + 15 = 46$
210–500	$13 + 18 = 31$
210–400	$2 + 11 = 13$
210–300	2

Solution:

a. On each line of the cumulative frequency table after the first, add the frequency from the corresponding line of the frequency table to the cumulative frequency on the preceding line, as shown in the table above.
b. See the accompanying figure.
c. Table 1 shows that 4 students scored above 700 and 2 students scored below 310. Thus $4 + 2 = 6$ of 60 students scored above 700 or below 310. Because $\frac{6}{60} = \frac{1}{10} = 10\%$, **10%** of the students scored 700 or below 310.

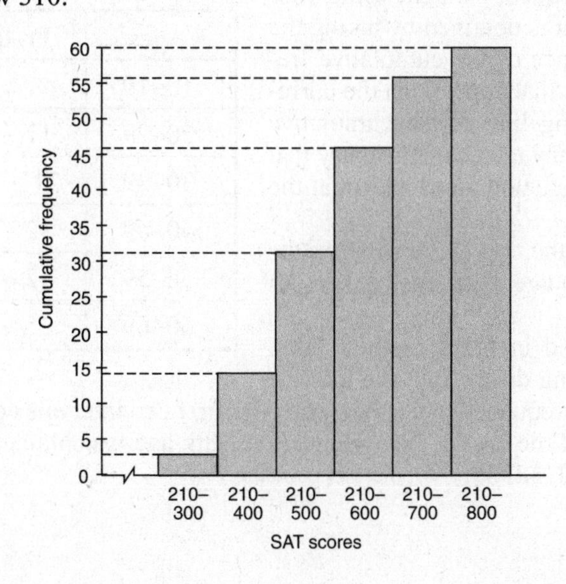

Example 2

Table 1 shows the cumulative frequency of the ages of 35 people standing on a movie theater line.

Table 1

Ages	Cumulative Frequency
10–19	2
10–29	17
10–39	27
10–49	32
10–59	32
10–69	35

Table 2

Ages	Frequency
10–19	2
20–29	
30–39	
40–49	
50–59	
60–69	

a. Using the data given in the cumulative frequency table (Table 1), copy and then complete the frequency table (Table 2).
b. Using the frequency table obtained in part a, determine the interval in which the median occurs.

Solution:

a. The frequency on each line after the first is obtained by taking the difference of the cumulative frequency that appears on the corresponding line of the cumulative frequency table and the entry that appears below it, as shown at the right.

Table 2

Ages	Frequency
10–19	2
20–29	$17 - 2 = 15$
30–39	$21 - 17 = 10$
40–49	$32 - 27 = 5$
50–59	$32 - 32 = 0$
60–69	$35 - 32 = 3$

b. Since there are 35 people, the median age is the age of the 18th person when the people are arranged in order of their ages. Counting down from the top line of the frequency table shows that the first two intervals contain $2 + 15 = 17$ of the lowest ages. Hence, the 18th age is contained in the next interval, **30–39**.

Check Your Understanding of Section 14.4

A. *Multiple Choice.*

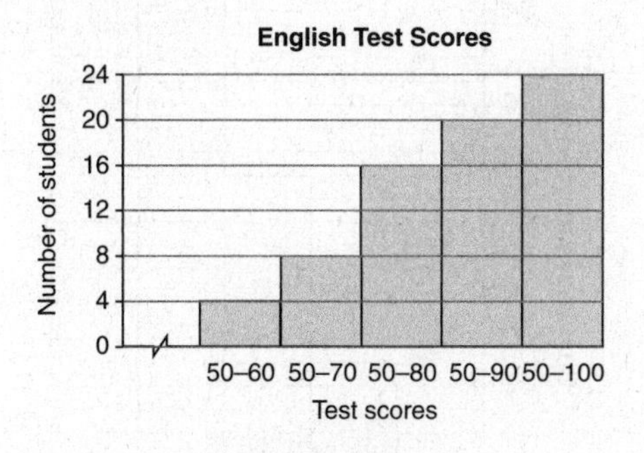

Exercise 1

1. The accompanying cumulative frequency histogram shows the distribution of scores that students received on an English test. How many students had scores between 71 and 80?
(1) 16 (3) 20
(2) 8 (4) 4

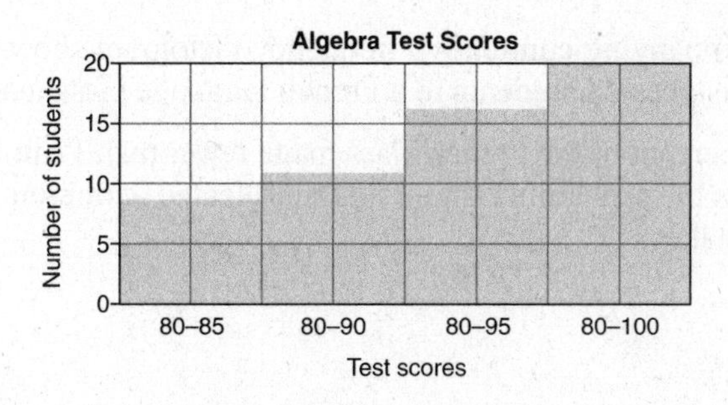

Exercise 2

2. The accompanying cumulative frequency histogram shows the distribution of scores of 20 students on an algebra test. What percent of the students had scores above 90?
(1) 20 (2) 25 (3) 45 (4) 55

B. *Show how you arrived at your answer.*

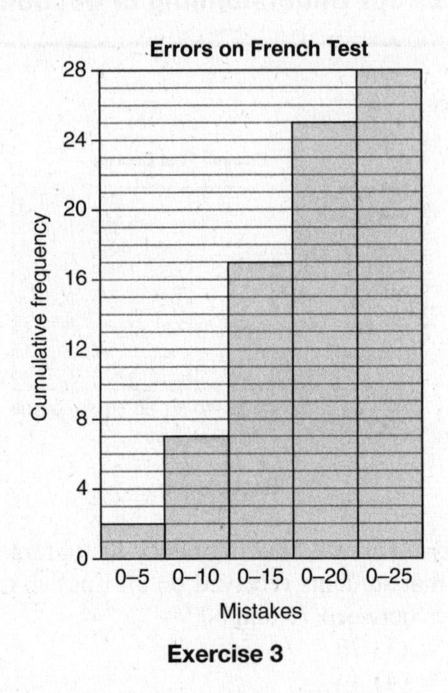

Exercise 3

3. The accompanying cumulative frequency histogram shows the distribution of mistakes 28 students in a French language class made on a test.

a. What percent of the French class made fewer than 11 mistakes?
b. What is the probability that a student selected at random made at least 16 mistakes?

4. The accompanying frequency table on the left below shows the distribution of scores that 100 students received on a standardized test.

a. Copy and complete the accompanying cumulative frequency table.

Scores	Frequency
91–100	15
81–90	26
71–80	23
61–70	15
51–60	11
41–50	5
31–40	3
21–30	2

Scores	Cumulative Frequency
21–100	
21–90	
21–80	
21–70	
21–60	
21–50	
21–40	
21–30	

b. Using the table completed in part a, draw a cumulative frequency histogram.

c. According to the frequency table, which interval contains the upper quartile?

(1) 31–40 (2) 61–70 (3) 71–80 (4) 81–90

d. What percent of the students scored less than 51?

5. The accompanying cumulative frequency histogram shows the distribution of the heights, in inches, of 24 high school students.

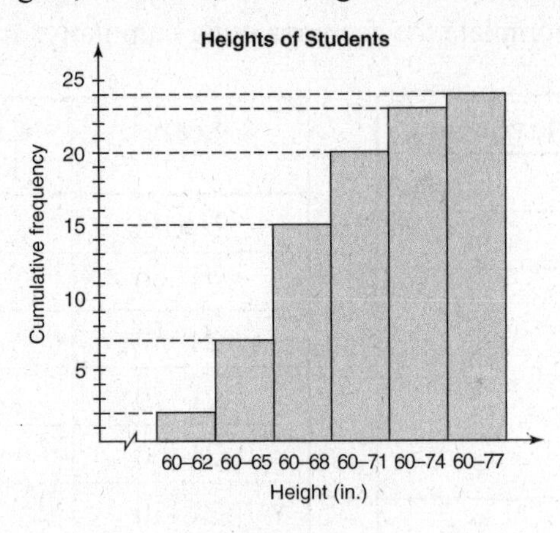

Exercise 5

a. Using the cumulative frequency histogram, copy and then complete the accompanying frequency table.

b. Using the table completed in part a:

Heights	Number of Students
60–62	
63–65	
66–68	
69–71	
72–74	
75–77	

(1) Draw a frequency histogram using graph paper.

(2) Determine the intervals that contain the lower and upper quartiles.

c. If a student is selected at random, what is the probability that the student is at most 68 inches tall?

6. The following data consists of the weights, in pounds, of 30 adults:

195, 206, 100, 98, 150, 210, 195, 106, 195, 168, 180, 212, 104, 195, 100, 216, 195, 209, 112, 99, 206, 116, 195, 100, 142, 100, 135, 98, 160, 155

Interval	Frequency	Cumulative Frequency
51–100		
101–150		
151–200		
201–250		

Exercise 6

a. Using the data, copy and complete the accompanying cumulative frequency table.
b. Construct a cumulative frequency histogram using graph paper.

7. The accompanying frequency histogram shows the weight, in pounds, of the students in a sixth grade class.

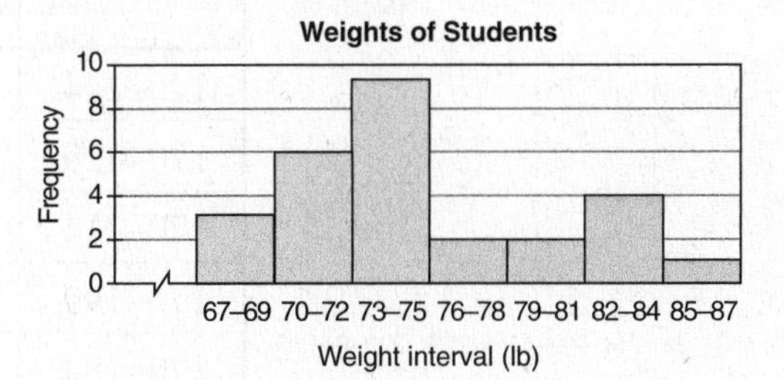

Exercise 7

a. According to the frequency histogram, in which interval does the median fall?
b. What is the probability that the weight of a student chosen at random will be greater than 75 pounds?
c. Copy and complete the accompanying cumulative frequency table.
d. Using the table completed in part c, construct a cumulative frequency histogram using graph paper.

Weight Interval (pounds)	Cumulative Frequency
67–69	
67–72	
67–75	
67–78	
67–81	
67–84	
67–87	

8. The scores of the teams in a local bowling league are listed in the accompanying table.

Teams

Aces	Bees	Cubs	Darts	Experts
186	177	199	197	193
224	207	212	196	214
216	235	188	226	231
207	223	239	205	200

Exercise 8

a. What is the mode of the 20 bowling scores listed?
b. Using the information provided in the accompanying histogram, copy and complete the cumulative frequency table.

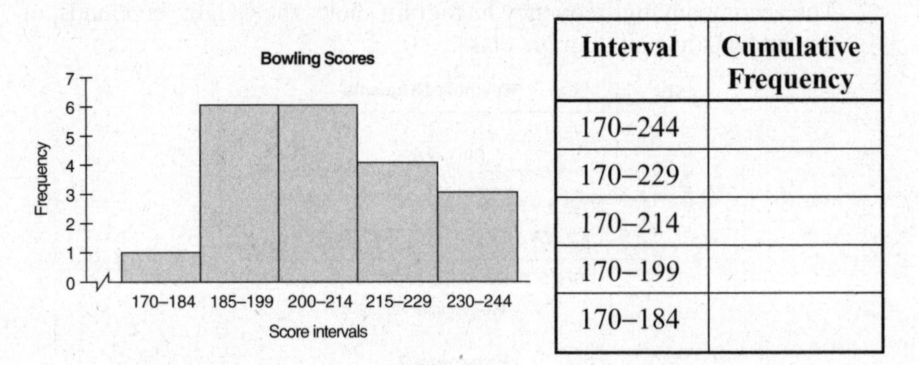

Interval	Cumulative Frequency
170–244	
170–229	
170–214	
170–199	
170–184	

c. On graph paper, construct a cumulative frequency histogram based on the table completed in part b.
d. Is a score of 214 below the 75th percentile? Justify your answer.

CHAPTER 15

COUNTING AND PROBABILITY OF COMPOUND EVENTS

15.1 COUNTING USING PERMUTATIONS

KEY IDEAS

According to the Fundamental Principle of Counting, if event A has a possible outcomes and event B has b possible outcomes, then there are $a \times b$ ways in which event A can be followed by event B. This principle can be used to find the number of ordered arrangements of two or more objects.

The Fundamental Principle of Counting

The Fundamental Principle of Counting works for a sequence of two or more events.

Example 1

If a man has five different shirts, four different neckties, and three different sport jackets, how many different possible outfits consisting of a shirt, tie, and sports jacket are possible?

Solution: The number of different possible outfits consisting of a shirt, tie, and sports jacket is the product of the number of possible choices for each item:

$$5 \times 4 \times 3 = 60$$

Hence, **60** different outfits consisting of one sport jacket, one shirt, and one tie are possible.

Example 2

Debbie orders from a lunch menu that has five appetizers, three soups, seven entrées, six vegetables, and four desserts. How many different meals consisting of either an appetizer *or* a soup, one entrée, one vegetable, and one dessert can Debbie order?

Solution:

- The number of different meals consisting of an *appetizer*, one entrée, one vegetable, and one dessert is the product of the number of different choices for each menu item:

$$5 \times 7 \times 6 \times 4 = 840$$

- The number of different meals consisting of a *soup*, one entrée, one vegetable, and one dessert is the product of the number of different choices for each menu item:

$$3 \times 7 \times 6 \times 4 = 504$$

- The number of different meals consisting of either an appetizer *or* a soup, one vegetable, and one dessert is the sum of the number of menu choices that include an appetizer and the number of menu choices that include a soup:

$$840 + 504 = \mathbf{1344}$$

Arranging *n* Objects in *n* Slots

The lock combination 5–19–34 is different than the lock combination 34–19–5. Each of these lock combinations represents a different arrangement or *permutation* of the same three numbers: 5, 19, and 34. A **permutation** is an arrangement of objects in which the order of the objects matters. To figure out how many three-number lock combinations can be formed using the numbers 5, 19, and 34, without repeating a number, reason as follows using the Fundamental Principle of Counting:

- Any one of the *three* numbers can fill the first position of the lock combination:

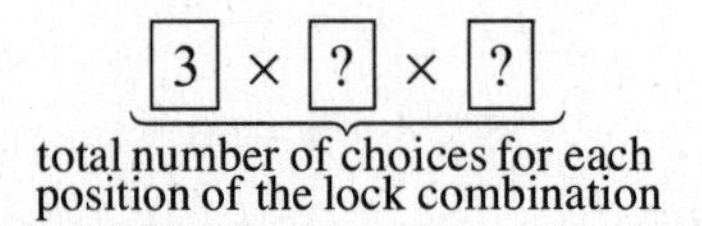

- Either one of the remaining *two* numbers can fill the second position of the lock combination:

$$\boxed{3} \times \boxed{2} \times \boxed{?}$$

- Since the remaining number must be used to fill the third position, there is only one possible choice for the third position of the lock combination:

$$\boxed{3} \times \boxed{2} \times \boxed{1} = 6$$

Thus, the *permutation of three objects* (the numbers 5, 19, and 34) *taken three at a time* (all three numbers are used in every lock combination) is 6. The shorthand notation ${}_nP_n$ represents the permutation of n objects taken n at a time. Thus,

$${}_3P_3 = 3! = 3 \times 2 \times 1 = 6$$

To illustrate further, the number of different ways in which four different books can be arranged on a shelf is

$${}_4P_4 = 4! = 4 \times 3 \times 2 \times 1 = 24$$

MATH FACTS

- When n objects are used to fill n available slots, the total number of different arrangements is given by ${}_nP_n$ where

  ${}_nP_n = n! =$ product of consecutive integers from n down to 1

- By definition, $0! = 1$.

Arranging *n* Objects in Fewer Than *n* Slots

The number of objects or people to be arranged may be greater than the available number of slots. For example, if 7 students run in a race in which there are no ties, then the number of possible arrangements of first, second, and third place is

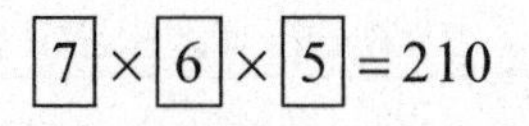

$$\boxed{7} \times \boxed{6} \times \boxed{5} = 210$$

The product $7 \times 6 \times 5$ represents the total number of possible arrangements of 7 students taken 3 at a time. This product can be expressed in permutation notation as ${}_7P_3$ where ${}_7P_3 = 7 \times 6 \times 5 = 210$. Thus, ${}_7P_3$ represents the product of the three greatest factors of 7!

MATH FACT

When n objects are used to fill r available slots, where r is less than or equal to n, the total number of different arrangements is given by ${}_nP_r$, where

${}_nP_r =$ the product of the r greatest factors of $n!$

Example 3

Shari knows the last four digits of a seven digit telephone number but only remembers that each of the first three digits of the telephone number is an odd number. Find the maximum number of telephone calls she must make before she dials the correct number when

a. The first three digits are different
b. Repetition of digits is allowed

Solution

a. There are five odd digits: 1, 3, 5, 7, and 9. The number of ordered arrangements of the five odd digits in the three available slots, without repetition of digits, is ${}_5P_3 = 5 \times 4 \times 3 = 60$. Shari would have to make a maximum of **60** telephone calls.
b. If a digit can be repeated, then the number of arrangements of the five odd digits in the three available slots is $5 \times 5 \times 5 = 125$. Shari would have to make a maximum of **125** telephone calls.

Sets of Objects with Some Identical

If in a set of n objects some are exactly alike, then the n objects can be arranged in fewer ways than n objects that are all different.

- If a out of n objects are identical, then the number of different arrangements of the n objects is $\frac{n!}{a!}$. The word "BETWEEN" has seven letters, and three of these letters (E) are identical. Thus, the number of different arrangements of these 7 letters is

$$\frac{7!}{3!} = \frac{7\times6\times5\times4\times \overset{1}{\cancel{3\times2\times1}}}{\cancel{3\times2\times1}}$$
$$= 7\times6\times5\times4$$
$$= 840$$

- If a set of n objects includes a identical objects, b identical objects, c identical objects, and so forth, then the number of different arrangements of the n objects is $\frac{n!}{a!\times b!\times c!\times\cdots}$. To find the number of ways in which four red flags, three blue flags, and one green flag can be arranged on a vertical flag pole, use the formula $\frac{n!}{a!\times b!\times c!}$, where $n = 4 + 3 + 1 = 8$, $a = 4$, $b = 3$, and $c = 1$:

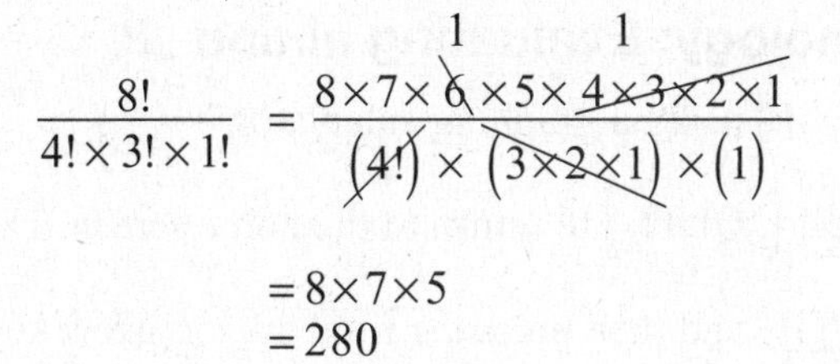

Counting Subject to Conditions

When counting arrangements of a set of objects, there may be conditions that require certain objects in the set to fill specific positions in each arrangement.

Example 4

How many *even* four-digit numbers can be formed using the digits 1, 2, 3, and 9 if no digit is repeated?

Solution: An even number ends in an even digit. Therefore, the last digit of the four digit number must be 2 since 1, 3, and 9 are odd integers. Once the last position of the four-digit number is filled, there are 3 digits remaining that can fill the first position. After that position is filled there are 2 digits left that can fill the second position, leaving the remaining digit for the third position. Hence, the total number of even four-digit numbers that can be formed is

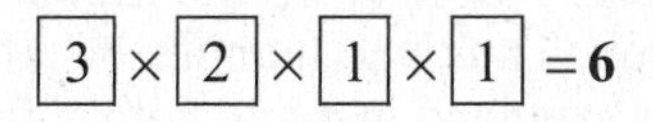

Example 5

How many three-digit numbers greater than 500 can be formed using the digits 1, 2, 3, 4, 5, and 6 if no digit is repeated?

Solution: Since the three-digit number must be greater than 500, there are two digits that can fill the first position, 5 or 6. After the first position is filled, there is no restriction on how to fill each of the remaining positions provided the same digit is not used more than once. Since five digits are available for the second position, any one of the remaining four digits can be used to fill the last position. Thus, the number of three-digit numbers greater than 500 that can be formed is

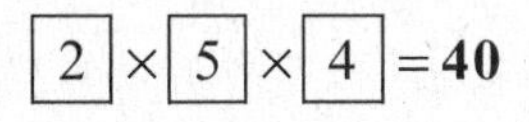

Using Technology: Evaluating $n!$ and $_nP_r$

To evaluate 7! or $_7P_3$ using a graphing calculator,

- Press [2nd] [QUIT] to return to the home screen. Then enter 7.
- Press [MATH] and then press the right cursor arrow key three times to access the MATH PRoBability menu.
- Press [4] [ENTER] to calculate 7! ; or calculate $_7P_3$ by pressing [2] to select $_nP_r$. Then press [3] [ENTER].

Probability and Arrangements

In order to write the numerator or denominator of a probability fraction, it may be necessary to first figure out the number of arrangements of objects that are possible according to the conditions of the problem.

Example 6

What is the probability that when Kevin, Maria, Ricardo, Mike, and Kimberly are arranged randomly in a line, a girl is first and last?

Solution: Of the five people, two are girls.

- There are two girls available to fill the first position, leaving one girl to fill the last position. Thus, the number of different ways in which the five people can be arranged in a line with a girl first and last is

$$\boxed{2} \times \underbrace{\boxed{3} \times \boxed{2} \times \boxed{1}}_{\text{Boys}} \times \boxed{1} = 12$$

- If there is no restriction on how the people line up, then the five people can be arranged in a line in $5! = 5 \times 4 \times 3 \times 2 \times 1 = 120$ different ways.
- The probability that a girl will be first and last when the five people are arranged randomly in a line is $\frac{12}{120}$ or $\mathbf{\frac{1}{10}}$.

Check Your Understanding of Section 15.1

A. *Multiple Choice.*

1. How many different six-letter arrangements can be made from the letters of the name "JENNIE"?
(1) 15 (2) 30 (3) 180 (4) 720

2. The *least* number of different 5-letter arrangements can be made from the letters in the word
(1) ANGLE (2) DADDY (3) ORDER (4) ADAPT

3. If the seven letters L, O, G, I, C, A, and L are randomly arranged in a row to form a seven-letter word, what is the probability that the word will be "LOGICAL?"

(1) $\frac{1}{7}$ (2) $\frac{2}{49}$ (3) $\frac{1}{2520}$ (4) $\frac{1}{5040}$

4. A locker combination system uses four digits from 0 to 9. How many different four-digit locker combinations are possible if the first and the last digit of the combination cannot be 0 and no digit can be repeated?
(1) 3024 (2) 4032 (3) 5040 (4) 7200

5. A history quiz consists of five true-false questions. If you guess on each question, then the probability of answering all five questions correctly is approximately
(1) 10% (2) 2% (3) 3% (4) 5%

6. If $\frac{5!}{6!} = \frac{1}{n+1}$, what is the value of n?

(1) 0 (2) $\frac{1}{5}$ (3) $\frac{1}{6}$ (4) 5

B. *Show how you arrived at your answer.*

7–10. *Find the value of each expression for $x = 5$.*

7. $\left(\frac{x+1}{2}\right)!$ **8.** $2(x-1)!$ **9.** $\frac{x!}{(2x)!}$ **10.** $\frac{(x+1)!}{(x-1)!}$

11. Marcy will be applying to three different colleges. She is choosing from among six four-year colleges in New York, three out-of-state colleges, and a certain number of two-year colleges. If Marcy found that there were 90 different ways in which she could choose to apply to one four-year college in New York, one two-year college, and one out-of-state college, how many choices of a two-year college did she have?

12. What is the number of different ways in which a chemistry book, a calculus book, a history book, a poetry book, and a dictionary can be arranged on a shelf so that the chemistry book or the history book appears first ?

13. How many four-digit odd numbers can be formed using the digits 0, 1, 2, 3, 4, and 5 if no digit is used in any number more than once?

14. In how many different ways can three black flags, four red flags, and two green flags be arranged on a vertical flagpole if a green flag is always on top?

15. a. In how many different ways can the letters of the word "COMMITTEE" be arranged?

b. If the letters of the word "COMMITTEE" are randomly arranged, what is the probability the letter C, O, or I appears first?

16. Five students, all of different heights, are to be randomly arranged in a line. What is the probability that the tallest student will be first and the shortest student will be last in line?

17. If the letters of the word "PARABOLA" are randomly arranged to form a new word, what is the probability that the three As will appear consecutively at the beginning or at the end of the new word?

18. In Jackson Country, Wyoming, license plates are made with two letters (A through Z) followed by three digits (0 through 9). The plates are made according to the following restrictions:

- The first letter must be J or W, and the second letter can be any of the 26 letters in the alphabet
- No digit can be repeated

How many different license plates can be made with these restrictions?

19. Three digits are selected at random and without repetition from the set $\{2, 4, 5, 6, 8\}$ and used to form a 3-digit whole number. Find the probability that the number formed will be

a. Less than 600 b. An odd number c. A number greater than 800

20. What is the probability that when Allan, Barbara, Jose, Steve, Charles, and Maria line up:

a. All four boys are before the two girls?
b. A girl is first and a boy is last?

21. The telephone company has run out of seven-digit telephone numbers for an area code. To fix this problem, the telephone company will introduce a new area code. Find the maximum number of new seven-digit telephone numbers that can be generated for the new area code if both of the following conditions must be met:

- The first digit cannot be 0 or 1.
- The first three digits cannot be the emergency number 911 or the number 411 used for information.

22. A certain state is considering two options for changing the arrangement of letters and numbers on its license plates:

OPTION 1: Three letters followed by a four-digit number with repetition of both letters and digits allowed
OPTION 2: Four letters followed by a three-digit number without repetition of either letters or digits

(Zero may be chosen as the first digit of the number in either option.)

Which option will enable the state to issue more license plates? How many *more* different license plates will that option yield?

15.2 PROBABILITY OF COMPOUND EVENTS

KEY IDEAS

A **compound event** is an event that is comprised of two or more other events. One way of finding the probability that a compound event will occur is to list all of the possible outcomes, count the favorable outcomes, and then find the ratio of the number of favorable outcomes to the total number of possible outcomes. A *tree diagram* may be helpful in identifying the set of all possible outcomes for a compound event.

Listing Outcomes

Consider a probability experiment that consists of two activities: tossing a fair coin and rolling a six-sided cube in which each side of the cube is labeled with a different number from 1 to 6. Each possible outcome represents a compound event. The set of possible outcomes can be described as a set of 12 ordered pairs:

$$\text{sample space} = \begin{Bmatrix} (\text{H},1), (\text{H},2), (\text{H},3), (\text{H},4), (\text{H},5), (\text{H},6), \\ (\text{T},1), (\text{T},2), (\text{T},3), (\text{T},4), (\text{T},5), (\text{T},6) \end{Bmatrix}$$

The first member of each ordered pair represents a possible outcome of tossing a coin where H is a head and T is a tail. The second member of each pair is a possible outcome of rolling the cube.

- What is the probability of getting a head and rolling a 4? Of the 12 possible outcomes, only one, (H,4), satisfies the given condition. Thus, $P(\text{H and } 4) = \frac{1}{12}$.
- What is the probability of getting a tail and rolling an odd number? Three of the 12 possible outcomes satisfy the given condition: (T,1), (T,3), and (T,5). Hence, $P(\text{T and rolling an odd number}) = \frac{3}{12}$.
- What is the probability of getting a head *or* rolling a 6? Of the 12 possible outcomes, there are seven favorable outcomes: (H,1), (H,2), (H,3), (H,4), (H,5), (H,6), and (T,6). Hence, $P(\text{H or rolling a } 6) = \frac{7}{12}$.

Example 1

In a certain class, there are four students in the first row: three girls, Ann, Barbara, and Cathy, and one boy, David. The teacher will select at random one of these students to solve a problem at the board. When the problem is completed, the teacher will select at random one of the remaining students in the first row to do a second problem at the board. Find the probability that

a. The teacher will select Ann first and Barbara second
b. David will be one of the two students selected

Solution: Because any one of 4 students can be selected first and any one of the remaining 3 students can be selected next, there are 4×3 or 12 possible outcomes in the sample space. Make a list in which each of these outcomes is represented by an ordered pair where A represents *A*nn being selected, B represents *B*arbara being selected and so forth:

$$\begin{array}{cccc} (A,B) & (B,A) & (C,A) & (D,A) \\ (A,C) & (B,C) & (C,B) & (D,B) \\ (A,D) & (B,D) & (C,D) & (D,C) \end{array}$$

a. Of the 12 ordered pairs in the sample space, only one, (A,B), corresponds to Ann selected first and Barbara second. Hence, the probability the teacher will select Ann first, and Barbara second is $\frac{1}{12}$.

b. Of the 12 ordered pairs in the sample space, six include David:

$$(A,D), (B,D), (C,D), (D,A), (D,B), \text{ and } (D,C)$$

Thus, the probability that David will be one of the two students selected is $\frac{6}{12}$.

Tree Diagrams

The sample space for a compound event can also be described using a **tree diagram** in which the path along the branches gives the set of all possible outcomes. For example, suppose Mr. and Mrs. Anderson plan on having three children born in different years. The tree diagram in Figure 15.3 shows all of the possible arrangements of boy and girl children in which B represents a boy and G represents a girl.

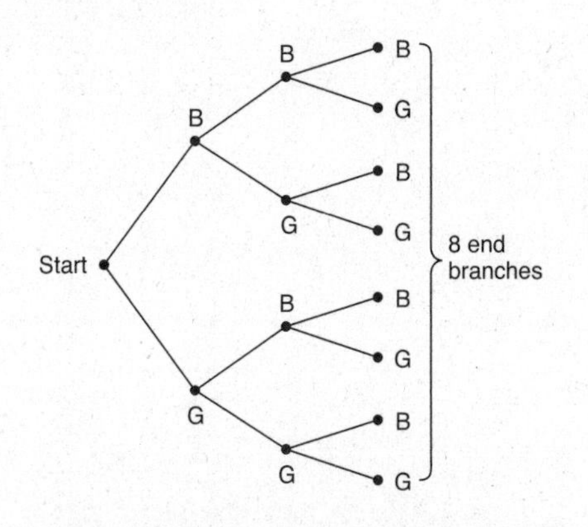

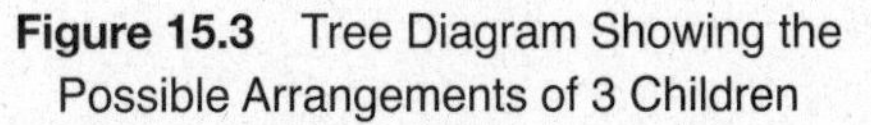
Figure 15.3 Tree Diagram Showing the Possible Arrangements of 3 Children

In Figure 15.3,

- There are two branches at the start since there are two possibilities for the first child. If the first child is a boy, there are two possibilities for the second child, and if the first child is a girl, there are two possibilities for the second child. For each of these four possibilities, branches are attached to represent the possible boy and girl outcomes for the third child.
- The tree ends in eight branches that corresponds to the eight possible arrangements of three children:

(B,B,B), (B,B,G), (B,G,B), (B,G,G), (G,B,B), (G,B,G), (G,G, B), and (G,G,G)

Although the Fundamental Principle of Counting could have been used to determine the *number* of possible outcomes ($2 \times 2 \times 2 = 8$), drawing a tree diagram describes how each of these outcomes are comprised.

Example 2

According to the tree diagram drawn in Figure 15.3, what is the probability that Mr. and Mrs. Anderson will have

a. All girls? b. *At least* two boys?

Solution:

a. Of the 8 possible arrangements of boy and girl children, only one has all girls (G, G, G). Hence, the probability of having all girls is $\frac{1}{8}$.
b. Of the 8 possible arrangements of boy and girl children, four have two boys or three boys: (B, B, B), (B, B, G), (B,G,B), and (G, B, B). Hence, the probability of having at least two boys is $\frac{4}{8}$ or $\frac{1}{2}$.

Example 3

At a local ice cream stand, Dawn has a choice of three flavors of ice cream—vanilla, chocolate, and strawberry; two types of cone—sugar and wafer; and three toppings—sprinkles, nuts, and cookie crumbs.

a. Draw a tree diagram or list the sample space showing all possible ice cream cones that Dawn can choose that have one flavor of ice cream, one type of cone, and one topping.

b. If Dawn does not order either vanilla ice cream or nuts, how many different choices can she make that have one flavor of ice cream, one type of cone, and one topping?

Solution: Dawn must select one flavor, one type of cone, and one topping from these choices:

Flavors: vanilla, chocolate, strawberry
Cone: sugar, wafer
Topping: sprinkles, nuts, cookie crumbs

a. Describe the sample space.

METHOD 1: List the sample space as a set of ordered triples of the form (flavor, cone, topping):

(vanilla, sugar, sprinkles)
(vanilla, sugar, nuts)
(vanilla, sugar, cookie crumbs)
(vanilla, wafer, sprinkles)
(vanilla, wafer, nuts)
(vanilla, wafer, cookie crumbs)

(chocolate, sugar, sprinkles)
(chocolate, sugar, nuts)
(chocolate, sugar, cookie crumbs)
(chocolate, wafer, sprinkles)
(chocolate, wafer, nuts)
(chocolate, wafer, cookie crumbs)

(strawberry, sugar, sprinkles)
(strawberry, sugar, nuts)
(strawberry, sugar, cookie crumbs)
(strawberry, wafer, sprinkles)
(strawberry, wafer, nuts)
(strawberry, wafer, cookie crumbs)

METHOD 2: Use a tree diagram to represent all possible choices:

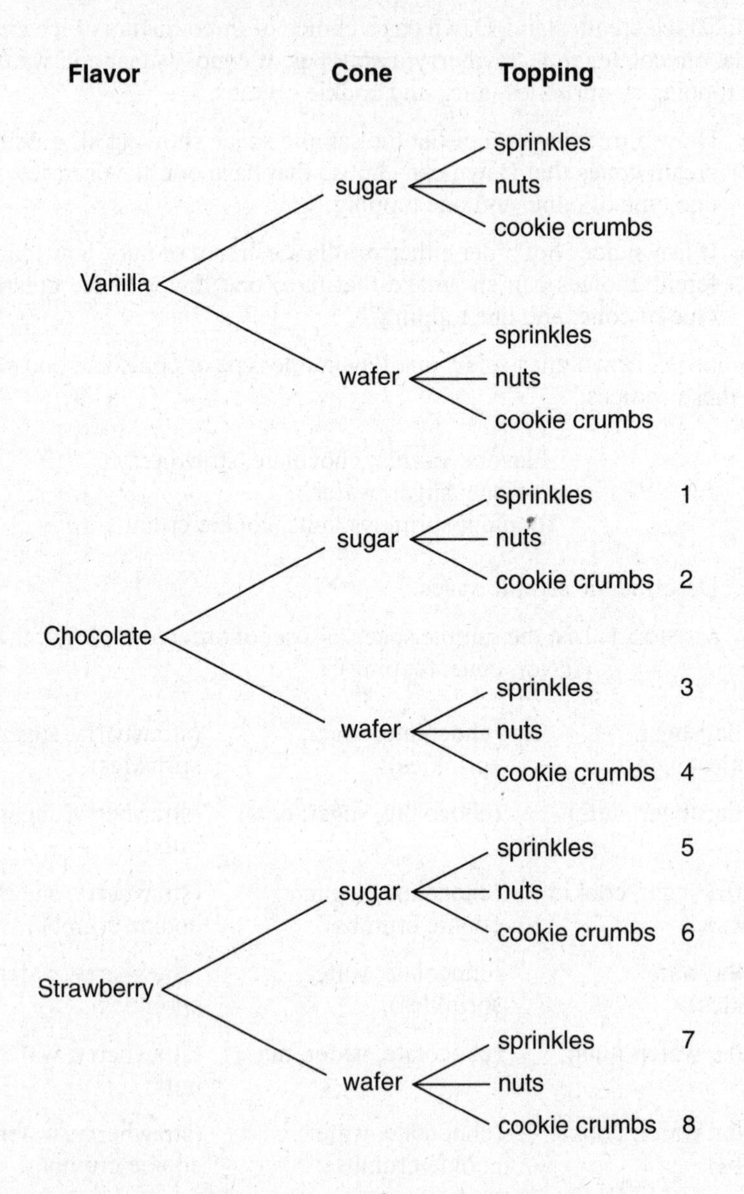

b. As indicated in the tree diagram above, 8 choices do *not* include vanilla or nuts.

Check Your Understanding of Section 15.2

Show how you arrived at your answer.

1. In a certain class, there are five students in the first row: three girls, Aeisha, Brooke, and Chloe, and two boys, Darrel and Eric. The teacher will select at random one of these five students to solve a problem at the board. When the problem is completed, the teacher will select at random one of the remaining four students in the first row to do a second problem at the board. What is the probability that

a. The teacher will call Aeisha first, and Eric second?
b. Both boys will be called?
c. A boy will *not* be called?
d. A boy and a girl will be called?

2. There are four coins in a jar: a penny, a nickel, a dime, and a quarter. One coin is removed at random. Without replacing the first coin, a second coin is removed. Draw a tree diagram or list the same space showing all the possible outcomes. Find the probability that the total value of the two coins selected will be

a. 11 cents
b. Greater than 35 cents
c. *At most* 30 cents

3. The assembly committee of the River High School student council consists of four students whose ages are 14, 15, 16, and 17. One student will be chosen at random to be chairperson, and then, from the remaining three, one will be chosen at random to be the recording secretary. Draw a tree diagram or list the same space showing all the possible outcomes. Find the probability that

a. The chairperson will be older than the recording secretary
b. Both students chosen will be under the age of 16
c. Both students chosen will be the same age

4. A book bag contains one novel, one biography, and one poetry book that all have the same size. Henry selects one book at random from the bag, looks at the book, and then places the book back in the bag. Henry then selects another book without looking. Draw a tree diagram or list the same space showing all the possible outcomes. Find the probability that

a. A novel will be selected both times
b. A poetry book will be selected *at least* once
c. A biography will *not* be selected
d. The same book will be selected both times

5. A coach has to purchase uniforms for a school team. A uniform consists of one pair of shorts and one shirt. The colors available for the shorts are black and white. The colors available for the shirt are green, orange, and yellow. Draw a tree diagram or list the same space showing all the possible outcomes. If each of the colors for the shorts and shirt have an equally likely chance of being selected, find the probability that in the uniform the coach will choose

a. The pants are black and the shirt is orange
b. The shirt is green.
c. The pants and shirt are different colors

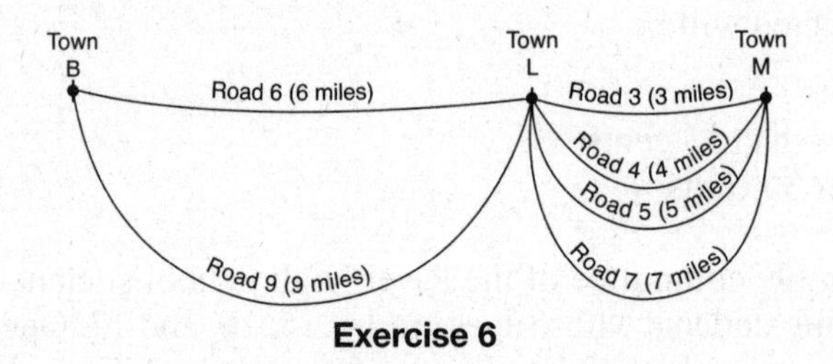

Exercise 6

6. The accompanying diagram shows two roads that lead from Town B to Town L and four roads that go from Town L to Town M. The numbers in parentheses show the distances between each of these towns.

a. Draw a tree diagram or list the sample space showing all possible routes from Town B to Town M.
b. Bonnie is planning a trip from Town B to Town M, passing through Town L. Find the probability that
(1) Both roads she chooses are odd-numbered roads
(2) The distance from Town B to Town M is at most 13 miles
(3) The distance from Town B to Town M is less than 9 miles

7. A restaurant sells kid's meals consisting of one main course, one side dish, and one drink as shown in the table below.

Kid's Meal Choices

Main Course	Side Dish	Drink
Hamburger	French fries	Milk
Chicken nuggets	Applesauce	Juice
Turkey sandwich		Soda

a. Draw a tree diagram or list the sample space showing all possible kid's meals. How many different kid's meals can a person order?
b. José does not drink juice. Determine the number of different kid's meals that do not include juice.
c. If José chooses a kid's meal at random for his sister, what is the probability that he chooses a kid's meal that includes chicken nuggets?

15.3 PROBABILITY FORMULAS FOR COMPOUND EVENTS

KEY IDEAS

Finding probability values for compound events by drawing tree diagrams or compiling lists of all possible outcomes can be time consuming and error prone, if not impractical. If two events have the same sample space, then simple formulas can be used to calculate the probability that:

- Either of the two events occurs.
- Both events occur.

Probability Involving "OR"

Figure 15.4 shows a spinner divided into eight equal regions. Because the spinner landing on yellow cannot happen at the same time as the spinner landing on white, these two events are *mutually exclusive*. The probability of the spinner landing on yellow or white is the sum of the individual probabilities:

$$\begin{aligned} P(\text{Yellow or White}) &= P(\text{Yellow}) + P(\text{White}) \\ &= \frac{3}{8} + \frac{2}{8} \\ &= \frac{5}{8} \end{aligned}$$

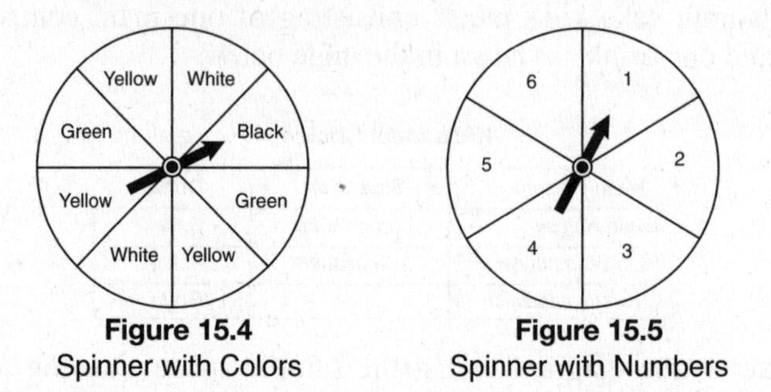

Figure 15.4
Spinner with Colors

Figure 15.5
Spinner with Numbers

Events may not be mutually exclusive. The spinner in Figure 15.5 is divided into six equal regions. Consider the events of spinning a number greater than 4 (event A) and spinning an odd number (event B). These events are not mutually exclusive since spinning a 5 is a favorable outcome for both events. Thus:

$$P(A \text{ or } B) = \frac{2}{6} + \frac{3}{6} - \frac{1}{6}$$
$$= \frac{4}{6}$$

Why is it necessary to subtract $\frac{1}{6}$? The value of $\frac{1}{6}$ corresponds to the probability of spinning a 5. Since spinning a 5 is counted as a favorable outcome when calculating $P(A)$, and again counted as a favorable outcome when calculating $P(B)$, it is being counted twice. Subtracting the extra probability value of $\frac{1}{6}$ makes the necessary adjustment.

MATH FACTS

Two events are **mutually exclusive** if they have the same sample space but cannot happen at the same time.

If events A and B are mutually exclusive:

$$P(A \text{ or } B) = P(A) + P(B)$$

If events A and B are not mutually exclusive:

$$P(A \text{ or } B) = P(A) + P(B) - P(A \text{ and } B)$$

where $P(A \text{ and } B)$ represents the probability of the favorable outcomes common to both events. If events A and B are mutually exclusive, then $P(A \text{ and } B) = 0$.

Independent Events

Two events are **independent events** if the outcome of one event does not affect the outcome of the second event. To find the probability that two or more independent events occur in succession, multiply their individual probability values together. For example:

- If a fair coin is tossed two times, the probability of tossing two heads is $\frac{1}{2} \times \frac{1}{2} = \frac{1}{4}$.
- If a fair coin is tossed three times, the probability of tossing three heads is $\frac{1}{2} \times \frac{1}{2} \times \frac{1}{2} = \frac{1}{8}$.

Suppose when a fair coin is tossed three times, it lands heads up on each toss. If the same coin is tossed again, the probability that it will land heads up on the fourth toss is $\frac{1}{2}$. Past results have no effect on a subsequent outcome provided the coin is fair.

Example 1

Brianna is using the two spinners shown below to play her new board game. The first spinner is divided into six equal sections, and the second spinner is divided into three equal sections. Brianna uses the first spinner to determine how many spaces to move and the second spinner to determine whether her move from the first spinner will be forward or backward, or whether she will lose her turn.

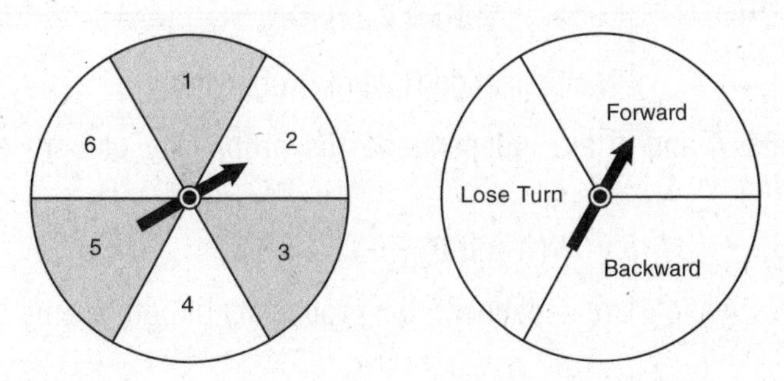

Find the probability that Maryann will move fewer than six spaces and will not lose her turn.

Solution: For the first event, five of the six outcomes are favorable. For the second event, two of the three outcomes are favorable.

$$P(\text{Move less than 6 spaces and Not lose turn}) = \frac{5}{6} \times \frac{2}{3} = \mathbf{\frac{10}{18}}$$

Dependent Events

When the outcome of one event affects the sample space of a second event, the two events are **dependent events**. For example, consider a jar of candy that contains 7 chocolates, 4 mints, and 2 hard candies. Each piece of candy has an equally likely chance of being picked from the jar. Two pieces of candy are selected from the jar, one after the other. If the second piece is selected without putting the first piece back into the jar, the sample space changes for the second pick so that the two picks represent dependent events.

Example 2

A jar of candy contains 7 chocolates, 4 mints, and 2 hard candies. What is the probability of choosing, at random, two chocolates without replacement?

Solution: Multiply together the probabilities of choosing a chocolate on each pick.

- The probability that the first candy picked is a chocolate is $\frac{7}{13}$.
- Assume the first pick is a chocolate and it is not replaced in the jar. Since 6 of the remaining 12 pieces in the jar are chocolates, the probability of picking a chocolate on the second pick is $\frac{6}{12}$.
- The probability of picking two chocolate pieces without replacement is:

$$\frac{7}{13} \times \frac{6}{12} = \mathbf{\frac{42}{156}}$$

MATH FACTS

Multiplication Rule of Probability

If events A and B are independent, the probability of both events occurring is:

$$P(A \text{ and } B) = P(A) \times P(B)$$

If events A and B are dependent, the probability of both events occurring is:

$$P(A \text{ and } B) = P(A) \times P(B) \text{ given event } A \text{ occurs first)}$$

When calculating the probability that the second of two dependent events occurs, adjust the sample space for the second event by assuming that the first event has already occurred successfully.

Example 3

In a certain algebra class, three girls and two boys sit in the first row. The teacher randomly selects one these students to solve a problem at the board. When the problem is completed, the teacher randomly selects one of the remaining students in the same row to do a second problem at the board. What is the probability that the teacher will call two girls to the board?

Solution: The probability that the first student called to the board will be a girl is $\frac{3}{5}$.

Assuming that the first student picked is a girl, 2 of the remaining 4 students are girls. The probability that the second student called to the board is a girl is $\frac{2}{4}$.

$$P(\text{Girl and Girl}) = \frac{3}{5} \times \frac{2}{4} = \mathbf{\frac{6}{20}}$$

Example 4

Two socks are picked at random from a drawer that contains 12 navy socks, 8 black socks, and no other socks. If the socks are taken one at a time and without replacement, what is the probability that the two socks will be the same color?

Solution: The probability of selecting two socks of the same color is the sum of the probabilities of picking two navy socks and of picking two black socks.

- The draw contains 12 + 8 = 20 socks. Of the 20 socks, 12 are navy. On the first pick, $P(\text{Navy}) = \frac{12}{20}$. Without replacement, 11 of the remaining 19 socks are navy. For the second pick, $P(\text{Navy}) = \frac{11}{19}$.

$$P(\text{Navy and Navy}) = \frac{12}{20} \times \frac{11}{19}$$
$$= \frac{132}{380}$$

- Of the 20 socks, 8 are black. On the first pick, $P(\text{Black}) = \frac{8}{20}$. Without replacement, 7 of the remaining 19 socks are black. For the second pick, $P(\text{Black}) = \frac{7}{19}$.

$$P(\text{Black and Black}) = \frac{8}{20} \times \frac{7}{19}$$
$$= \frac{56}{380}$$

- Add the probabilities of picking two navy socks or two black socks:

$$P(\text{Same color}) = P(\text{Navy and Navy}) + P(\text{Black and Black})$$
$$= \frac{132}{380} + \frac{56}{380}$$
$$= \mathbf{\frac{188}{380}}$$

Check Your Understanding of Section 15.3

A. Multiple Choice.

1. Selena and Tracy play on a softball team. Selena has 8 hits out of 20 times at bat, and Tracy has 6 hits out of 16 times at bat. Based on their past performance, what is the probability that both girls will get a hit next time at bat?

(1) $\frac{1}{2}$ (2) $\frac{14}{36}$ (3) $\frac{31}{40}$ (4) $\frac{48}{320}$

2. A bag of marbles contains two green, one blue, and three red marbles. If two marbles are chosen at random without replacement, what is the probability that both will be red?

(1) $\frac{1}{5}$ (2) $\frac{1}{6}$ (3) $\frac{1}{10}$ (4) $\frac{1}{12}$

3. A fair coin is tossed four times and lands head up on each toss. What is the probability that if the coin is tossed again it will *not* land head up?

(1) 0 (2) $\frac{1}{2}$ (3) $\left(\frac{1}{2}\right)^5$ (4) $1-\left(\frac{1}{2}\right)^4$

4. There are 11 girls and 19 boys in a mathematics class. If the teacher randomly selects two students to come to the board to show their work, what is the probability both students will be girls?

(1) $\frac{11}{90}$ (2) $\frac{121}{900}$ (3) $\frac{11}{87}$ (4) $\frac{121}{870}$

5. A pencil holder contains six blue pencils and three red pencils. If two pencils are picked at random and without replacement, what is the probability that both are blue?

(1) $\frac{2}{9}$ (2) $\frac{6}{9}$ (3) $\frac{30}{72}$ (4) $\frac{30}{81}$

6. A whole number from 1 to 12, inclusive, is picked at random. What is the probability that the number is less than 7 or prime?

(1) $\frac{1}{2}$ (2) $\frac{7}{12}$ (3) $\frac{2}{3}$ (4) $\frac{11}{12}$

***B.** Show how you arrived at your answer.*

7. John's sock drawer contains 10 identical navy blue socks, 14 identical black socks, and 4 identical white socks with no other socks. If John selects 2 socks at random from the drawer without replacement, what is the probability that the socks will *not* be the same color?

S	M	T	W	TH	F	S
					1	2
3	4	5	6	7	8	9
10	11	12	13	14	15	16
17	18	19	20	21	22	23
24	25	26	27	28	29	30
31						

Exercise 8

8. Assume the accompanying table shows the calendar dates for the month of August. Aisha needs to pick two days from this month so that she can go shopping with her mother before the new school year begins. If Aisha picks two days at random from the calendar, find the probability that the dates she selects will be

a. A Wednesday and a Friday
b. Both weekend days (Saturday or Sunday)
c. Both before August 16
d. A Sunday and a weekday (M–F)

9. A softball team plays two games each weekend, one on Saturday and the other on Sunday. The probability of winning the game scheduled for next Saturday is $\frac{3}{5}$ and the probability of winning the following game scheduled for Sunday is $\frac{4}{7}$.

a. What is the probability that the team will *lose* both games?
b. What is the probability that the team will win *at least* one of the two games ?

10. From a jar that contains three orange marbles, four brown marbles, and two red marbles, three marbles are drawn in sequence and without replacement. Find the probability that

a. Two orange marbles and a red marble will be drawn in that order
b. Three marbles having the same color will be drawn

11. A bookshelf contains six mysteries and three biographies. Two books are selected at random without replacement.

a. What is the probability that both books are biographies?
b. What is the probability that one book is a mystery and the other is a biography?

12. A penny, a nickel, and a dime are in a box. Bob randomly selects a coin, notes its value, and returns it to the box. He then randomly selects another coin from the box.

a. What is the probability that a nickel will be drawn *at least* once?
b. What is the probability that the total value of both coins that are selected will exceed 6¢?

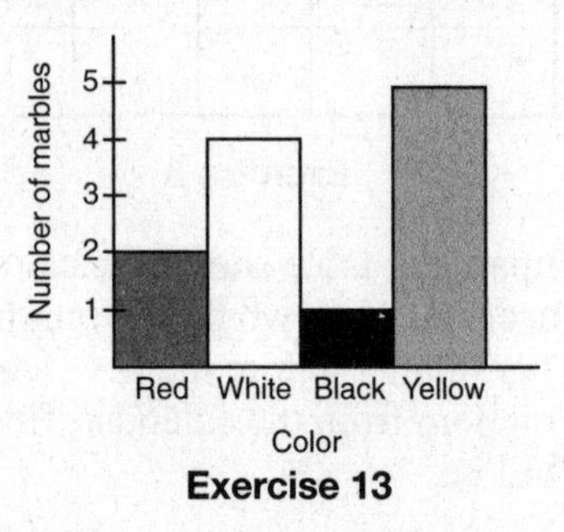

Exercise 13

13. Adam bought a package of marbles and sorted all of them by color as shown by the accompanying graph.

a. If one marble is selected at random, what is the probability it will be red, black, or yellow?
b. If two marbles are selected at random, without replacement, what is the probability that a black marble will be selected?
c. If two marbles are selected at random, without replacement, what is the probability that neither marble selected will be red or black?

14. A jar contains 26 marbles. Each marble is red, white, or blue. The number of white marbles is three times the number of red marbles, and the number of blue marbles is 1 more than twice the number of red marbles.

a. Find the number of marbles of each color in the jar.
b. One marble is selected at random, its color is noted, and then it is replaced in the jar. A second marble is then selected at random. Find the probability that
(1) The first marble will be white and the second marble will be red
(2) Both marbles will be the same color

15. Joan has a hat in which she puts exactly one \$20 bill and a number of \$1 bills. Joan tells Vincent that if he picks two bills from the hat without replacement, the probability that the two bills will add up to more than \$20 is $<\frac{1}{2}$. How many \$1 bills did Joan put in the hat? Explain how you arrived at your answer.

16. There are only three flavors of gumdrops in a jar containing 40 gumdrops. There are 3 times as many cherry gumdrops as lemon gumdrops. There are 4 more than twice as many orange gumdrops as lemon gumdrops.

a. How many of each flavor gumdrop are in the jar?
b. Two gumdrops are drawn at random and without replacement. Find the probability that both are the same flavor.

17. Elton bought a pack of 16 baseball cards. He sorted them by position and noted that he had pictures of 4 pitchers, 5 outfielders, and 7 infielders. Two cards are randomly chosen from the pack of 16 cards without replacement. Find the probability that

a. At least one of the two cards shows a pitcher
b. Neither card shows a pitcher

18. Alexi's wallet contains four \$1 bills, three \$5 bills, and one \$10 bill. If Alexi randomly removes two bills without replacement, determine whether the probability that the bills will total \$15 is greater than the probability that the bills will total \$2.

19. From a jar of candy that contains 7 chocolates, 4 mints, and 2 hard candies, three pieces are selected at random, without replacement. Find the probability of picking three of the same type of candies.

Answers and Solution Hints to Practice Exercises

CHAPTER 1

Section 1.1

1. (4)
2. (2)
3. (4)
4. (3)
5. (3)
6. (3)
7. (3)
8. (1)
9. (4)
10. (3)
11. $26 \le x < 35$
12. 15
13. 30

Section 1.2

1. (3)
2. (2)
3. (1)

Section 1.3

1. (2)
2. (2)
3. (3)
4. (3)
5. (2)
6. (1)
7. {5, 6, 7, 8, 9}
8. {–3, –2, –1, 0, 1, 2}
9. {1, 3, 5, 7, 9, 11, 13, 15, 17, 19}
10. {–3, –2, 2, 3}
11. {3, 6, 12}
12. {1, 4, 6, 8, 9, 10, 12, 14, 15, 16, 18, 20}
13. {8}
14. {31, 37, 43, 47}
15. a. {1, 2, 3, 5, 7, 15}
    b. {3, 5}
    c. {7, 9, 11, 13, 17, 19}
16. a. 24
    b. 7
    c. 37
    d. 2
17. a. {2, 3}
    b. {2, 5, 6}
    c. {2}
    d. {1, 3, 8, 10}

Section 1.4

1. (1)
2. (4)
3. (1)
4. –7
5. 5
6. –2.5
7. $\frac{1}{6}$
8. – 4
9. 4
10. –2
11. 13
12. 12
13. 60
14. –2
15. –1
16. – 4
17. –0.45
18. 2
19. –21
20. –3
21. 4
22. – 4
23. –5
24. {– 4}
25. –8,5
26. –24, –1
27. 8
28. –18
29. 30
30. –3
31. – 4

Section 1.5

1. (3)
2. (3)
3. (4)
4. (2)
5. (1)
6. (1)
7. (2)
8. (4)
9. (2)
10. (2)
11. (2)
12. (3)
13. $-x + 3$
14. $2n + 1$
15. $14x - 1$
16. $7y - 4$
17. $11 - 4x$
18. $9m + 5$

Section 1.6

1. (1)
2. (1)
3. (3)
4. (1)
5. (4)
6. (2)
7. (2)
8. (1)
9. (3)
10. (3)
11. (3)
12. (1)
13. (3)
14. (4)
15. (4)
16. (1)
17. 1.05×10^{12}
18. 1.457×10^{-8}
19. 1.25×10^{-13}
20. 4.0×10^{8}
21. 1.2×10^{-4}
22. 7.5×10^{4}
23. 128
24. $-32x^{10}y^{15}$
25. 1
26. $\frac{11}{4}$
27. {1, 2, 3, 4, 5}
28. 16

Section 1.7

1. (3)
2. (4)
3. (1)
4. (2)
5. (3)
6. (2)
7. (1)
8. (3)
9. 81
10. −1
11. 6

Section 1.8

1. (2)
2. (2)
3. (3)
4. (1)
5. (2)
6. (4)
7. (4)
8. (1)
9. (1)
10. (1)
11. $2(x-5) = 18$
12. $2x - 4 < 11$
13. $3x + 6 \leq 39$
14. $2\,(x+3) \geq 12$
15. $x - 2 = \frac{1}{2}x + 1$
16. $2(x-5) \leq 24$
17. $2x - 3 = x + 7$

CHAPTER 2

Section 2.1

1. −14
2. − 4
3. −6
4. 70
5. −6
6. −15
7. −9.2
8. −2.3
9. $-\frac{2}{3}$
10. 24
11. −24
12. −8.5
13. $-\frac{5}{3}$
14. −25.2
15. 6
16. $\frac{3}{8}$
17. $\frac{28}{3}$
18. 7
19. − 4.8
20. $\frac{11}{6}$
21. −5.7

Section 2.2

1. −10
7. $\frac{5}{3}$
13. 8
19. 28
25. −10

2. –5	**8.** $-\frac{27}{2}$	**14.** 50	**20.** 22	**26.** 270
3. 10	**9.** 0	**15.** $-\frac{1}{3}$	**21.** $\frac{23}{5}$	**27.** 9
4. 13	**10.** 9	**16.** –2	**22.** {8}	**28.** 15
5. 10	**11.** $\frac{5}{7}$	**17.** 5	**23.** { }	**29.** 7
6. 39	**12.** 8	**18.** –5	**24.** {–7, 6}	**30.** 43

Section 2.3

1. (3)	**11.** 9	**21.** $\frac{5}{4}$	**31.** 18, 20, 22
2. (4)	**12.** –7	**22.** 3570	**32.** 13, 15, 17, 19
3. (4)	**13.** 2	**23.** 18	
4. (4)	**14.** –13	**24.** 20	
5. (3)	**15.** 2.1	**25.** 8	
6. (2)	**16.** –6	**26.** 3.6	
7. (2)	**17.** 10	**27.** 21	
8. (2)	**18.** 3	**28.** 23, 25, 27, 29	
9. (2)	**19.** 18	**29.** 21, 22, 23, 24	
10. – 4	**20.** 8	**30.** 22, 24, 26	

Section 2.4

1. (3)
2. (3)
3. 1232
4. 266
5. 1,560
6. 7
7. 26
8. 36
9. 5 nickels, 2 quarters, 1 dime
10. 315
11. 7
12. 80
13. 9
14. Seth, 101; Jason, 51; Raoul, 104
15. 34 cheesecakes and 17 apple pies
16. $167.50

Section 2.5

1. (2)	**3.** (3)	**5.** (3)	**7.** (2)
2. (3)	**4.** (2)	**6.** (4)	**8.** (1)

9. (1)
10. (2)
11. (2)
12. $\frac{A-p}{pr}$
13. $\frac{b-x}{a}$
14. $\frac{J-28}{4}$
15. $\frac{b-c}{a}$
16. 50
17. 64
18. 9
19. 26
20. 5
21. 3, 8, 10
22. $p = 50; A = 108$
23. 15

Section 2.6

1. (4)
2. (2)
3. (3)
4. (1)
5. (4)
6. (4)
7. (1)
8. (4)
9. (4)
10. (4)
11. (4)
12. {0, 1, 2, 3, 4}
13. {–5, – 4 ,–3, –2, –1}
14. $x > -3$
15. $x \leq 2$
16. $x \geq 3$
17. $x > -5$
18. $28
19. 486 girls, 439 boys
20. 5
21. 14
22. 342
23. $1095 \leq x \leq 1209$
24. 206

CHAPTER 3

Section 3.1

1. (4)
2. (2)
3. (2)
4. (1)
5. (3)
6. (3)
7. (2)
8. $37.20
9. 18, 14, 10, 6, 2
10. 40
11. 72
12. 2.5
13. 32
14. 250,000
15. $120
16. Rick, 10; Mark, 14; Vanessa, 9; Sandy, 5; Ariela, 6
17. 7
18. 18
19. $32,000
20. Disagree. If $x = -2$ and $y = -3$, $x > y$ but $x^2 \not> y^2$.
21. Use $n = 40$ as a counterexample.
22. 15
23. 335
24. 5
25. 6

Section 3.2

1. (2) **2.** (2) **3.** (3) **4.** (4) **9.** (4,7) **10.** –3

Section 3.3

1. –2
2. 3
3. 1
4. –3
5. 4
6. 20
7. 8
8. 74
9. a. 3
b. $24
10. 85
11. a. 9
b. 40.5%

CHAPTER 4

Section 4.1

1. (2)
2. (4)
3. (3)
4. (3)
5. (2)
6. (1)
7. (4)
8. (3)
9. (4)
10. 1225
11. 2
12. $68,000
13. 42
14. 1280
15. 184
16. 8
17. a. 24
b. 3:7

Section 4.2

1. (1)
2. (1)
3. (4)
4. $-\frac{1}{2}$
5. –6
6. 20
7. 14
8. –22
9. 5
10. 18
11. 8
12. 275
13. 227.2
14. 16
15. 216
16. 36
17. $\frac{11}{18}$
18. 4.03
19. 20
20. 12

Section 4.3

1. (2)
2. (3)
3. (2)
4. (1)
5. (3)
6. (3)
7. 50 mph
8. 60 mph
9. 3:30 PM
10. 72
11. 40
12. 2.5
13. 9.6
14. 40 mph
15. 12:50 PM
16. 10

Section 4.4

1. (2)
2. (4)
3. (2)
4. (3)
5. (3)
6. (1)
7. (3)
8. (4)
9. (2)
10. (3)
11. (1)
12. 20.7%
13. 600
14. 560
15. 145
16. $195
17. 75%
18. 6
19. 13
20. 28%
21. 18
22. $1600
23. $800
24. 9
25. $40
26. 12.5
27. a. 50
b. 7.5%
28. $45
29. 28
30. $167.50
31. $47.08 in either case.
32. No; 40%
33. $148.54

Section 4.5

1. (2)
2. (2)
3. (1)
4. (4)
5. (2)
6. (4)
7. (3)
8. (3)
9. $\frac{1}{5}$
10. $\frac{2}{5}$
11. $\frac{13}{20}$
12. $\frac{7}{10}$
13. 200
14. $\frac{3}{5}$
15. a. 17.8%
b. 89
16. 8
17. 24
18. 16
19. a. 3
b. $\frac{8}{15}$

CHAPTER 5

Section 5.1

1. (1)
2. (3)
3. (1)
4. (1)
5. (4)
6. $-2x + 4$
7. $8y - 10$
8. $-3n^2 - 5$
9. $2x^3 + 8x^2 - 6x - 9$
10. $5x - 7y + 7z$
11. $-a^2 - 3.5a - 2$
12. $1.5b - 4.1c + \frac{5}{6}e$
13. $\frac{5}{6}m - n - \frac{7}{12}p$
14. $-4x^3 + 5x^2 - 13$
15. $-2x^2 - 2x - 6$
16. $4x^3 + 6x^2 - 14x + 6$
17. $a = 9, b = -9$

Section 5.2

1. (3)
2. (3)
3. (2)
4. (1)
5. (2)
6. (1)
7. (1)
8. (4)
9. (1)
10. (4)
11. $4a^3b^4$
12. $-24x^6$
13. $-0.06xy^5$
14. $\frac{y}{x^2}$
15. $\frac{-2}{3b^2}$
16. $\frac{3x^3}{y}$
17. $3x^2y$
18. $5y^4 - 40y^2 - 20y$
19. $n^2 + 3n - 40$
20. $3x^2 - 20x - 63$
21. $-y^2 - y + 20$
22. $9r^2 - 25$
23. $0.24m^2 - 4m - 50$
24. $\frac{2}{5}k^2 - 11k - 200$
25. $3m^2 + 5mn - 2n^2$
26. $\frac{3}{4}a^2 + \frac{11}{24}ab - b^2$
27. $4x^2 - 12x + 9$
28. $8c^2 - 2c - 3$
29. $32b^6c^7$
30. $\frac{-7p}{q}$
31. $\frac{9}{4}x^{10}y^6$
32. $\frac{2r^3}{s} - 3r^2s$
33. $-7c^2 + 4c - 1$
34. $2a^2 - 1.5ab$
35. True because $\frac{n+(n+1)+...+(n+4)}{5} = \frac{5n+10}{5} = n+2$ with a 0 remainder.
36. a. $-x + 9x + 15$

36. b. 2

37. 8 by 5

38. 9

Section 5.3

1. (2)
2. (2)
3. (3)
4. (1)
5. $3x(5x - 2)$
6. $7(p^2 + q^2)$
7. $x(x^2 + x - 1)$
8. $3y^3(y^4 - 2y^2 + 4)$
9. $-4(a + b)$

10. $4(2b^7 - 5a^4)$
11. $5xy(2x^2 - 3y)$
12. $12x(0.02x - 0.03y)$
13. $6u^3w^2(3u^2 - 5w^5)$
14. $4x^8y^3(4x^4y - 3)$
15. $\frac{1}{4}ab\left(a - 3b^2\right)$
16. $h^2k\left(6 + \frac{1}{2}h^2k\right)$
17. $\dfrac{3x}{x+2}$
18. $\dfrac{a+bx}{a-b}$

Section 5.4

1. $x^2 - \frac{1}{4}$
2. $25w^2 - 64$
3. $\frac{x^2}{4} - 16y^2$
4. $0.09y^4 - 1$
5. $9x^2 - 4y^4$
6. $9h^6 - 4k^4$
7. $(y + 12)(y - 12)$
8. $(9 + x)(9 - x)$
9. $\left(p + \frac{1}{3}\right)\left(p - \frac{1}{3}\right)$
10. $\left(1 + \frac{4}{5}x\right)\left(1 - \frac{4}{5}x\right)$
11. $(3e + 0.2)(3e - 0.2)$
12. $(7e + 11f)(7e - 11f)$
13. $(10a^3 + 7b)(10a^3 - 7b)$
14. $\left(\frac{1}{2}r + \frac{1}{3}s\right)\left(\frac{1}{2}r - \frac{1}{3}s\right)$
15. $\left(\frac{2}{3}c + 0.1\right)\left(\frac{2}{3}c - 0.1\right)$
16. $(2ab + 3c^3)(2ab - 3c^3)$
17. $(4y^2 + 7x)(4y^2 - 7x)$
18. $(0.3w + 0.8z^3)(0.3w - 0.8z^3)$
19. $(x^n + 1)(x^n - 1)$
20. $(a^x + b^y)(a^x - b^y)$
21. $\left(\frac{3}{x} + \frac{5}{y^2}\right)\left(\frac{3}{x} - \frac{5}{y^2}\right)$

Section 5.5

1. (2)
2. (3)
3. (3)
4. (2)
5. (2)
6. (4)
7. (1)
8. (2)
9. (3)
10. (4)

11. $y(a + b)$
12. $(x + 3)(x + 5)$
13. $(x - 3)(x - 7)$
14. $(y + 3)(y + 3)$
15. $(p - 4)(p - 4)$
16. $(b - 5)(b + 8)$
17. $(w - 7)(w - 6)$
18. $(a - 5)(a + 9)$
19. $(n - 20)(n + 3)$
20. $(x^2 - 2)(x^2 - 3)$
21. $2y(y + 5)(y - 5)$
22. $-5(t + 1)(t - 1)$
23. $4(m + n)(m - n)$
24. $4(y + 1)(y + 4)$
25. $8xy(y + 3)(y - 3)$
26. $-2(y + 2)(y + 5)$
27. $2x(x - 7)(x + 8)$
28. $4y^3(y - 8)(y + 5)$

29. $2x(3x + 5y)(3x - 5y)$
30. $(m + 4)(m - 4)(m^2 + 16)$
31. $\frac{1}{2}x(x + 6)(x - 6)$
32. $(y^2 + 3x)(y^2 - 3x)(y^4 + 9x^2)$
33. $(3x - 7)(x + 3)$
34. $(4n - 1)(n + 3)$
35. $(5s + 1)(s - 3)$

Section 5.6

1. (3)
2. (1)
3. (1)
4. (1)
5. (4)
6. $\{-2, -1\}$
7. $\{-3, 2\}$
8. $\{-3, 8\}$
9. $\{-2, 5\}$
10. $\{0, 13\}$
11. $\{5\}$
12. $\{0, 6\}$
13. $\{-3, 1\}$
14. $\{3, 7\}$
15. $\{4, 5\}$
16. $\{-15, 6\}$
17. $\{-4, 7\}$
18. $\{-3, 2\}$
19. $\{-2, 9\}$
20. $\{-3, 5\}$
21. $\{-3, 2\}$
22. $\{-3, 7\}$
23. $\{1, 6\}$
24. 6
25. 5
26. $h = 3, w = 5$
27. a. 120
b. 5
28. 37
29. $\left\{-\frac{1}{2}, 3\right\}$
30. $\left\{-5, \frac{1}{3}\right\}$
31. $\left\{-2, \frac{4}{3}\right\}$
32. 2.5

Section 5.7

1. 7,8
2. 7
3. 7,8,9
4. 7
5. 10, 11, 12
6. 7
7. 13, 15, 17
8. 1, 2, 3, 4
9. 3 cm
10. 24 cm by 18 cm
11. 2 in.
12. 54
13. 13, 8
14. 10 ft by 12 ft
15. 81, 117, 162
16. 10 in. by 20 in.
17. 6 ft
18. 10 in. by 3 in.
19. 5 in. by 9 in. or 3 in. by 11 in.

CHAPTER 6

Section 6.1

1. (1)
2. (2)
3. (4)
4. (1)
5. (3)
6. 4
7. $-\frac{3}{7}$
8. $0, \frac{1}{2}$

9. −3, 2

10. $-\frac{1}{2}$

11. $\frac{2}{x+8}$

12. $\frac{a-4}{3}$

13. $\frac{b-a}{2}$

14. $0.6x - 0.2$

15. $-\frac{b}{5a}$

16. $\frac{r(3-r)}{2}$

17. $\frac{10}{y-3x}$

18. $\frac{3}{y+5}$

19. $\frac{-5}{x+1}$

20. $\frac{x+y}{x-y}$

21. $\frac{p-7}{p+2}$

22. $\frac{n+6}{n+1}$

23. $\frac{x-5}{x+2}$

24. $\frac{2p-3q}{q}$

25. $\frac{-s(s+2r)}{4}$

26. $\frac{2x-3y}{5xy}$

27. $\frac{m-2n}{m+2n}$

Section 6.2

1. (2)

2. (4)

3. $\frac{2a-1}{3b}$

4. $\frac{y}{2x}$

5. $\frac{8x}{k(x-k)}$

6. 4

7. $\frac{x-3}{x}$

8. $2m$

9. $\frac{-25r^2}{r+2}$

10. $\frac{3}{y-2}$

11. 1

12. $\frac{n-7}{n-5}$

Section 6.3

1. (2)

2. (3)

3. (4)

4. (4)

5. (2)

6. (3)

7. $\frac{b}{2x}$

8. $\frac{y-2}{2xy}$

9. $\frac{b+1}{3b}$

10. $\frac{x+1}{2x}$

11. $\frac{4d-1}{6d}$

12. −1

13. $\frac{8c+7}{14c}$

14. $\frac{1}{a+3}$

15. $\frac{8y-3}{18y}$

16. $\frac{3-5a^2b}{15a^2b}$

17. $\frac{y-4x}{2xy^2}$ **18.** $\frac{-2}{p+1}$ **19.** $\frac{1}{3x-2y}$ **20.** $\frac{5m+21}{4m}$ **21.** $\frac{k-2}{k+5}$

Section 6.4

1. 3	**6.** 6	**11.** 9	**16.** $1.8C + 32$	**21.** 6
2. −12	**7.** 30	**12.** 3	**17.** $x > 6$	**22.** −3
3. 3	**8.** 6	**13.** 3	**18.** $x \geq 3$	**23.** $\frac{1}{2}$
4. 2	**9.** −2	**14.** $\frac{1}{2}$	**19.** $x < 27$	**24.** 11
5. −36	**10.** 3	**15.** −5, 6	**20.** 48	**25.** 147

CHAPTER 7

Section 7.1

1. (1)	**4.** (4)	**7.** (4)	**10.** $-\frac{14}{5}$	**13.** 6
2. (3)	**5.** (3)	**8.** 0.5	**11.** 2	**14.** $550
3. (4)	**6.** (3)	**9.** $\frac{2}{3}$	**12.** 10	

Section 7.2

1. (1)	**7.** (1)	**13.** $\frac{3}{2}$	**19.** $8\sqrt{5}$
2. (1)	**8.** 0.8	**14.** $126\sqrt{3}$	**20.** $6y+4\sqrt{3}$
3. (3)	**9.** $20\sqrt{3}$	**15.** 45	**21.** $\pm 2\sqrt{3}$
4. (2)	**10.** $\frac{\sqrt{7}}{3}$	**16.** $6\sqrt{7}$	**22.** 112
5. (1)	**11.** $4\sqrt{3}$	**17.** $\frac{9\sqrt{2}}{2}$	**23.** $\frac{9}{2}$
6. (1)	**12.** $3\sqrt{10}$	**18.** $9\sqrt{2}$	**24.** 88

Section 7.3

1. (3)
2. (1)
3. (3)
4. (2)
5. (2)
6. $-4\sqrt{11}$
7. $12\sqrt{2}$
8. $8\sqrt{3}$
9. $6+2\sqrt{5}$
10. $19-8\sqrt{3}$
11. $2\sqrt{5}$
12. 3
13. 2
14. $43-30\sqrt{2}$
15. $10\sqrt{3}$
16. 5
17. $\frac{8\sqrt{3}}{3}$
18. -1
19. $4\sqrt{3}$
20. 1

Section 7.4

1. (4)
2. (3)
3. (2)
4. (4)
5. (3)
6. (1)
7. (4)
8. (1)
9. (2)
10. (3)
11. (3)
12. 16
13. 80.8
14. 127.3
15. 108
16. 14
17. $9\sqrt{2}$
18. 36,48
19. 5
20. 6.7
21. 2.8
22. Martin first, then Mary 40 minutes later.

Section 7.5

1. (2)
2. (2)
3. 3.5
4. $1 \pm \sqrt{2}$
5. $-5 \pm 2\sqrt{7}$
6. $\frac{5\pm\sqrt{33}}{4}$
7. $\frac{5\pm\sqrt{61}}{2}$
8. $\frac{1\pm\sqrt{13}}{3}$

CHAPTER 8

Section 8.1

1. (1)
2. (1)
3. (2)
4. (1)
5. (3)
6. $\sin J = \frac{24}{25}$, $\cos J = \frac{7}{25}$, $\tan J = \frac{24}{7}$
7. 18.47
8. 14.11
9. 2.91
10. 56.3°
11. 41.4°
12. 11.5°
13. 68.7°
14. 28°
15. a. 15.6
16. a. 58.7° b. 122
17. a. 77° b. 13.6

Section 8.2

1. 185
2. 109
3. 81.3
4. 34°
5. 224
6. a. 56°
b. 12.5
7. 72.0
8. 754
9. 148.6
10. 3 feet 1 inch
11. 116
12. 32°

CHAPTER 9

Section 9.1

1. (3)
2. (2)
3. (3)
4. 92
5. 45.5
6. $24.5\sqrt{3}$
7. 90
8. 270
9. $27,342
10. 260
11. 162

Section 9.2

1. (3)
2. (2)
3. (1)
4. 210
5. 12
6. 18
7. a. 5.2
b. 13%

Section 9.3

1. (2)
2. (3)
3. (1)
4. (1)
5. (1)
6. (1)
7. (1)
8. (3)
9. (2)
10. 8.6
11. 15
12. 178.3
13. 2.7

Section 9.4

1. (1)
2. (1)
3. (2)
4. $256 - 64\pi$
5. $392 - 98\pi$
6. a. 66.0
b. 67%
7. 36
8. 60.7
9. 256
10. $2950.33
11. a. 103
b. 40%
12. 39.4

Section 9.5

1. (1)
2. (4)
3. (3)
4. (3)
5. (4)
6. (2)
7. (1)
8. (3)
9. (4)
10. 35
11. 7
12. $\frac{2197\pi}{8}$
13. 1715
14. 0.025
15. 0.026
16. 0.102
17. 0.029
18. 0.7%
19. 8.2%
20. 47
21. a. 23 cm by 21 cm
b. 702

CHAPTER 10

Section 10.1

1. (2) **4.** (3) **7.** (1) **10.** 9 **13.** –4

2. (1) **5.** (4) **8.** $-\frac{1}{2}$ **11.** $-\frac{2}{3}$ **14.** y is changing at a constant rate of 3.5

3. (4) **6.** (3) **9.** $\frac{5}{3}$ **12.** $-\frac{3}{4}$

15. 0, –3

16. Graph ordered pairs of the form (year, dollar value): (2005, $24,000), (2006, $22,500), (2007, $21,000), (2008, $19,500), (2009, $18,000), and (2010, $16,500).

17. $\frac{1}{a+b}$

Section 10.2

1. (3)

2. (3)

3. (1)

4. (1)

5. (4)

6. (3)

7. (3)

8. (1)

9. (4)

10. (3)

11. (3)

12. (1)

13. (1)

14. $m=-\frac{1}{2}, b=5$

15. $m=\frac{2}{3}, b=5$

16. $m=\frac{2}{3}, b=-\frac{8}{3}$

17. $y = 3x - 7$

18. $y = -3x + 9$

19. $y = 6x - 13$

20. $y = -4x - 1$

21. $y=-\frac{3}{2}x-7$

22. $y = 6x + 3$

23. a. $v = -32t + 112$
b. 3.5

24. a. $\ell: y=\frac{5}{3}x+2$
$m: y=-x+2$

25. a. $y=\frac{2}{3}x-2$
b. Yes

26. a. 10°C
b. $C=\frac{5}{9}(F-32)$

Section 10.3

1. (1) **3.** (1) **5.** (1) **7.** (4)

2. (3) **4.** (4) **6.** (2) **8.** (3)

9. (1) **11.** 56 **13.** $-\frac{9}{2}$ **15.** 7

10. (2) **12.** $y = 2x - 9$ **14.** $y = 4x + 5$ **16.** a. 24
b. $y = \frac{3}{4}x$

Section 10.4

1–3. Draw the line through:
1. (–1,–3), (0,–1), and (1,1)
2. (–2,2), (0,3), and (2,4)
3. (–1,3), (0,1), and (1,–1)
4. Graph the points (–3,5), (–2,2), (–1,–1), (0,– 4), (1,–7), and (2,–10)

5–10. Draw the line through:
5. (0,– 4) and (2,0) **7.** (0,–6) and (3,0) **9.** (0,–6) and (4,0)

6. (0,4) and (8,0) **8.** (0,3) and (2,0) **10.** $\left(0,-\frac{5}{2}\right)$ and (1,0)

11–13. Graph the line such that:

11. $m = -2, b = 3$ **12.** $m = \frac{1}{2}, b = 3$ **13.** $m = -\frac{1}{3}, b = 4$

14. a. $y = 5x - 300$
b. $x = 60$

Section 10.5

1. (2) **3.** (3) **5.** 1 **7.** 340

2. (2) **4.** 14 **6.** $\frac{4}{3}$ **8.** 45

Section 10.6

1. $y - 2 = 2(x - 8)$ **4.** $y = \frac{1}{2}x + \frac{3}{2}$ **7.** –6

2. $y + 1 = \frac{1}{2}(x + 5)$ **5.** $y = -\frac{1}{2}x + \frac{1}{2}$ **8.** $y = -x + 3$

3. $y + 3 = -\frac{1}{2}(x - 7)$ **6.** $y + 5 = 4(x - 7)$ **9.** a. $D = 9h + 12$
b. $9

CHAPTER 11

Section 11.1

1. (3) **2.** (1) **3.** (3) **4.** (1) **5.** (2) **6.** (1)

7. a. $y=\frac{4}{3}x-4$
b. $x=\frac{3}{4}y+3$

8. $A = 10x - x^2$

9. a. $y=\frac{36}{x}$
b. $p=2x+\frac{72}{x}$

10. a. $V=x^3-x$
b. 60

11. Table:

x	1	2	3	...	8
y	100	160	220	...	520

Equation: $y = 40 + 60x$, where $x \in \{1, 2, 3, \ldots, 8\}$
Graph: Plot the set of eight ordered pairs in the table.

Section 11.2

1. (1)
2. (4)
3. (3)
4. (3)
5. (1)
6. a. $P = 2d + 10$
b. 26
7. a. $y = 10 + 8x$
b. $46
8. a. 8
b. 6
c. 4
9. a. $25
b. May
c. $15
d. $y = 75 + 10x$

10. a. 10
b. A : $y = 50x$; B : $y = 40x$
c. 80
11. a. Cost of electric increases at the greatest rate since the slope of its line is greater than the slope of the line that represents oil.
b. Electric by $7000.
c. $C = \$4000 + \$1000Y$

Section 11.3

1. (2)
2. (2)
3. (3)
4. (2)
5. (1)
6. (1)
7. (4)
8. $\{-5, 3\}$
9. No
10. Domain is the set of all nonzero real numbers. The range is ± 1.

11. b. The graph of $y = |x + c|$ is the graph of $y = |x|$ shifted c units to left when $c > 0$ and c units to right when $c < 0$.

Section 11.4

1.

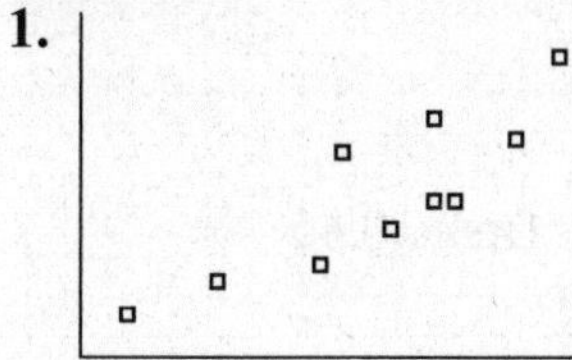

2.

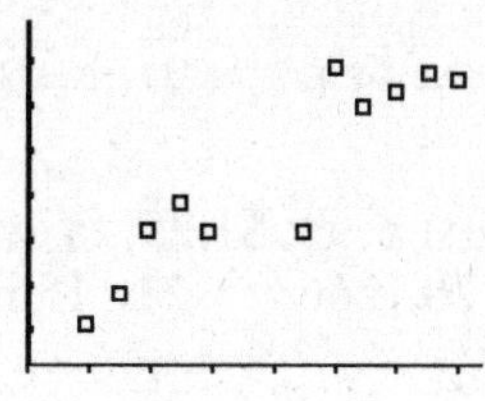

Section 11.5

1. (1)
2. (4)
3. (1)
4. (1)
5. (4)
6. (3)
7. (4)
8. (1)
9. (4)

10. a.

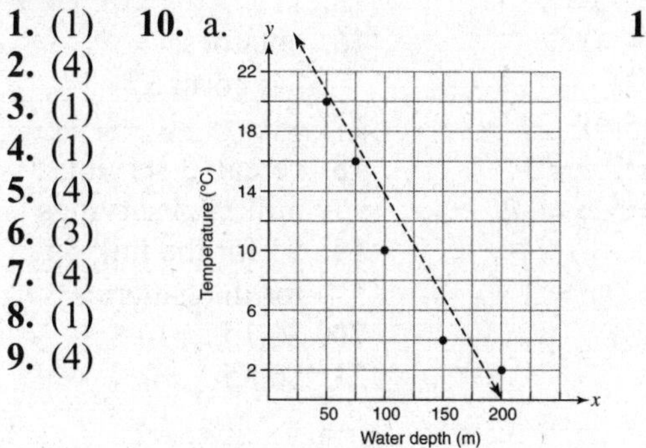

b. $y \approx -0.12x + 25$
c. –9

11. a.

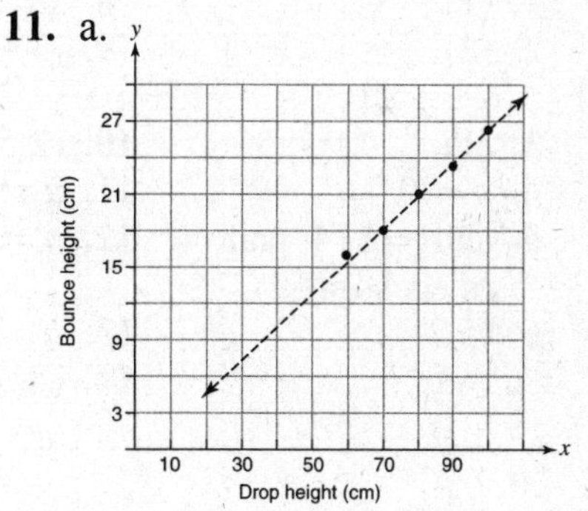

b. $\overline{x} = 80$, $y = 20.8$
c. $y = 0.25x + 0.8$

Chapter 12

Section 12.1

1. (3)
2. (3)
3. (1)
4. (3)
5. (2)
6. (2)
7. (2)
8. (–2,3)
9. (–1,–7)
10. (–1,2)
11. (1,–6)
12. (3,–1)
13. (–6,6) and (3,3)
14. (3,1)
15. (3,– 4)
16. (–16,–5)
17. $x = 3$
18. (1,3), (–2, –3), and (3,–3)
19. 12
20. 25
21. 10
22. 24
23. b. 16
24. 48

25. a. store A: $C = 15 + 2n$; store B: $C = 3.50n$
b. 10

Section 12.2

1. (4)
2. (2)
3. (1)
4. $y = -39$
5. $s = 5$
6. $x = 2, y = -10$
7. $x = -2, y = -8$
8. $a = -1, b = -1$
9. $x = 3, y = 4$
10. $x = 40, y = -30$
11. $x = 3, y = 9$
12. \$0.70
13. 74°
14. $\frac{5}{12}$
15. 82

16. 15 liters of 15% solution and 35 liters of 25% solution

17. a. $\overleftrightarrow{AB}: y=-x+7$; $\overleftrightarrow{CD}: y=2x-8$

b. (5,2)

18. faxing first page: \$1.25; faxing each page after first: \$0.45

19. 60 **20.** 266 **21.** 15 hats and 45 T-shirts

Section 12.3

1. (1)
2. (2)
3. (1)
4. $n=0$
5. shirt = \$14; pair of socks = \$9
6. (5,2)
7. (−3,2)
8. $s=9, t=7$
9. (3,− 4)
10. (2.5,4)
11. $u=2, v=3$
12. $a=5, b=-2$
13. $b=9, c=2$
14. (4,3)
15. 115
16. potatoes: \$4; corn: \$7
17. \$10
18. cereal: 3 servings; milk: 2.5 servings
19. \$3 for the foil; \$5 for the batteries
20. \$6.15
21. \$6.85

Section 12.4

1. (3)
2. (3)
3. (4)
4. (2)
5. (3)
6. (4)
7. (3)

8. Use a Decimal window:

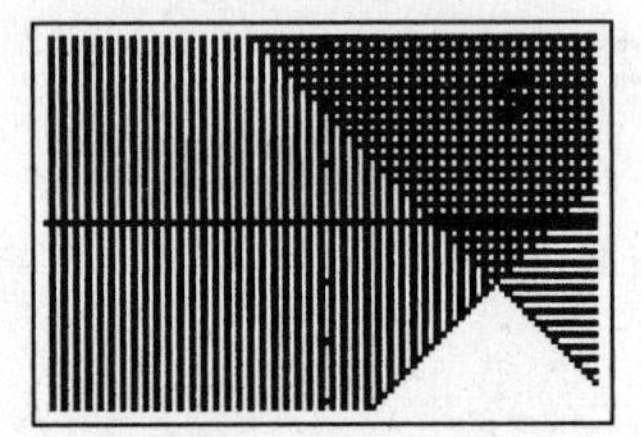

9. a. Use a standard window:

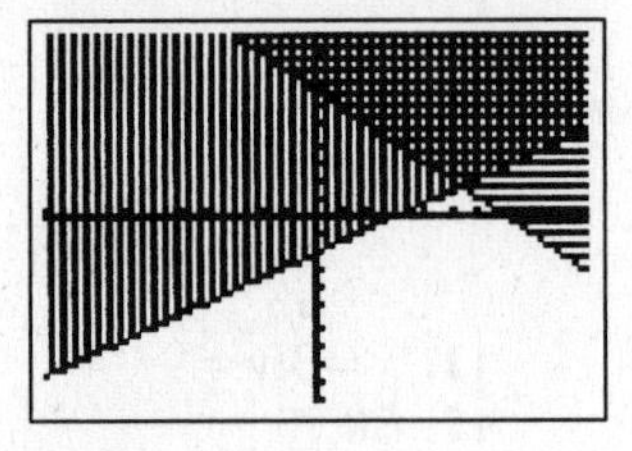

b. (3)

10. a. Use a Decimal window:

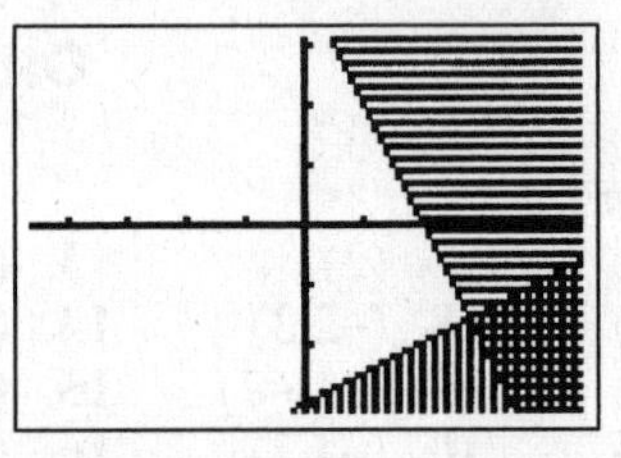

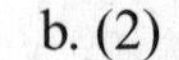

b. (2)

11. a. Use a standard window:

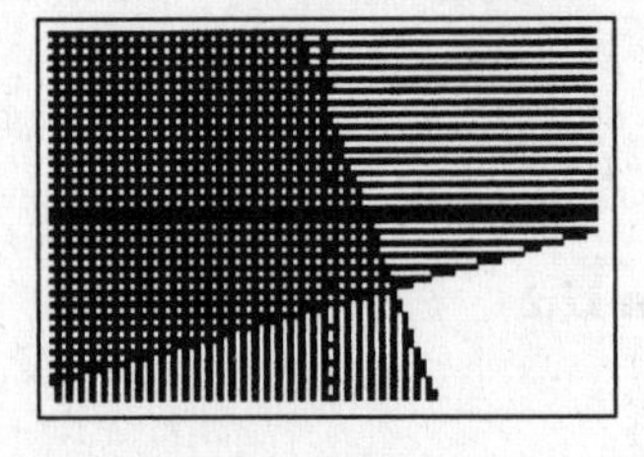

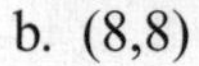

b. (8,8)

12. Graph $x \leq 14, y \leq 8$, and $x+y \leq 18$ on the same set of axes.

CHAPTER 13

Section 13.1

1. (2) **5.** (4) **9.** (1) **13.** (4)
2. (4) **6.** (2) **10.** (3) **14.** $y + 5 = 3(x + 2)$
3. (2) **7.** (3) **11.** (1) **15.** 48
4. (2) **8.** (4) **12.** (2) **16.** $y = 0.4x + 5.8$

17. a. See graph. b. $y = -4$

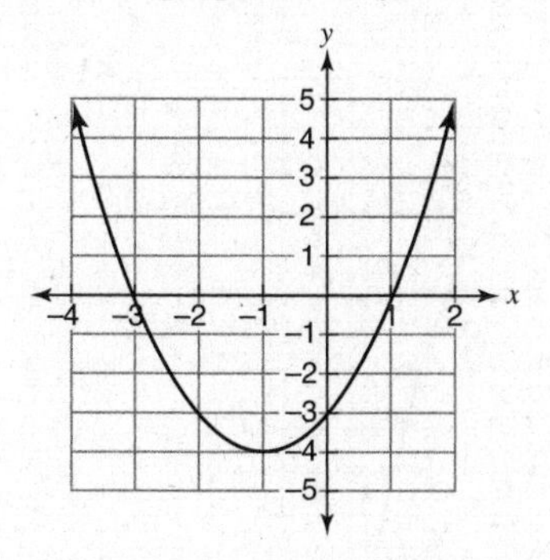

18. a. See graph. b. 9

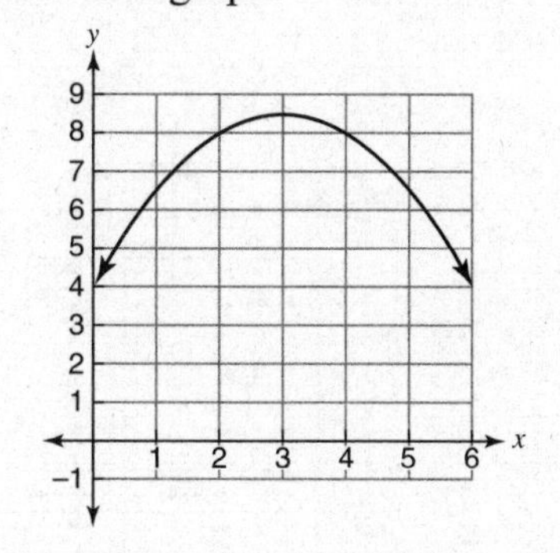

19. a. See graph. b. 3

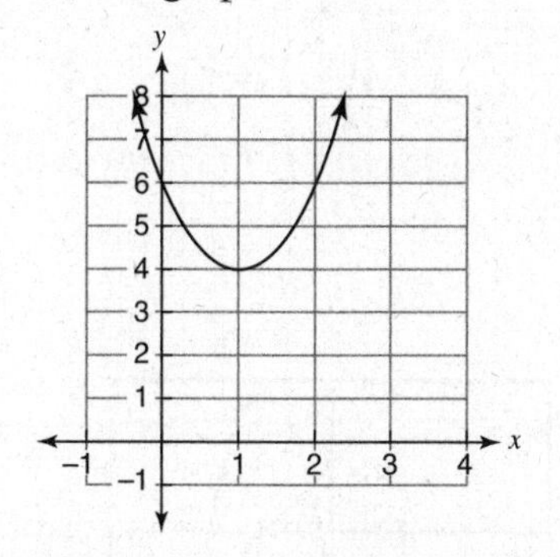

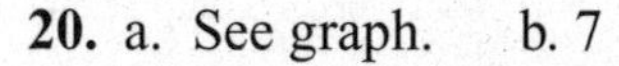

20. a. See graph. b. 7

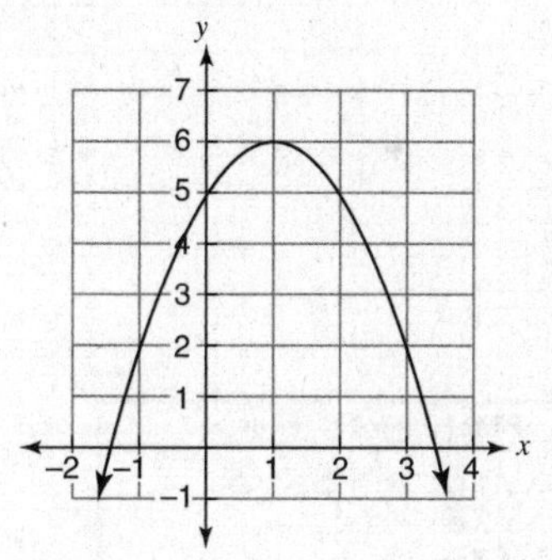

21. a. \$37.50 b. \$5062.50

Section 13.2

1. (2)
2. (2)
3. (1)
4. (1)
5. $y = x^2 - 3x - 10$
6. $y = -3x^2 - 12x$

7. b. 3,5
c. $y = -1$
d. -2

8. b. $-6,1$
c. 13

9. b. 9 ft in 3 sec

10. a. 1.5, 7.5
c. At $t = \frac{1.5+7.5}{2} = 4.5$, $h = 324$

11. a. 324 ft in 4.5 sec
b. 9

12. a. 3
b. 49

13. a. 7
b. 36.75

Section 13.3

1. (1)
2. (2)
3. (4)
4. (3)
5. (1)
6. (1)
7. (−2,−5), (−1,−4)
8. (1,6), (4,9)
9. (−4,0), (1,5)
10. (2,−2), (5,1)
11. (2,−3), (6,5)
12. (4,7)

13. a. (1,2) and (6,−3). See accompanying graph.

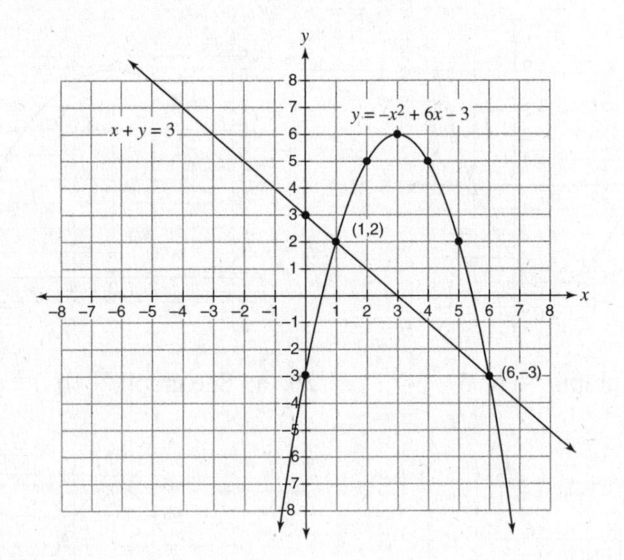

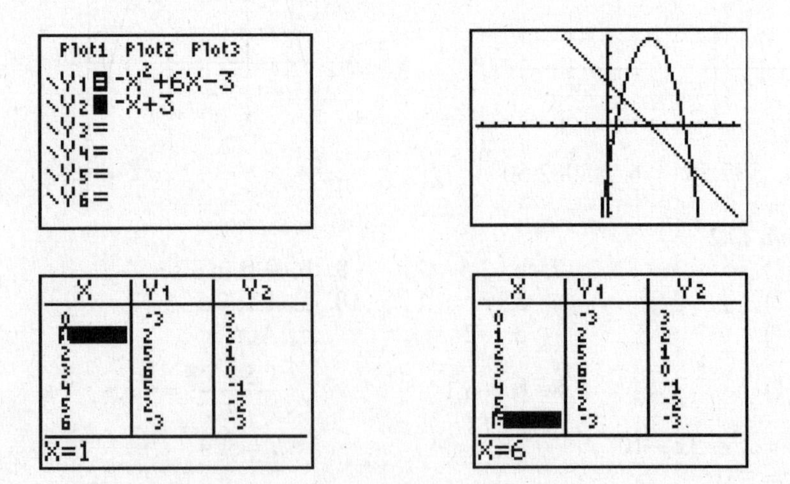

14. a. (2,13) and (–1,–2). See accompanying graph.

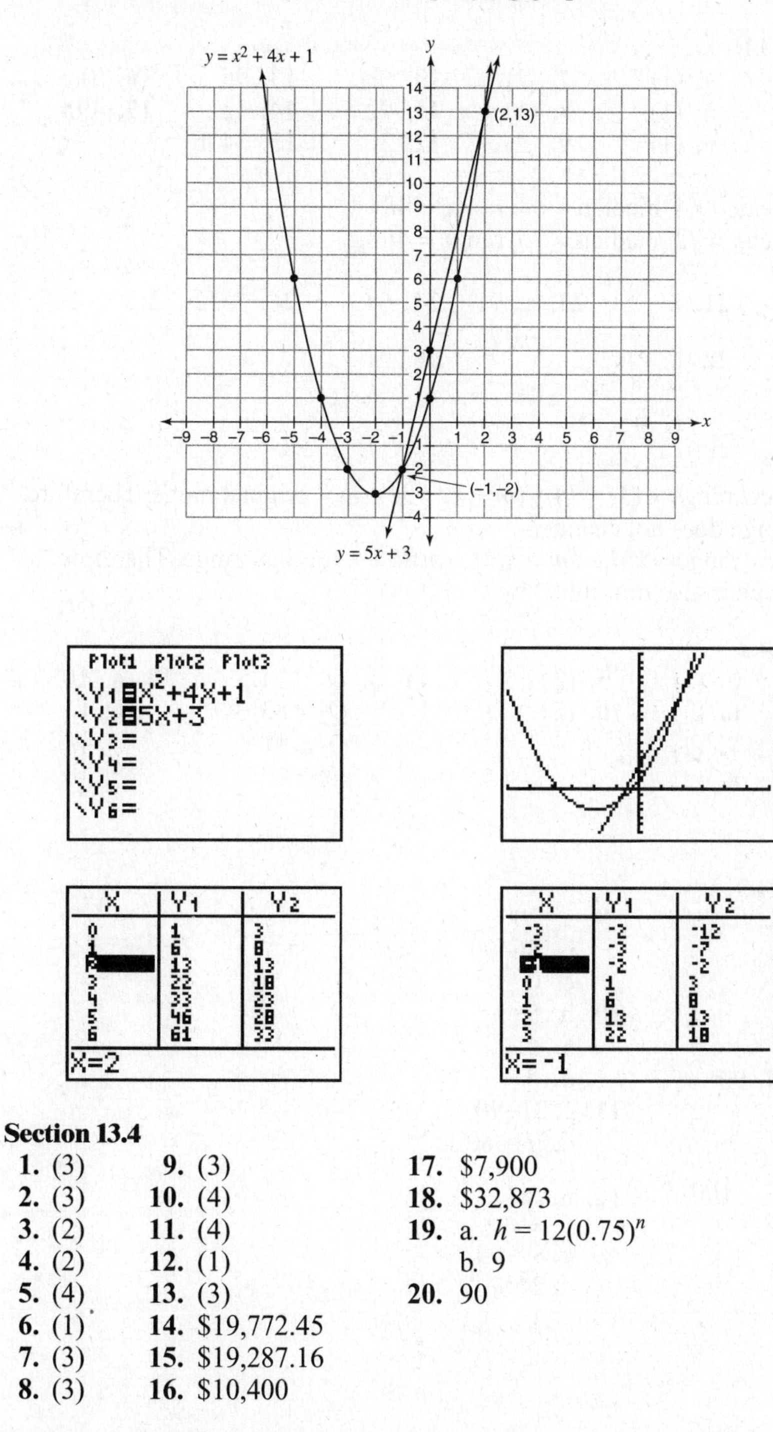

Section 13.4

1. (3)
2. (3)
3. (2)
4. (2)
5. (4)
6. (1)
7. (3)
8. (3)
9. (3)
10. (4)
11. (4)
12. (1)
13. (3)
14. \$19,772.45
15. \$19,287.16
16. \$10,400
17. \$7,900
18. \$32,873
19. a. $h = 12(0.75)^n$
b. 9
20. 90

CHAPTER 14

Section 14.1

1. (4)	**4.** (1)	**7.** (2)	**10.** (2)	**13.** 95	**16.** 91
2. (2)	**5.** (1)	**8.** (1)	**11.** 92	**14.** 42	**17.** 49π
3. (3)	**6.** (1)	**9.** (2)	**12.** 20	**15.** 34 lb	

18. a. mean = 64, median = 68, range = 48
b. mean = 72, median = 76, range = 48
c. 51

19. 96
20. 46
21. 8
22. a. $\frac{1}{8}$
b. 0
23. a. 91
b. 74
24. 64
25. 91
26. 7, 12, 22

27. a. New range = $(M + k) - (m + k) = M - m$ = original range. Therefore, range does not change.
b. New range = $kM - km = k(M - m) = k \times$ original range. Therefore, range is also multiplied by k.

Section 14.2

1. (3)
2. (2)
3. (1)
4. (1)
5. (3)
6. (3)
7. (2)
8. (4)
9. (2)
10. (2)
11. a. $Q_1 = 71.5$, $Q_2 = 83$, $Q_3 = 91$
b. 88
12. a. 10
b. I
c. $\frac{1}{4}$

Section 14.3

1. (3)
2. (2)
3. (3)
4. (1)
5. (3)
6. a. 81–90
b. $\frac{12}{20}$
7. c. 91 – 100
8. a. 78.5
b. 75
9. a. 20
b. 86–90
c. 0
10. b. 5
c. 4
d. 4
11. c. 81–90
d. 71–80
12. b. $\frac{21}{25}$
c. 89–94
d. 24%

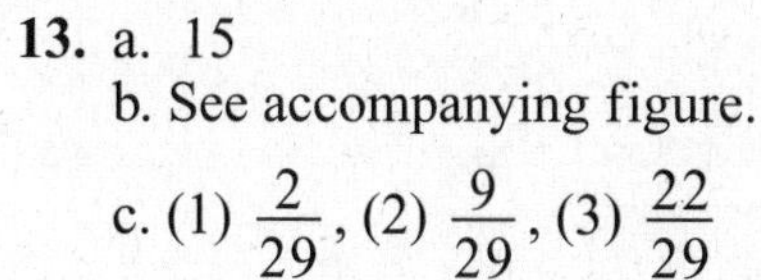

13. a. 15

b. See accompanying figure.

c. (1) $\frac{2}{29}$, (2) $\frac{9}{29}$, (3) $\frac{22}{29}$

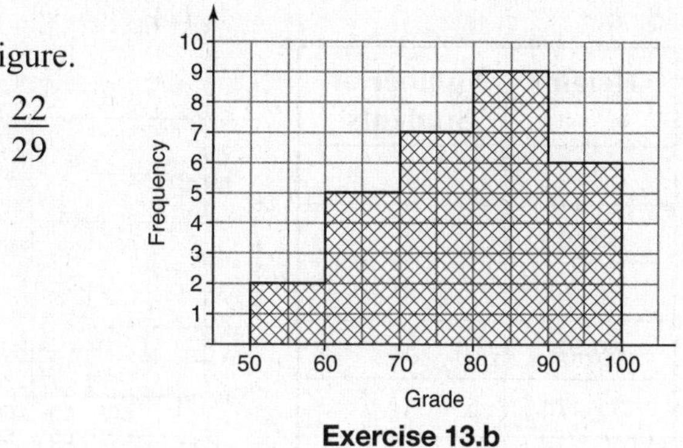

Exercise 13.b

Section 14.4

1. (2) **2.** (3) **3.** a. 25

b. $\frac{11}{28}$

4. a.

Scores	Cumulative Frequency
20–100	100
21–90	85
21–80	59
21–70	36
21–60	21
21–50	10
21–40	5
21–30	2

Exercise 4.a

b.

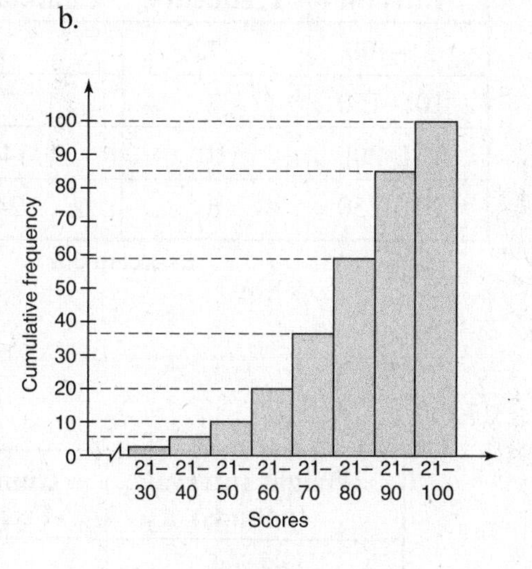

Exercise 4.b

c. (4)

d. 10%

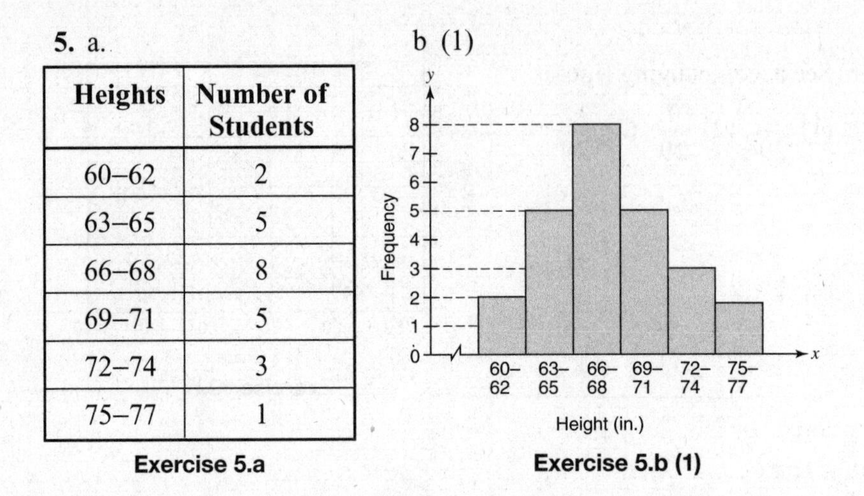

5. a.

Heights	Number of Students
60–62	2
63–65	5
66–68	8
69–71	5
72–74	3
75–77	1

Exercise 5.a

b (1)

Exercise 5.b (1)

5. b. (2) Q_1, 63–65; Q_3, 69–71

5. c. 0.625

6. a.

Interval	Frequency	Cumulative Frequency
51–100	7	**7**
101–150	7	7 + 7 = **14**
151–200	10	14 + 10 = **24**
201–250	6	24 + 6 = **30**

Exercise 6.a

7. a. 73–75

b. $\frac{9}{27}$

c.

Weight Interval (pounds)	Cumulative Frequency
67–69	**3**
67–72	3 + 6 = **9**
67–75	9 + 9 = **18**
67–78	18 + 2 = **20**
67–81	20 + 2 = **22**
67–84	22 + 4 = **26**
67–87	26 + 1 = **27**

8. a. 207

d. Because 75% of 20 is 15 and there are 13 scores at or below 214, 214 is below the 75th percentile.

b.

Interval	Cumulative Frequency
170–244	17 + 3 = **20**
170–229	13 + 4 = **17**
170–214	7 + 6 = **13**
170–199	1 + 6 = **7**
170–184	**1**

Exercise 8.b

CHAPTER 15

Section 15.1

1. (3)

2. (2)

3. (3)

4. (2)

5. (3)

6. (4)

7. 6

8. 48

9. 30,240

10. 30

11. 5

12. 48

13. 144

14. 280

15. a. 45,360

b. $\frac{1}{3}$

16. $\frac{6}{120}$

17. $\frac{240}{6720}$

18. 37,440

19. a. $\frac{36}{60}$

b. $\frac{12}{60}$

c. $\frac{12}{60}$

20. a. $\frac{48}{720}$

b. $\frac{192}{720}$

21. 7.98×10^6

22. Option 2 will yield 82,576,000 more possibilities.

Section 15.2

1. a. $\frac{1}{20}$ b. $\frac{2}{20}$ c. $\frac{6}{20}$ d. $\frac{12}{20}$

2. a. $\frac{2}{12}$ b. 0 c. $\frac{10}{12}$

3. a. $\frac{6}{12}$ b. $\frac{2}{12}$ c. 0

4. a. $\frac{1}{9}$ b. $\frac{5}{9}$ c. $\frac{4}{9}$ d. $\frac{3}{9}$

5. a. $\frac{1}{6}$ b. $\frac{2}{6}$ c. 1

6. a. (6,3), (6,4), (6,5), (6,7), (9,3), (9,4), (9,5), (9,7)

b. (1) $\frac{3}{8}$, (2) $\frac{6}{8}$, (3) 0

7. a. 18

(hamburger, french fries, milk)
(hamburger, french fries, juice)
(hamburger, french fries, soda)
(hamburger, applesauce, milk)
(hamburger, applesauce, juice)
(hamburger, applesauce, soda)
(chicken nuggets, french fries, milk)
(chicken nuggets, french fries, juice)
(chicken nuggets, french fries, soda)
(chicken nuggets, applesauce, milk)
(chicken nuggets, applesauce, juice)
(chicken nuggets, applesauce, soda)
(turkey, french fries, milk)
(turkey, french fries, juice)
(turkey, french fries, soda)
(turkey, applesauce, milk)
(turkey, applesauce, juice)
(turkey, applesauce, soda)

b. 12

c. $\frac{6}{18}$ *or* $\frac{1}{3}$

Section 15.3

1. (4)

2. (1)

3. (2)

4. (3)

5. (3)

6. (3)

7. $\frac{472}{756}$

8. a. $\frac{20}{930}$
   b. $\frac{90}{930}$
   c. $\frac{210}{930}$
   d. $\frac{105}{930}$

9. a. $\frac{6}{35}$
   b. $\frac{29}{35}$

10. a. $\frac{1}{110}$
    b. 0

11. a. $\frac{6}{72}$
    b. $\frac{36}{72}$

12. a. $\frac{5}{9}$
    b. $\frac{6}{9}$

13. a. $\frac{8}{12}$
    b. $\frac{22}{132}$
    c. $\frac{72}{132}$

14. a. 4 red, 12 white, 10 blue
    b. (1) $\frac{48}{676}$,(2) $\frac{260}{676}$

15. 3

16. a. 18 cherry, 6 lemon, 16 orange
    b. $\frac{576}{1560}$

17. a. $\frac{108}{240}$
    b. $\frac{132}{240}$

18. No, because $P(15)=\frac{6}{56}<P(2)=\frac{12}{56}$

19. $\frac{234}{1716}$

Glossary of Integrated Algebra Terms

A

Abscissa See *x-coordinate*.

Absolute value The absolute value of a number x, denoted by $|x|$, is its distance from 0 on the number line. To remove the absolute value sign, write x without its sign, as in $|-3| = -(-3) = 3$ and $|+3| = 3$.

Additive inverse The opposite of a real number. The sum of a number and its additive inverse is 0. The additive inverse of +3 is −3 as $(+3) + (-3) = 0$.

Algebraic expression Any combination of numbers and variables connected by one or more arithmetic operations.

Altitude A line segment that is perpendicular to the side to which it is drawn.

Angle of depression The angle through which the horizontal line of sight must be lowered in order to view an object below it.

Angle of elevation The angle through which the horizontal line of sight must be raised in order to view an object above it.

Area The number of square units a region encloses.

Associative property The sum or product of three numbers is the same regardless of which two numbers are chosen first to be added or multiplied together:

$$(a+b)+c=a+(b+c)$$
$$(ab)c=a(bc)$$

where a, b and c are real numbers.

Average See *Mean*.

Axis of symmetry The line of symmetry of a parabola that contains the turning point (vertex). For the parabola $y = ax^2 + bx + c$, an equation of the axis of symmetry is $x = -\frac{b}{2a}$.

B

Base of a power The number that is used as the factor when a product is expressed in exponential form. In 3^4, the *exponent*, 4, tells the number of times the *base*, 3, is to be used as a factor in the expanded product, as in $3^4 = 3 \times 3 \times 3 \times 3$.

Binary operation An operation that works on two elements of a set at a time. Addition, subtraction, multiplication, and division are examples of binary operations.

Binomial The sum or difference of two unlike terms, as in $2x + 3y$.

Box-and-whisker plot A diagram that shows how spread out data are about the median using five key values: the lowest score, the lower quartile, the median, the upper quartile, and the highest score.

C

Circumference The distance around a circle. The circumference C of a circle with radius r is given by the formula, $C = 2\pi r$.

Closure property A set is closed under an operation if the result of performing that operation on members of the set is also a member of the same set. For example, the set of integers is closed under addition but not division since the quotient of two integers is not necessarily an integer.

Coefficient The number that multiplies one or more variable factors in a monomial. The coefficient of $-5xy^2$ is -5.

Coefficient of linear correlation A number from -1 to $+1$, denoted by r, that indicates the strength and direction of the relationship between two sets of data. If both experimental variables move in the same direction, then r will be positive. The closer $|r|$ is to 1, the stronger the relationship between the variables that represents the two sets of data.

Collinear points Points that lie on the same line.

Commutative property The order in which two numbers are added or multiplied together does not matter:

$$a + b = b + a$$
$$a \times b = b \times a$$

where a and b are real numbers.

Complementary angles Two angles whose degree measures add up to 90.

Congruent figures Figures that have the same size and shape.

Constant A quantity that does not change in value such as π.

Coordinate plane A plane divided into four parts, called *quadrants*, by a horizontal number line and vertical number line intersecting at their zero points, called the *origin*.

Cosine ratio In a right triangle, the ratio of the length of the leg adjacent to a given acute angle to the length of the hypotenuse.

Counterexample A single, specific instance that contradicts a proposed generalization.

Counting Principle If event A can occur in m ways and event B can occur in n ways, then the number of ways in which both events can occur is m times n.

Cumulative frequency The sum of all frequencies from a given data point up to and including another data point.

Cumulative frequency histogram A histogram whose bar heights represent the cumulative frequency at stated intervals.

D

Degree of a monomial The sum of the exponents of its variable factors. The degree of $7xy^2$ is 3 since the exponent of x is 1, the exponent of y is 2, and $1 + 2 = 3$.

Degree of a polynomial For a polynomial in one variable, the greatest degree of its terms. The degree of $3x^2 + 5x + 11$ is 2.

Direct variation When two variables are related so that their ratio remains the same.

Distributive property A sum may be multiplied by a number by multiplying each addend separately by that number and then adding the products:

$$a(b + c) = ab + ac$$

where a, b, and c are real numbers.

Domain The set of all first members or x-values of the ordered pairs that comprise a relation.

E

Empirical probability An estimate of the probability that an event will happen using sample data or the results of performing repeated trials of a probability experiment.

Empty set A set that has no elements denoted by { } or ∅. Also called the *Null set*.

Equation A mathematical statement where two expressions have the same value. An equal sign is used to separate the two expressions, as in $3x - 7 = x + 1$, thereby indicating the left side has the same value as the right side.

Equivalent equations Two equations that have the same solution set, as with $2x = 6$ and $x + 2 = 5$.

Event A subset of the sample space of a probability experiment.

Exponent A number written to the right and a half line above another number, called the *base*, that tells the number of times the base is used as a factor in a product. Thus, $5^3 = 5 \times 5 \times 5 = 125$.

Exponential decay When a quantity decreases by a fixed percent of its current value over successive time periods. If A is the initial amount and r is the rate of decay per time period, then after n successive time periods the amount that is left, y, is given by $y = A(1 - r)^n$ where $0 < r < 1$.

Exponential function A function of the form $y = b^x$ where b is a positive number other than 1.

Exponential growth When a quantity increases by a fixed percent of its current value over successive time periods. If A is the initial amount and r is the rate of growth per time period, then after n successive time periods the amount that is present, y, is given by $y = A(1 + r)^n$ where $r > 0$.

F

Factor An exact divisor of a given expression. For example, 3 is a factor of 12 ; because $2x(3x + 2) = 6x^2 + 4x$, $2x$ and $3x + 2$ are factors of $6x^2 + 4x$.

Factorial *n* Denoted by $n!$ and, for any positive integer n, is the product of the consecutive whole numbers from n to 1. By definition, $0! = 1$. Thus, $4! = 4 \times 3 \times 2 \times 1 = 24$.

Factoring The process that reverses multiplication. Factoring a polynomial breaks down the polynomial into the product of two or more lower degree polynomials.

Factoring completely Factoring so that each of the factors of a given expression cannot be factored further.

Favorable outcomes The set of outcomes for which an event occurs.

FOIL An abbreviation of a rule for multiplying two binomials horizontally by forming the sum of the products of: the first terms (F), the outer terms (O), the inner term (I), and the last terms (L) of the binomial factors, as in

$$(x+5)(x-2) = \overbrace{x \cdot x}^{F} + \overbrace{-2x}^{O} + \overbrace{5x}^{I} + \overbrace{-2 \cdot 5}^{L}$$
$$= x^2 + 3x - 10$$

Frequency The number of times that a score appears in a set of data values.

Function A set of ordered pairs in which no two ordered pairs have the same first member and different second members. A function may be represented as a table, graph, or equation.

G

Greatest common factor (GCF) For a given polynomial, the greatest monomial that is an exact divisor of each of its terms. The GCF of $8a^2b^2 + 20ab^3$ is $4ab^2$.

H

Half-life The time needed for a substance to decay to one-half of its original amount.

Histogram A vertical bar graph whose bars are adjacent to each other. The height of each rectangular bar shows an amount or frequency of the quantity that is indicated at the base of the bar.

Hypotenuse In a right triangle, the side opposite the right angle.

I

Independent events When the outcome of one event does not affect the outcome of a second event.

Inequality A sentence that uses an inequality symbol to compare two expressions.

Inequality relation A comparison using an inequality symbol: $<$ means "is less than"; $\leq$ means "is less than or equal to"; $>$ means "is greater than"; $\geq$ means "is greater than or equal to."

Integer An element of the set

$\{\ldots, -4, -3, -2, -1, 0, 1, 2, 3, 4, \ldots\}$.

Irrational number A number that cannot be expressed as the quotient of two integers, such as $\sqrt{3}$ and π. Non-ending decimals with no repeating pattern represent irrational numbers.

Isosceles trapezoid A trapezoid whose nonparallel sides have the same length.

L

Line of best fit The line that can be drawn "closest" to a set of data points. The stronger the linear relationship between the sets of data that variables x and y represent, the more closely the line will fit the data. Because an equation of this line can be determined using a statistical procedure called *regression analysis*, it is sometimes referred to as a *regression line*.

Linear function A function whose graph is a nonvertical line.

Literal equation An equation in which one or more of the coefficients of the variables are letters, as in $ax + by = c$.

M

Mathematical model A quantitative tool such as a graph, table, or equation that shows the relationship between real-world variables.

Mean For a given set of n numbers, their sum divided by n.

Measure of central tendency A single number that represents a central or middle value about which a set of data values are clustered. The mean, median, and mode are measures of central tendency.

Median The middle score when a set of data values are arranged in size order. If the set of data values has an even number of scores, then the median is the average of the two middle scores.

Midpoint The point on a line segment that divides it into two segments that have the same length.

Mode The score in a set of data values that has the greatest frequency.

Monomial A single term that consists of a number, a variable, or the product of a number and one or more variables.

Multiplication property of 0 If the product of two expressions is 0, then at least one of these is equal to 0.

Multiplicative identity The element of a set that, when multiplied by any member of the same set, results in that same member. The multiplicative identity for the set of real numbers is 1.

Multiplicative inverse For any given member of a set, the element that must multiply it to produce the multiplicative identity for that set. For the set of real numbers, the multiplicative inverse of each nonzero number is its reciprocal because the product is 1.

Mutually exclusive events Events from the same sample space that have no outcomes in common.

N

Negative number A number less than 0.

Null set See *Empty set.*

O

Open sentence A sentence whose truth value cannot be determined until each variable is replaced with a specific number.

Ordered pair Two numbers that are written in a definite order.

Ordinate See *y-coordinate.*

Origin The zero point on a number line.

P

Parabola The graph of a quadratic equation in which either x or y, but not both, are squared. The graph of $y = ax^2 + bx + c$ $(a \neq 0)$ is a parabola that has a vertical line of symmetry, an equation of which is $x = -\frac{b}{2a}$.

Parallel lines Lines that lie in the same plane that do not intersect.

Parallelogram A quadrilateral in which both pairs of opposite sides are parallel. The diagonals of a parallelogram bisect each other.

Perfect square A rational expression whose square root is also rational, such as 4 $\left(\sqrt{4} = 2\right)$ and $\frac{9}{25}\left(\sqrt{\frac{9}{25}} = \frac{3}{5}\right)$.

Perimeter The distance around a figure.

Permutation An ordered arrangement of objects.

Permutation notation The notation ${}_nP_n$ represents the arrangement of n objects in n positions so ${}_nP_n = n!$. The arrangement of n objects when fewer than n positions are available is denoted by ${}_nP_r$ where $r \leq n$. To evaluate ${}_nP_r$, multiply together the r greatest factors of n, as in ${}_5P_3 = 5 \times 4 \times 3 = 60$.

Perpendicular lines Two lines that intersect at right angles.

Point-slope form An equation of a line of the form $y - b = m(x - a)$, where m is the slope of the line and (a,b) is a point on the line.

Polynomial A monomial or the sum of two or more monomials.

Positive number A number greater than 0.

Power A number written with an exponent, as in 2^4 which is read "2 raised to the fourth power."

Prime number A positive integer greater than 1 whose only positive factors are itself and 1. The number 2 is the only even integer that is a prime number.

Principal square root The positive square root of a positive number. Although 49 has two square roots, +7 and −7, its principal square root is +7. The radical sign, $\sqrt{\ }$, indicates the principal square root of the number underneath it. Thus, $\sqrt{49} = 7$.

Probability of an event A measure of the likelihood that an event will occur expressed as a number from 0 to 1. If all the outcomes of an event are equally likely to occur, then the probability that the event will occur is the number of ways in which it can occur divided by the total number of equally likely outcomes.

Proportion An equation that states that two ratios are equal.

***p*th percentile** The score at or below which p% of the scores in a set of data values lie.

Pythagorean Theorem The relationship that states that in a right triangle, the square of the length of the hypotenuse is equal to the sum of the squares of the lengths of the two legs.

Q

Quadrant One of four rectangular regions into which the coordinate plane is divided.

Quadratic equation An equation in which a variable has an exponent of 2 and no variable has an exponent greater than 2.

Quadratic polynomial A polynomial whose degree is 2.

Quadrilateral A polygon with four sides.

R

Radical sign The symbol $\sqrt{\ }$ that denotes the principal root of the number underneath it, as in $\sqrt{49} = 7$.

Radicand The number that appears underneath a radical sign. In $\sqrt{49}$, the radicand is 49.

Radius of a circle Any segment whose endpoints are the center and a point on the circle. The radius of a circle also refers to the distance from the center to any point on the circle.

Range The set of all second members or y-values of the ordered pairs that belong to a relation.

Range of a set of scores In statistics, the difference between the highest and the lowest scores.

Rate A comparison of two quantities measured in different units by division.

Ratio A comparison of two quantities measured in the same units by division. The ratio of a to b ($a : b$) is the fraction $\frac{a}{b}$, provided $b \neq 0$.

Rational number A number that can be written as a fraction having an integer in the numerator and a nonzero integer in the denominator. The set of rational numbers includes decimals in which a set of digits endlessly repeat, as in $0.25000... \left(=\frac{1}{4}\right)$ and $0.333... \left(=\frac{1}{3}\right)$.

Real number A number that is either rational or irrational. The set of all points on the number line corresponds in one-to-one fashion to the set of real numbers.

Reciprocal For a nonzero number, the number by which it must be multiplied to produce 1. The reciprocal of $\frac{a}{b}$ is $\frac{b}{a}$ provided $a,b \neq 0$.

Rectangle A parallelogram with four right angles. The diagonals of a rectangle have the same length.

Regression line See *Line of best fit*.

Regular polygon A polygon in which all of the sides have the same length and all of the angles have the same degree measure.

Relation A set of ordered pairs.

Replacement set The set whose elements may be substituted for a variable. Also referred to as the *domain* of the variable.

Rhombus A parallelogram whose four sides have the same length. The diagonals of a rhombus intersect at right angles.

Right angle An angle that measures 90°.

Right triangle A triangle that contains a right angle.

Root A number from the replacement set that when substituted for a variable makes an equation a true statement. An equation may have more than one root or may have no roots.

S

Sample space The set of all possible outcomes of a probability experiment.

Scientific notation A number expressed as the product of a number from 1 to 10 and a power of 10. In scientific notation, 81,000 is written as 8.1×10^4 and 0.0072 is written as 7.2×10^{-3}.

Set A collection of objects whose members are described within braces, { }. The set of integers greater than 2 and less than 6 is {3, 4, 5}.

Similar triangles Two triangles are similar when two angles of one triangle have the same degree measures as the corresponding angles of the other triangle. The lengths of corresponding sides of similar triangles are in proportion.

Sine ratio In a right triangle, the ratio of the length of the leg that is opposite a given acute angle to the length of the hypotenuse.

Slope A measure of steepness. A line that rises as x increases has a positive slope. A line that falls as x increases has a negative slope. The slope of a horizontal line is 0, and the slope of a vertical line is undefined.

Slope formula The formula

$$m = \frac{y_B - y_A}{x_B - x_A}$$

where m is the slope of a nonvertical line that contains the points $A(x_A, y_A)$ and $B(x_B, y_B)$.

Slope-intercept form An equation of a line of the form $y = mx + b$, where m is the slope of the line and b is the y–intercept.

Solution set The set consisting of those members of the replacement set that, when substituted for the variable in an equation or inequality, results in a true statement.

Square A parallelogram with four right angles and whose sides have the same length. The diagonals of a square have the same length and intersect at right angles.

Square root One of two identical factors of a nonnegative number. Every positive number has two square roots. The square root of 9 is +3 or −3.

Statement A sentence that can be judged as true or false, but not both.

Successes See *Favorable outcomes.*

Supplementary angles Two angles whose degree measures add up to 180.

Surface area The sum of the areas of each of the surfaces of a solid figure.

System of equations A set of equations with the same variables whose solution is the set of values that make each of the equations true at the same time.

T

Tangent ratio In a right triangle, the ratio of the length of the leg that is opposite a given acute angle to the length of the leg that is adjacent to the same angle.

Terms The parts of an algebraic expression that are separated by addition or subtraction signs.

Theorem A generalization that can be proved.

Theoretical probability A probability ratio that can be calculated based on a logical analysis of a probability experiment without the need to actually perform it.

Trapezoid A quadrilateral with exactly one pair of parallel sides.

Tree diagram A diagram whose branches describe the different possible outcomes in a probability experiment.

Trinomial A polynomial with three unlike terms.

Turning point of a parabola See *Vertex of a parabola.*

V

Variable A symbol, usually a single lowercase letter, that represents an unspecified member of a given set called the *replacement set* or *domain* of the variable.

Venn diagram A diagram in which circles are used to represent the logical relationships between two or more sets that may have members in common.

Vertex of a parabola The point at which the axis of symmetry intersects a parabola. At this point, the parabola reaches either its maximum or minimum y-value. Also referred to as the "turning point."

Vertical angles The opposite pairs of equal angles that are formed when two lines intersect.

Vertical line test A graph represents a function if it is *not* possible to draw a vertical line that intersects the graph in more than one point.

Volume A measure of capacity that gives the number of unit cubes a solid figure can hold.

X

***x*-axis** The horizontal axis in the coordinate plane.

***x*-coordinate** The first member (abscissa) in an ordered pair of numbers that indicates the location of a point in the coordinate plane. The x-coordinate of (3,2) is 3.

***x*-intercept** The x–coordinate of the point at which a graph intersects the x–axis.

Y

***y*-axis** The vertical axis in the coordinate plane.

***y*-coordinate** The second member (ordinate) in an ordered pair of numbers that indicates the location of a point in the coordinate plane. The y-coordinate of (3,2) is 2.

***y*-intercept** The y-coordinate of the point at which a graph intersects the y-axis.

Z

Zero product rule If the product of two expressions is 0, then at least one of these is equal to 0.

THE INTEGRATED ALGEBRA REGENTS EXAMINATION

The Regents Examination in Integrated Algebra is a three-hour exam that is divided into four parts with a total of 39 questions. All 39 questions must be answered. Part I consists entirely of regular multiple-choice questions. Parts II, III, and IV each contain a set of questions that must be answered directly in the question booklet. You are required to show how you arrived at the answers for the questions in Parts II, III, and IV. The accompanying table shows how the exam breaks down.

Question Type	Number of Questions	Credit Value
Part I: Multiple choice	30	$30 \times 2 = 60$
Part II: 2-credit open ended	3	$3 \times 2 = 6$
Part III: 3-credit open ended	3	$3 \times 3 = 9$
Part IV: 4-credit open ended	3	$3 \times 4 = 12$
	Total = 39 questions	Total = 87 points

How Are Credits Distributed by Topic Area?

The table below shows the percentage of total credits that will be aligned to each major topic area.

Major Topic Area	% of Total Credits
I. Number Sense and Operations	6–10%
II. Algebra	50–55%
III. Geometry and Coordinates	14–19%
IV. Measurement	3–8%
V. Probability and Statistics	14–19%

How Is the Exam Scored?

- Each of your answers to the 30 multiple-choice questions in Part I will be scored as either right or wrong.

- Solutions to questions in Parts II, III, and IV that are not completely correct may receive partial credit according to a special rating guide that is provided by the New York State Education Department. In order to receive full credit for a correct answer to a question in Parts II, III, or IV, you must show or explain how you arrived at your answer by indicating the key steps taken, including appropriate formula substitutions, diagrams, graphs, and charts. A correct numerical answer with no work shown will receive only 1 credit.
- The raw scores for the four parts of the test are added together. The maximum total raw score for the Integrated Algebra Regents Examination is 87 points. Using a special conversion chart that is provided by the New York State Education Department, your total raw score will be equated to a final test score that falls within the usual 0 to 100 scale.

What Type of Calculator Is Required?

Graphing calculators are *required* for the Integrated Algebra Regents Examination. During the administration of the Regents exam, schools are required to make a graphing calculator available for the exclusive use of each student. You will need to use your calculator to work with trigonometric functions of angles, find roots of numbers, and perform routine calculations.

Knowing how to use a graphing calculator gives you more options when deciding how to solve a problem. Rather than solving a problem algebraically with pen and paper, it may be easier to solve the same problem using a graph or table created by a graphing calculator. A graphical or numerical solution using a calculator can also be used to help confirm an answer obtained by solving the problem algebraically.

Are Any Formulas Provided?

The Integrated Algebra Regents Examination test booklet will include a reference sheet containing the formulas in the accompanying table. Keep in mind that you may be required to know other formulas that are not included in this sheet.

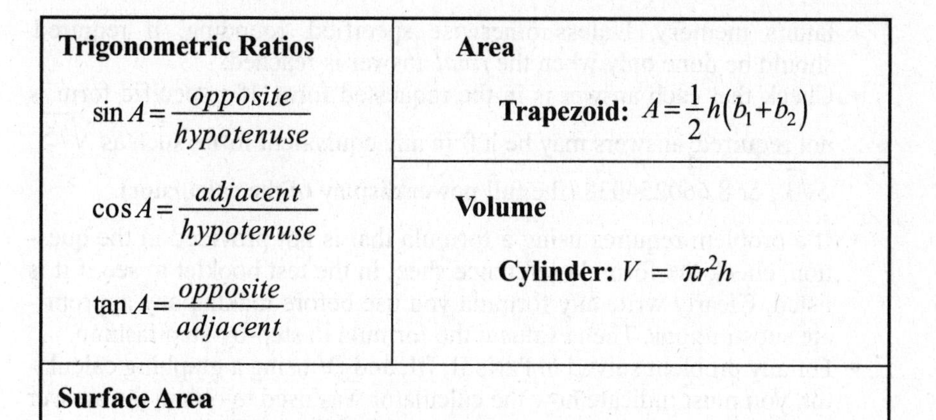

Trigonometric Ratios	Area
$\sin A = \dfrac{opposite}{hypotenuse}$	**Trapezoid:** $A = \frac{1}{2}h(b_1 + b_2)$
$\cos A = \dfrac{adjacent}{hypotenuse}$	**Volume**
$\tan A = \dfrac{opposite}{adjacent}$	**Cylinder:** $V = \pi r^2 h$

Surface Area
Rectangular prism (closed box): $SA = 2\ell w + 2hw + 2\ell h$
Cylinder: $SA = 2\pi r^2 + 2\pi rh$

Coordinate Geometry
Slope: $m = \dfrac{\Delta y}{\Delta x} = \dfrac{y_2 - y_1}{x_2 - x_1}$

What Else Should I Know?

- Do not omit any questions from Part I. Since there is no penalty for guessing, make certain that you record an answer for each of the 30 multiple-choice questions.
- If the method of solution is not stated in the problem, choose an appropriate method (numerical, graphical, or algebraic) with which you are most comfortable.
- If you solve a problem in Parts II, III, or IV using a trial-and-error approach, show the work for at least *three* guesses with appropriate checks. Should the correct answer be reached on the first trial, you must further illustrate your method by showing that guesses below and above the correct guess do not work.
- Avoid rounding errors when using a calculator. Unless otherwise directed, the (pi) key on a calculator should be used in computations involving the constant π rather than the common rational approximation of 3.14 or $\frac{22}{7}$. When performing a sequence of calculations in which the result of one calculation is used in a second calculation, do not round off. Instead, use the full power/display of the calculator by performing a "chain" calculation, saving intermediate results in the calcu-

lator's memory. Unless otherwise specified, rounding, if required, should be done only when the *final* answer is reached.

- Check that each answer is in the requested form. If a specific form is not required, answers may be left in any equivalent form, such as $\sqrt{75}$, $5\sqrt{3}$, or 8.660254038 (the full power/display of the calculator).
- If a problem requires using a formula that is not provided in the question, check the formula reference sheet in the test booklet to see if it is listed. Clearly write any formula you use before making any appropriate substitutions. Then evaluate the formula in step-by-step fashion.
- For any problem solved in Parts II, III, and IV using a graphing calculator, you must indicate how the calculator was used to obtain the answer such as by copying graphs or tables created by your calculator together with the equations used to produce them. When copying graphs, label each graph with its equation, state the dimensions of the viewing window, and identify the intercepts and any points of intersection with their coordinates. Whenever appropriate, indicate the rationale of your approach.

Examination June 2012
Integrated Algebra

FORMULAS

Trigonometric ratios

$$\sin A = \frac{\text{opposite}}{\text{hypotenuse}}$$

$$\cos A = \frac{\text{adjacent}}{\text{hypotenuse}}$$

$$\tan A = \frac{\text{opposite}}{\text{adjacent}}$$

Area Trapezoid $A = \frac{1}{2}h(b_1 + b_2)$

Volume Cylinder $V = \pi r^2 h$

Surface area Rectangular prism $SA = 2lw + 2hw + 2lh$

Cylinder $SA = 2\pi r^2 + 2\pi rh$

Coordinate geometry $m = \frac{\Delta y}{\Delta x} = \frac{y_2 - y_1}{x_2 - x_1}$

PART I

Answer all 30 questions in this part. Each correct answer will receive 2 credits. No partial credit will be allowed. For each question, write in the space provided the numeral preceding the word or expression that best completes the statement or answers the question. [60 credits]

1 In a baseball game, the ball traveled 350.7 feet in 4.2 seconds. What was the average speed of the ball, in feet per second?

(1) 83.5
(2) 177.5
(3) 354.9
(4) 1,472.9

1 _____

2 A survey is being conducted to determine if a cable company should add another sports channel to their schedule. Which random survey would be the *least* biased?

(1) surveying 30 men at a gym
(2) surveying 45 people at a mall
(3) surveying 50 fans at a football game
(4) surveying 20 members of a high school soccer team

2 _____

3 The quotient of $\frac{8x^5 - 2x^4 + 4x^3 - 6x^2}{2x^2}$ is

(1) $16x^7 - 4x^6 + 8x^5 - 12x^4$
(2) $4x^7 - x^6 + 2x^5 - 3x^4$
(3) $4x^3 - x^2 + 2x - 3x$
(4) $4x^3 - x^2 + 2x - 3$

3 _____

4 Marcy determined that her father's age is four less than three times her age. If x represents Marcy's age, which expression represents her father's age?

(1) $3x - 4$
(2) $3(x - 4)$
(3) $4x - 3$
(4) $4 - 3x$

4 ____

5 A set of data is graphed on the scatter plot below.

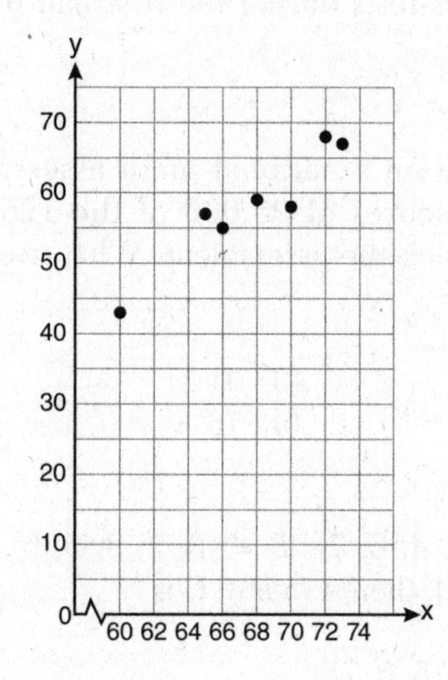

This scatter plot shows

(1) no correlation
(2) positive correlation
(3) negative correlation
(4) undefined correlation

5 ____

6 Which situation is an example of bivariate data?

(1) the number of pizzas Tanya eats during her years in high school
(2) the number of times Ezra puts air in his bicycle tires during the summer
(3) the number of home runs Elias hits per game and the number of hours he practices baseball
(4) the number of hours Nellie studies for her mathematics tests during the first half of the school year

6 _____

7 Brianna's score on a national math assessment exceeded the scores of 95,000 of the 125,000 students who took the assessment. What was her percentile rank?

(1) 6
(2) 24
(3) 31
(4) 76

7 _____

8 If $A = \{0, 1, 3, 4, 6, 7\}$, $B = \{0, 2, 3, 5, 6\}$, and $C = \{0, 1, 4, 6, 7\}$, then $A \cap B \cap C$ is

(1) $\{0, 1, 2, 3, 4, 5, 6, 7\}$
(2) $\{0, 3, 6\}$
(3) $\{0, 6\}$
(4) $\{0\}$

8 _____

9 Which graph represents a function?

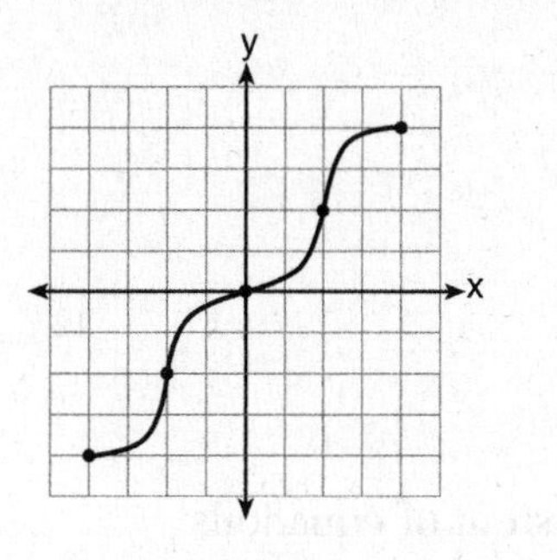

(1)

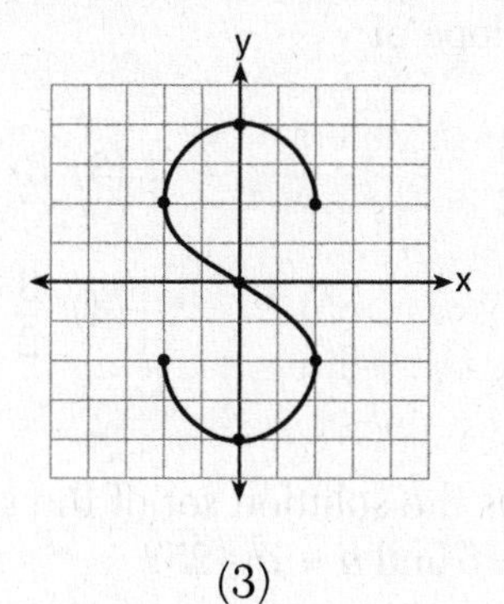

(3)

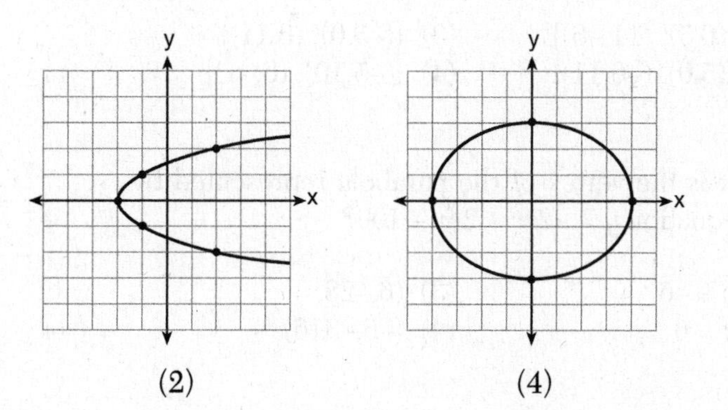

(2) (4)

9 ____

10 What is the product of $(3x + 2)$ and $(x - 7)$?

(1) $3x^2 - 14$
(2) $3x^2 - 5x - 14$
(3) $3x^2 - 19x - 14$
(4) $3x^2 - 23x - 14$

10 ____

11 If five times a number is less than 55, what is the greatest possible integer value of the number?

(1) 12
(2) 11
(3) 10
(4) 9

11 ____

12 The line represented by the equation $2y - 3x = 4$ has a slope of

(1) $-\frac{3}{2}$ (3) 3

(2) 2 (4) $\frac{3}{2}$ 12 _____

13 What is the solution set of the system of equations $x + y = 5$ and $y = x^2 - 25$?

(1) $\{(0,5), (11,-6)\}$ (3) $\{(-5,0), (6,11)\}$

(2) $\{(5,0), (-6,11)\}$ (4) $\{(-5,10), (6,-1)\}$ 13 _____

14 What is the vertex of the parabola represented by the equation $y = -2x^2 + 24x - 100$?

(1) $x = -6$ (3) $(6,-28)$

(2) $x = 6$ (4) $(-6,-316)$ 14 _____

15 If $k = am + 3mx$, the value of m in terms of a, k, and x can be expressed as

(1) $\frac{k}{a+3x}$ (3) $\frac{k-am}{3x}$

(2) $\frac{k-3mx}{a}$ (4) $\frac{k-a}{3x}$ 15 _____

16 Which expression represents $\frac{x^2-3x-10}{x^2-25}$ in simplest form?

(1) $\frac{2}{5}$ (3) $\frac{x-2}{x-5}$

(2) $\frac{x+2}{x+5}$ (4) $\frac{-3x-10}{-25}$ 16 _____

17 Which interval notation describes the set $S = \{x \mid 1 \le x < 10\}$?

(1) [1,10]
(2) (1,10]
(3) [1,10)
(4) (1,10)

17_____

18 The bull's-eye of a dartboard has a radius of 2 inches and the entire board has a radius of 9 inches, as shown in the diagram below.

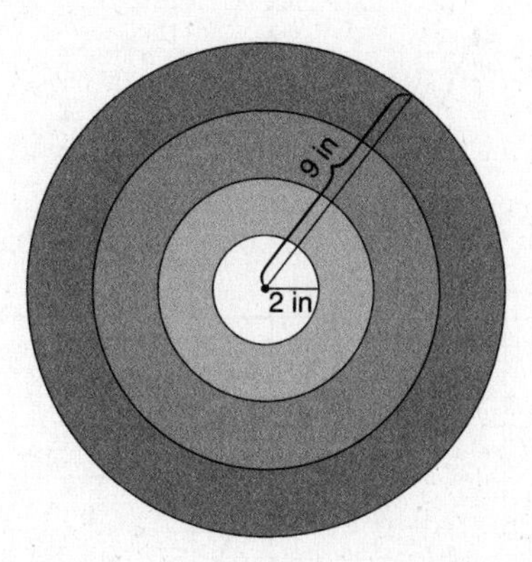

If a dart is thrown and hits the board, what is the probability that the dart will land in the bull's-eye?

(1) $\frac{2}{9}$
(2) $\frac{7}{9}$
(3) $\frac{4}{81}$
(4) $\frac{49}{81}$

18_____

19 What is one-third of 3^6?

(1) 1^2
(2) 3^2
(3) 3^5
(4) 9^6

19_____

20 The expression $\frac{2x+13}{2x+6}-\frac{3x-6}{2x+6}$ is equivalent to

(1) $\frac{-x+19}{2(x+3)}$

(2) $\frac{-x+7}{2(x+3)}$

(3) $\frac{5x+19}{2(x+3)}$

(4) $\frac{5x+7}{4x+12}$

20 _____

21 Which equation is represented by the graph below?

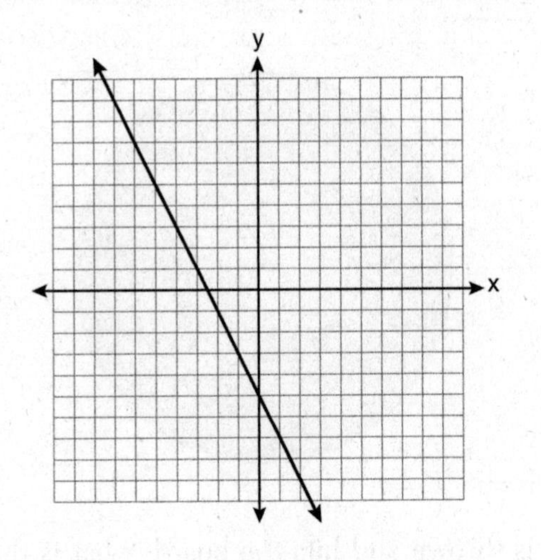

(1) $2y + x = 10$

(2) $y - 2x = -5$

(3) $-2y = 10x - 4$

(4) $2y = -4x - 10$

21 _____

22 Which coordinates represent a point in the solution set of the system of inequalities shown below?

$$y \le \frac{1}{2}x + 13$$

$$4x + 2y > 3$$

(1) (–4,1) (3) (1,–4)
(2) (–2,2) (4) (2,–2)

22 _____

23 The length of one side of a square is 13 feet. What is the length, to the *nearest foot*, of a diagonal of the square?

(1) 13 (3) 19
(2) 18 (4) 26

23 _____

24 In $\triangle ABC$, $m\angle C = 90$. If $AB = 5$ and $AC = 4$, which statement is *not* true?

(1) $\cos A = \frac{4}{5}$ (3) $\sin B = \frac{4}{5}$

(2) $\tan A = \frac{3}{4}$ (4) $\tan B = \frac{5}{3}$

24 _____

25 If n is an odd integer, which equation can be used to find three consecutive odd integers whose sum is –3?

(1) $n + (n + 1) + (n + 3) = -3$
(2) $n + (n + 1) + (n + 2) = -3$
(3) $n + (n + 2) + (n + 4) = -3$
(4) $n + (n + 2) + (n + 3) = -3$

25 _____

26 When $8x^2 + 3x + 2$ is subtracted from $9x^2 - 3x - 4$, the result is

(1) $x^2 - 2$
(2) $17x^2 - 2$
(3) $-x^2 + 6x + 6$
(4) $x^2 - 6x - 6$

26 _____

27 Factored completely, the expression $3x^3 - 33x^2 + 90x$ is equivalent to

(1) $3x(x^2 - 33x + 90)$
(2) $3x(x^2 - 11x + 30)$
(3) $3x(x + 5)(x + 6)$
(4) $3x(x - 5)(x - 6)$

27 _____

28 Elizabeth is baking chocolate chip cookies. A single batch uses $\frac{3}{4}$ teaspoon of vanilla. If Elizabeth is mixing the ingredients for five batches at the same time, how many tablespoons of vanilla will she use?

3 teaspoons = 1 tablespoon

(1) $1\frac{1}{4}$
(2) $1\frac{3}{4}$
(3) $3\frac{3}{4}$
(4) $5\frac{3}{4}$

28 _____

29 A car depreciates (loses value) at a rate of 4.5% annually. Greg purchased a car for \$12,500. Which equation can be used to determine the value of the car, V, after 5 years?

(1) $V = 12{,}500(0.55)^5$
(2) $V = 12{,}500(0.955)^5$
(3) $V = 12{,}500(1.045)^5$
(4) $V = 12{,}500(1.45)^5$

29 ____

30 The cumulative frequency table below shows the length of time that 30 students spent text messaging on a weekend.

Minutes Used	Cumulative Frequency
31–40	2
31–50	5
31–60	10
31–70	19
31–80	30

Which 10-minute interval contains the first quartile?

(1) 31–40
(2) 41–50
(3) 51–60
(4) 61–70

30 ____

PART II

Answer all 3 questions in this part. Each correct answer will receive 2 credits. Clearly indicate the necessary steps, including appropriate formula substitutions, diagrams, graphs, charts, etc. For all questions in this part, a correct numerical answer with no work shown will receive only 1 credit. [6 credits]

31 Solve the following system of equations algebraically for y:

$$2x + 2y = 9$$

$$2x - y = 3$$

32 Three storage bins contain colored blocks. Bin 1 contains 15 red and 14 blue blocks. Bin 2 contains 16 white and 15 blue blocks. Bin 3 contains 15 red and 15 white blocks. All of the blocks from the three bins are placed into one box.

If one block is randomly selected from the box, which color block would most likely be picked? Justify your answer.

33 Students calculated the area of a playing field to be 8,100 square feet. The actual area of the field is 7,678.5 square feet. Find the relative error in the area, to the *nearest thousandth*.

PART III

Answer all 3 questions in this part. Each correct answer will receive 3 credits. Clearly indicate the necessary steps, including appropriate formula substitutions, diagrams, graphs, charts, etc. For all questions in this part, a correct numerical answer with no work shown will receive only 1 credit. [9 credits]

34 On the set of axes below, graph the equation $y = x^2 + 2x - 8$.

Using the graph, determine and state the roots of the equation $x^2 + 2x - 8 = 0$.

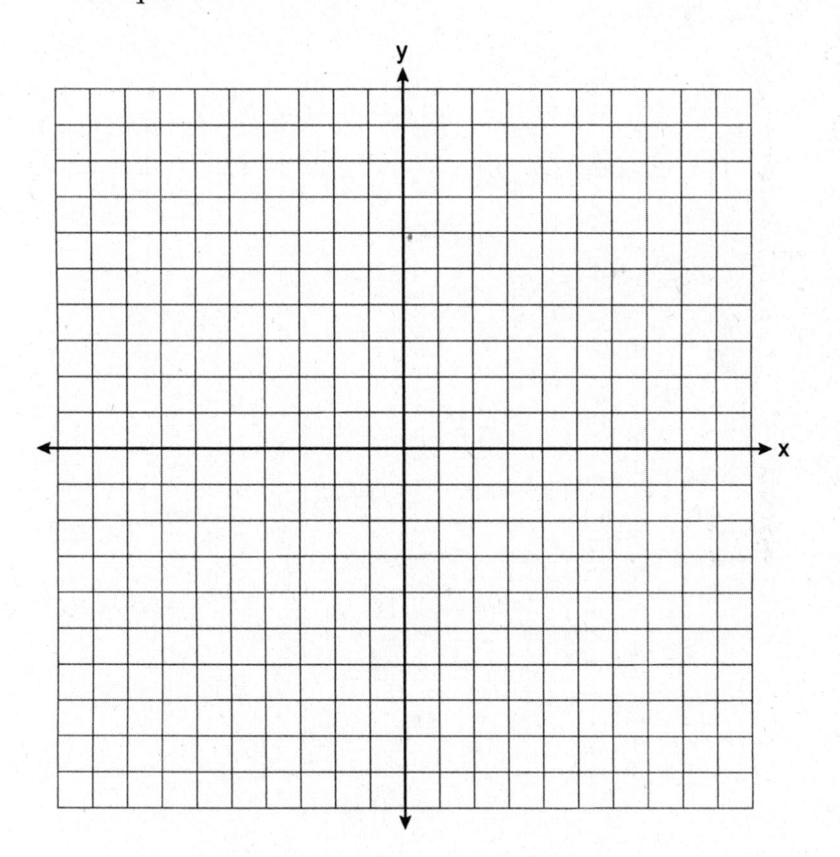

35 A 28-foot ladder is leaning against a house. The bottom of the ladder is 6 feet from the base of the house. Find the measure of the angle formed by the ladder and the ground, to the *nearest degree*.

36 Express $\frac{3\sqrt{75}+\sqrt{27}}{3}$ in simplest radical form.

PART IV

Answer all 3 questions in this part. Each correct answer will receive 4 credits. Clearly indicate the necessary steps, including appropriate formula substitutions, diagrams, graphs, charts, etc. For all questions in this part, a correct numerical answer with no work shown will receive only 1 credit. [12 credits]

37 Mike buys his ice cream packed in a rectangular prism-shaped carton, while Carol buys hers in a cylindrical-shaped carton. The dimensions of the prism are 5 inches by 3.5 inches by 7 inches. The cylinder has a diameter of 5 inches and a height of 7 inches.

Which container holds more ice cream? Justify your answer.

Determine, to the *nearest tenth of a cubic inch*, how much *more* ice cream the larger container holds.

38 Solve algebraically for x: $3(x + 1) - 5x = 12 - (6x - 7)$

39 A large company must choose between two types of passwords to log on to a computer. The first type is a four-letter password using any of the 26 letters of the alphabet, without repetition of letters. The second type is a six-digit password using the digits 0 through 9, with repetition of digits allowed.

Determine the number of possible four-letter passwords.

Determine the number of possible six-digit passwords.

The company has 500,000 employees and needs a different password for each employee. State which type of password the company should choose. Explain your answer.

Answers June 2012

Integrated Algebra

Answer Key

PART I

1. (1)	**7.** (4)	**13.** (2)	**19.** (3)	**25.** (3)
2. (2)	**8.** (3)	**14.** (3)	**20.** (1)	**26.** (4)
3. (4)	**9.** (1)	**15.** (1)	**21.** (4)	**27.** (4)
4. (1)	**10.** (3)	**16.** (2)	**22.** (4)	**28.** (1)
5. (2)	**11.** (3)	**17.** (3)	**23.** (2)	**29.** (2)
6. (3)	**12.** (4)	**18.** (3)	**24.** (4)	**30.** (3)

PART II

31. 2

32. White

33. 0.055

PART III

34. The equation is graphed correctly and –4 and 2 are stated.

35. 78

36. $6\sqrt{3}$

PART IV

37. Cylinder or Carol's, and 14.9 in^3

38. 4

39. 358,800 and 1,000,000, and six-digit or numeric password

In **PARTS II–IV** you are required to show how you arrived at your answers. For sample methods of solutions, see Barron's *Regents Exams and Answers* book for Integrated Algebra.

Examination
June 2013
Integrated Algebra

FORMULAS

Trigonometric ratios

$$\sin A = \frac{\text{opposite}}{\text{hypotenuse}}$$

$$\cos A = \frac{\text{adjacent}}{\text{hypotenuse}}$$

$$\tan A = \frac{\text{opposite}}{\text{adjacent}}$$

Area Trapezoid $A = \frac{1}{2}h(b_1 + b_2)$

Volume Cylinder $V = \pi r^2 h$

Surface area Rectangular prism $SA = 2lw + 2hw + 2lh$

Cylinder $SA = 2\pi r^2 + 2\pi rh$

Coordinate geometry $m = \frac{\Delta y}{\Delta x} = \frac{y_2 - y_1}{x_2 - x_1}$

PART I

Answer all 30 questions in this part. Each correct answer will receive 2 credits. No partial credit will be allowed. For each question, write in the space provided the numeral preceding the word or expression that best completes the statement or answers the question. [60 credits]

1 Which expression represents "5 less than twice x"?

(1) $2x - 5$
(2) $5 - 2x$
(3) $2(5 - x)$
(4) $2(x - 5)$

1 ____

2 Gabriella has 20 quarters, 15 dimes, 7 nickels, and 8 pennies in a jar. After taking 6 quarters out of the jar, what will be the probability of Gabriella randomly selecting a quarter from the coins left in the jar?

(1) $\frac{14}{44}$ (3) $\frac{14}{50}$

(2) $\frac{30}{44}$ (4) $\frac{20}{50}$

2 ____

3 Based on the line of best fit drawn below, which value could be expected for the data in June 2015?

(1) 230
(2) 310
(3) 480
(4) 540

3 _____

4 If the point $(5,k)$ lies on the line represented by the equation $2x + y = 9$, the value of k is

(1) 1
(2) 2
(3) −1
(4) −2

4 _____

5 A soda container holds $5\frac{1}{2}$ gallons of soda. How many ounces of soda does this container hold?

1 quart = 32 ounces
1 gallon = 4 quarts

(1) 44
(2) 176
(3) 640
(4) 704

5 _____

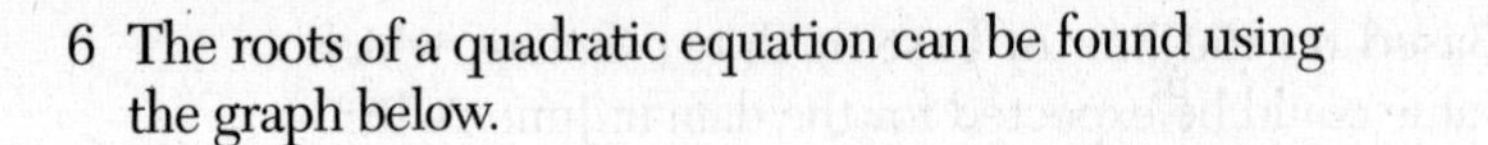

6 The roots of a quadratic equation can be found using the graph below.

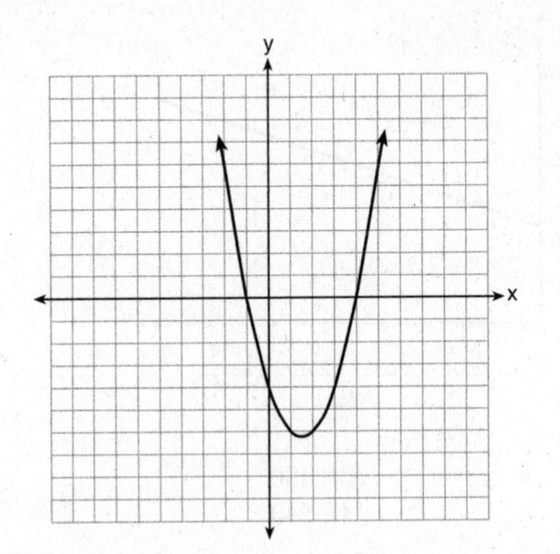

What are the roots of this equation?

(1) –4, only
(2) –4 and –1
(3) –1 and 4
(4) –4, –1, and 4

6 _____

7 If the area of a rectangle is represented by $x^2 + 8x + 15$ and its length is represented by $x + 5$, which expression represents the width of the rectangle?

(1) $x + 3$
(2) $x - 3$
(3) $x^2 + 6x + 5$
(4) $x^2 + 7x + 10$

7 _____

8 Which set of data describes a situation that would be classified as qualitative?

(1) the colors of the birds at the city zoo
(2) the shoe size of the zookeepers at the city zoo
(3) the heights of the giraffes at the city zoo
(4) the weights of the monkeys at the city zoo

8 ____

9 The value of the expression $6! + \dfrac{5!(3!)}{4!} - 10$ is

(1) 50
(2) 102
(3) 740
(4) 750

9 ____

10 Which interval notation represents $-3 \le x \le 3$?

(1) $[-3, 3]$
(2) $(-3, 3]$
(3) $[-3, 3)$
(4) $(-3, 3)$

10 ____

11 The solutions of $x^2 = 16x - 28$ are

(1) −2 and −14
(2) 2 and 14
(3) −4 and −7
(4) 4 and 7

11 ____

12 If the expression $(2y^a)^4$ is equivalent to $16y^8$, what is the value of a?

(1) 12
(2) 2
(3) 32
(4) 4

12 ____

13 Which table shows bivariate data?

Age (yr)	Frequency
14	12
15	21
16	14
17	19
18	15

(1)

Time Spent Studying (hr)	Test Grade (%)
1	65
2	72
3	83
4	85
5	92

(3)

Type of Car	Average Gas Mileage (mpg)
van	25
SUV	23
luxury	26
compact	28
pickup	22

(2)

Day	Temperature (degrees F)
Monday	63
Tuesday	58
Wednesday	72
Thursday	74
Friday	78

(4)

13 ____

14 The box-and-whisker plot below represents the results of test scores in a math class.

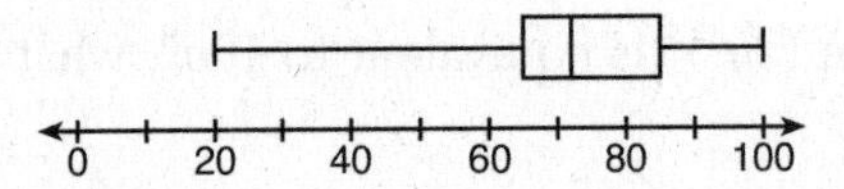

What do the scores 65, 85, and 100 represent?

(1) Q_1, median, Q_3
(2) Q_1, Q_3, maximum
(3) median, Q_1, maximum
(4) minimum, median, maximum

14 ____

15 The expression $\frac{x-3}{x+2}$ is undefined when the value of x is

(1) –2, only
(2) –2 and 3
(3) 3, only
(4) –3 and 2

15 _____

16 If $rx - st = r$, which expression represents x?

(1) $\frac{r+st}{r}$
(2) $\frac{r}{r+st}$
(3) $\frac{r}{r-st}$
(4) $\frac{r-st}{r}$

16 _____

17 What is the solution of the equation $\frac{x+2}{2} = \frac{4}{x}$?

(1) 1 and –8
(2) 2 and –4
(3) –1 and 8
(4) –2 and 4

17 _____

18 Which type of function is graphed below?

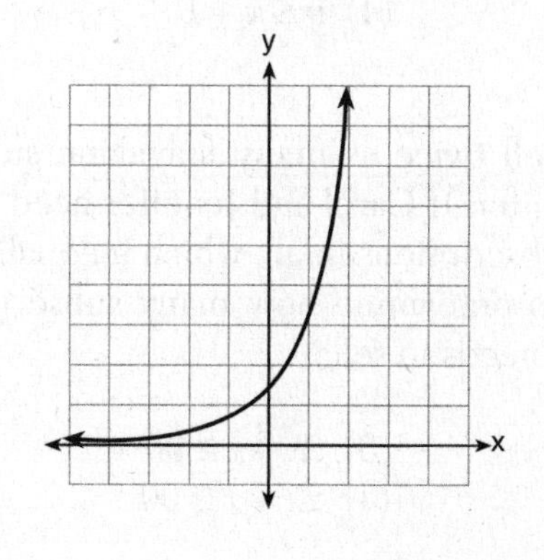

(1) linear
(2) quadratic
(3) exponential
(4) absolute value

18 _____

19 What is the slope of the line represented by the equation $4x + 3y = 12$?

(1) $\frac{4}{3}$ (3) $-\frac{3}{4}$

(2) $\frac{3}{4}$ (4) $-\frac{4}{3}$ 19_____

20 The diagram below shows the graph of which inequality?

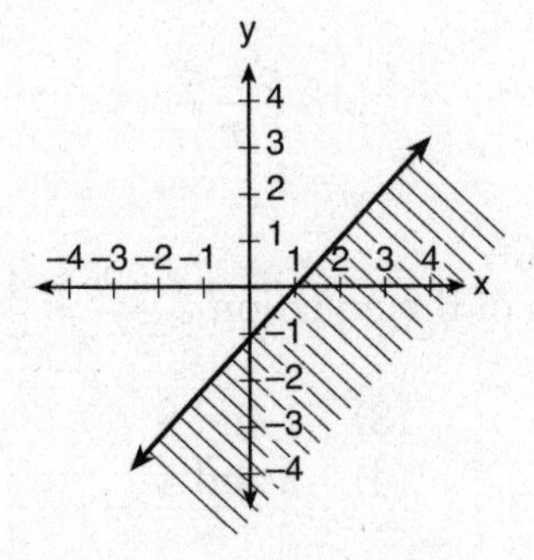

(1) $y > x - 1$ (3) $y < x - 1$
(2) $y \geq x - 1$ (4) $y \leq x - 1$ 20_____

21 Carol plans to sell twice as many magazine subscriptions as Jennifer. If Carol and Jennifer need to sell at least 90 subscriptions in all, which inequality could be used to determine how many subscriptions, x, Jennifer needs to sell?

(1) $x \geq 45$ (3) $2x - x \geq 90$
(2) $2x \geq 90$ (4) $2x + x \geq 90$ 21_____

22 When $2x^2 - 3x + 2$ is subtracted from $4x^2 - 5x + 2$, the result is

(1) $2x^2 - 2x$
(2) $-2x^2 + 2x$
(3) $-2x^2 - 8x + 4$
(4) $2x^2 - 8x + 4$

22 _____

23 Which expression represents the number of hours in w weeks and d days?

(1) $7w + 12d$
(2) $84w + 24d$
(3) $168w + 24d$
(4) $168w + 60d$

23 _____

24 Given:

$R = \{1, 2, 3, 4\}$
$A = \{0, 2, 4, 6\}$
$P = \{1, 3, 5, 7\}$

What is $R \cap P$?

(1) $\{0, 1, 2, 3, 4, 5, 6, 7\}$
(2) $\{1, 2, 3, 4, 5, 7\}$
(3) $\{1, 3\}$
(4) $\{2, 4\}$

24 _____

25 Which equation could be used to find the measure of angle D in the right triangle shown in the diagram below?

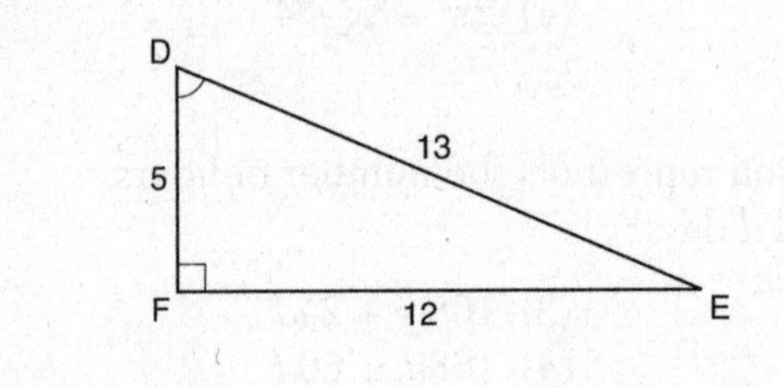

(1) $\cos D = \frac{12}{13}$ (3) $\sin D = \frac{5}{13}$

(2) $\cos D = \frac{13}{12}$ (4) $\sin D = \frac{12}{13}$ 25_____

26 If the roots of a quadratic equation are −2 and 3, the equation can be written as

(1) $(x - 2)(x + 3) = 0$
(2) $(x + 2)(x - 3) = 0$
(3) $(x + 2)(x + 3) = 0$
(4) $(x - 2)(x - 3) = 0$ 26_____

27 Which equation represents a line that is parallel to the y-axis and passes through the point (4,3)?

(1) $x = 3$ (3) $y = 3$
(2) $x = 4$ (4) $y = 4$ 27_____

28 There are 18 students in a class. Each day, the teacher randomly selects three students to assist in a game: a leader, a recorder, and a timekeeper. In how many possible ways can the jobs be assigned?

(1) 306
(2) 816
(3) 4896
(4) 5832

28 ____

29 In triangle *RST*, angle *R* is a right angle. If $TR = 6$ and $TS = 8$, what is the length of $\overline{RS}$?

(1) 10
(2) 2
(3) $2\sqrt{7}$
(4) $7\sqrt{2}$

29 ____

30 How many solutions are there for the following system of equations?

$$y = x^2 - 5x + 3$$
$$y = x - 6$$

(1) 1
(2) 2
(3) 3
(4) 0

30 ____

PART II

Answer all 3 questions in this part. Each correct answer will receive 2 credits. Clearly indicate the necessary steps, including appropriate formula substitutions, diagrams, graphs, charts, etc. For all questions in this part, a correct numerical answer with no work shown will receive only 1 credit. [6 credits]

31 Solve the inequality $-5(x - 7) < 15$ algebraically for x.

32 Oatmeal is packaged in a cylindrical container, as shown in the diagram below.

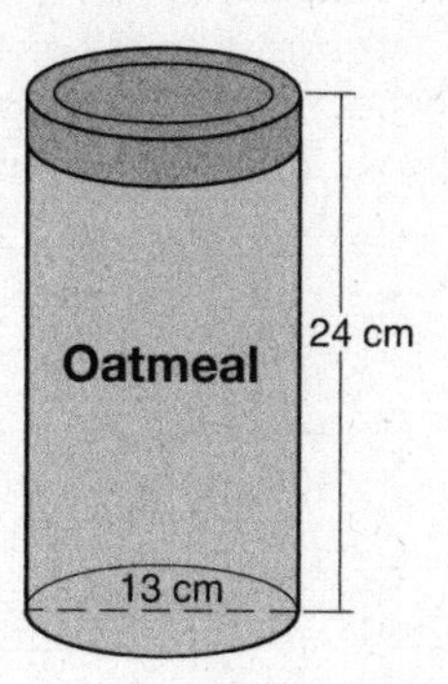

The diameter of the container is 13 centimeters and its height is 24 centimeters. Determine, in terms of π, the volume of the cylinder, in cubic centimeters.

33 The distance from Earth to Mars is 136,000,000 miles. A spaceship travels at 31,000 miles per hour. Determine, to the *nearest day*, how long it will take the spaceship to reach Mars.

PART III

Answer all 3 questions in this part. Each correct answer will receive 3 credits. Clearly indicate the necessary steps, including appropriate formula substitutions, diagrams, graphs, charts, etc. For all questions in this part, a correct numerical answer with no work shown will receive only 1 credit. [9 credits]

34 The menu for the high school cafeteria is shown below.

Main Course	Vegetable	Dessert	Beverage
veggie burger	corn	gelatin	milk
pizza	green beans	fruit salad	juice
tuna sandwich	carrots	yogurt	bottled water
frankfurter		cookie	
chicken tenders		ice cream cup	

Determine the number of possible meals consisting of a main course, a vegetable, a dessert, and a beverage that can be selected from the menu.

Determine how many of these meals will include chicken tenders.

If a student chooses pizza, corn or carrots, a dessert, and a beverage from the menu, determine the number of possible meals that can be selected.

35 A man standing on level ground is 1000 feet away from the base of a 350-foot-tall building. Find, to the *nearest degree*, the measure of the angle of elevation to the top of the building from the point on the ground where the man is standing.

36 Express $\sqrt{25} - 2\sqrt{3} + \sqrt{27} + 2\sqrt{9}$ in simplest radical form.

PART IV

Answer all 3 questions in this part. Each correct answer will receive 4 credits. Clearly indicate the necessary steps, including appropriate formula substitutions, diagrams, graphs, charts, etc. For all questions in this part, a correct numerical answer with no work shown will receive only 1 credit. [12 credits]

37 Solve algebraically: $\frac{2}{3x} + \frac{4}{x} = \frac{7}{x+1}$

[Only an algebraic solution can receive full credit.]

38 A jar contains five red marbles and three green marbles. A marble is drawn at random and not replaced. A second marble is then drawn from the jar.

Find the probability that the first marble is red and the second marble is green.

Find the probability that both marbles are red.

Find the probability that both marbles are the same color.

39 In the diagram below of rectangle $AFEB$ and a semicircle with diameter $\overline{CD}$, $AB = 5$ inches, $AB = BC = DE = FE$, and $CD = 6$ inches. Find the area of the shaded region, to the *nearest hundredth of a square inch.*

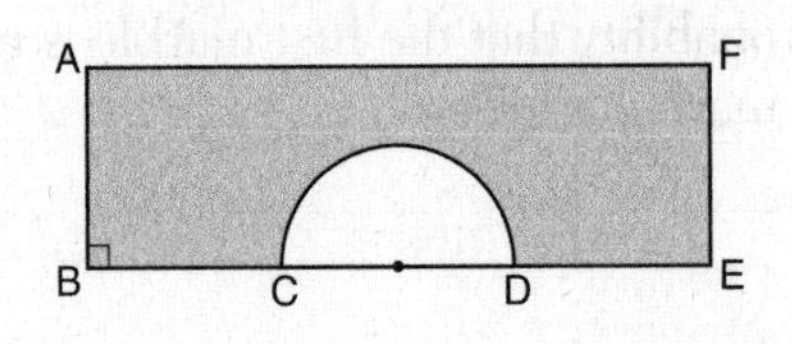

Answers June 2013

Integrated Algebra

Answer Key

PART I

1. (1)	**7.** (1)	**13.** *	**19.** (4)	**25.** (4)
2. (1)	**8.** (1)	**14.** (2)	**20.** (4)	**26.** (2)
3. (3)	**9.** (3)	**15.** (1)	**21.** (4)	**27.** (2)
4. (3)	**10.** (1)	**16.** (1)	**22.** (1)	**28.** (3)
5. (4)	**11.** (2)	**17.** (2)	**23.** (3)	**29.** (3)
6. (3)	**12.** (2)	**18.** (3)	**24.** (3)	**30.** (1)

*Due to lack of specificity in the wording of Question 13, all students were given credit for this question.

PART II

31. $x > 4$

32. 1014π

33. 183

PART III

34. 225, 45, and 30

35. 19

36. $11 + \sqrt{3}$

PART IV

37. 2

38. $\frac{15}{56}$, $\frac{20}{56}$, and $\frac{26}{56}$

39. 65.86

In **PARTS II–IV** you are required to show how you arrived at your answers. For sample methods of solutions, see Barron's *Regents Exams and Answers* book for Integrated Algebra.

Index